JOSEPH H. JACKSON

PICTORIAL
GUIDE
TO THE
PLANETS

Second Edition

THOMAS Y. CROWELL

COMPANY NEW YORK

ESTABLISHED 1834

End papers:

Comet Ikeya (1963a), one of the brighter comets of recent years, was discovered on January 2, 1963, by the Japanese amateur astronomer Kaoru Ikeya. This 15-minute exposure was made with the 40-inch Naval Observatory reflector on March 20, 1963, the day before the comet passed through perihelion.
—Elizabeth Roemer, U.S. Naval Observatory, Flagstaff, Ariz.

Title Page:

Mars, about 250,000 miles out, from Mariner 7.
—Jet Propulsion Laboratory

Designed by Eugene P. Schlerman

Manufactured in the United States of America

3 4 5 6 7 8 9 10

Library of Congress Cataloging in Publication Data

Jackson, Joseph Hollister.
    Pictorial guide to the planets.

    Bibliography: p.
      1. Planets. I. Title.
QB601.J3 1973   523.4   72-7573
ISBN 0-690-62443-3

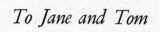

To Jane and Tom

ACKNOWLEDGMENTS

Telling the story of the sun, the moon, and the planets as they are known today has been as absorbing as studying their features through a telescope, watching a radar echo trace the record it has carried back from the moon's surface, or examining a meteorite from interplanetary space. Even more gratifying, perhaps, has been the opportunity to talk with astronomers and space scientists about their work and to experience at first hand their desire to share their knowledge without reserve.

Although I remain solely responsible for the contents of this book and any errors it may contain, I am particularly indebted to Carl Sagan, Laboratory for Planetary Studies, Cornell University, for his encouragement and advice; to Simone Gossner, formerly of the U.S. Naval Observatory, for many worthwhile suggestions; and to Hugh Rawson, Thomas Y. Crowell Company, and to my wife, Jane, for valuable editorial guidance.

Among the many scientists who generously donated ideas and illustrations, I wish especially to express appreciation for the contributions of the late Dinsmore Alter of the Griffith Observatory, Audouin Dollfus of the Observatoire de Paris, Brian Mason of the American Museum of Natural History, Peter Millman of the National Research Council of Canada, Elizabeth Roemer of Flagstaff Station, U.S. Naval Observatory, E. K. Bigg, Commonwealth Scientific and Industrial Research Organization, Australia, George C. Atamian, Talcott Mountain Science Center, Avon, Connecticut, Robert S. Dietz, Altantic Oceanographic and Meteorological Laboratories, Miami, George R. Carruthers, U.S. Naval Research Laboratory, and Clyde T. Holliday, The Johns Hopkins University.

Observatories that supplied photographs to round out this pictorial guide include especially the Dominion Astrophysical Observatory (Victoria, B.C.), the Lick Observatory, the Lowell Observatory, the Meudon and Pic-du-Midi observatories (France), the Mount Wilson and Palomar Observatories, and the Mount Stromlo and Siding Spring Observatories (Australia).

Other individuals and organizations who contributed in various ways include the Astronomical Society of the Pacific, the Commission for Physical Study of the Planets and Satellites of the International Astronomical Union, Donald Bradley of the Research Division of New York University College of Engineering, B. J. Levin of the O. Schmidt Institute of Physics of the Earth (Moscow), and Patrick Moore, the British amateur astronomer; the United States Air Force, Army, and Navy; the Atomic Energy Commission, the Coast and Geodetic Survey, the National Academy of Sciences–National Research Council, the National Aerotautics and Space Administration, the National Weather Service, National Oceanic and Atmospheric Administration, Lawrence Radiation Laboratory, and U.S. Geological Survey; and J. B. Kendrick of Thompson Ramo Wooldridge Technology Laboratories, William H. Pickering, Jet Propulsion Laboratory, and the California Institute of Technology and Carnegie Institution of Washington for the astronomical color photographs copyrighted. Many publishers have cooperated, in particular Houghton Mifflin Company, Boston, on the color photographs, as have such periodicals as *Astronautics & Aeronautics*, the *Journal of Geophysical Research*, the *Journal of the Royal Astronomical Society of Canada*, *New Scientist*, *Science*, *Nature*, and *Nature Physical Science*.

It was my good fortune at an early age to be taught the use of the *Ephemeris* and the operation of a 10-inch refractor by its designer, the late Governor James Hartness of Springfield, Vermont. This book ultimately derives from him and from the late Mrs. Hartness, who patiently imparted to a young boy all she knew about the heavens.

New York, N.Y.   Joseph H. Jackson

FOREWORD

Five hundred years ago, near the beginning of the Renaissance, European civilization stood on the threshold of discovery. The plucky caravels of Prince Henry the Navigator were being urged for a variety of reasons—economic, political, and for the love of adventure—down the west coast of Africa, around the Cape of Good Hope, and on toward the Indian subcontinent. An extraordinary account of distant travel, published a century before by a Venetian nobleman named Marco Polo, was being reread. Sailing ships had reached a stage in design sophistication where extended voyages on the open seas seemed possible. Then, in the course of a single century, came the establishment of an eastward sea route to India, the apparent discovery of a westward route to the South Seas, the realization that a new continent had been discovered, a preliminary exploration, and the initial steps at colonization of the New World. What a ring those words had then! The New World! A distant land, populated by strange men, where your fortune could be made, and your life begun again.

The Age of Discovery was partly a cause and partly an effect of the great cultural outburst that has led to the values and achievements of our own times. The many who stayed behind acquired a legacy from the deeds and thoughts of those few who sailed beyond the western seas to the New World. The exploration of the Americas provided a sense of perspective. There were other plants and animals in the New World, other peoples, other societies, other ways of life. In a devastating social criticism of contemporary European society, Voltaire imagined a Huron Indian brought to France, and viewed the social landscape through his eyes. The opening of the New World brought out much of the best, and some of the worst, in human beings; it provided an escape from stagnation and the prospect of a second chance. On the whole, the results have been salutary, and perhaps essential for the evolution of our civilization.

But today, the exploration of the earth's surface has been completed. For the joys of discovery and the lure of the frontier we must turn to the oceans and the skies. More than one new world now awaits exploration. There are eight other planets and thirty-one satellites in our solar system—and tens of thousands of smaller bodies. We are alive at the first moment in history when man steps forth beyond the ancient confines of the planet of his birth. Beyond lie other worlds.

The moon is a somber, airless, desolate place. But because of the absence of atmosphere and water, its surface is largely unweathered. Thus, clues to the origin and evolution of the solar system may reward the first lunar explorers.

Venus is hot, dark, and arid. Because of its dense cloud layer, no man has ever seen its surface—our information comes from infrared and microwave observations. What is the surface like? Why is it so hot? Where did all the water go?

Mars is chilly; its air is thin; oxygen seems to be absent; and water is scarce. Yet there is some evidence of life. Are there really organisms on Mars? How do they manage? Is the vegetation sparse, or may there be jungles? Is it quite certain that no higher animal life can exist on Mars?

As different as the earth, the moon, Mars, and Venus are from each other, Jupiter is another kind of place entirely. We cannot see beneath its brightly colored, banded and belted clouds. But the atmosphere abounds in hydrogen and helium. Organic molecules must be present. We do not know how far below the clouds the surface lies. Worse yet, we do not even know whether Jupiter has a surface, in the ordinary sense. The range of pressures that operate there is outside ordinary terrestrial experience.

And on it goes. Each world has its own kind of differentness, its unique problems and challenges. But we shall seek them out. The step we make now is irrevocable. The eight planets, thirty-one moons, and miscellaneous debris that Dr. Jackson introduces on these pages will occupy us for centuries: Langrenus, Pandorae Fretum, Sinus Aestuum, Tithonius Lacus, Io, Enceladus. . . . The strange names of those distant places will come readily to our children's lips. There may our grandchildren be born. But even centuries hence, when every planet, moon, and asteroid has been trod, when all that is habitable has been colonized, planetary exploration need not cease. For beyond this provincial solar system, light-years distant in the immensity of space, beckon, vast and imponderable, a hundred billion stars.

Carl Sagan
Director, Laboratory for
Planetary Studies,
Cornell University,
Ithaca, New York

CONTENTS

I   COUNTDOWN TO THE PLANETS   1

II   OUR POSITION IN THE UNIVERSE   3
     Appearance and Reality; Motion of the Earth; Milky Way Galaxy;
     Beyond the Galaxy

III THE SOLAR SYSTEM   11
     The Sun; Senior Members of the System; Search for New Members;
     Planetary Orbits; Elements of Orbits

IV   THE PLANETS AND THE MOON   23
     Birth of the Planets; Scale of the Planetary System; Revolution of
     the Planets; Motion of the Moon; Other Earth Satellites; Size and
     Shape of the Planets; Planetary Rotation

V   PLANET EARTH   34
     The Great Globe Itself; Views of the Earth; Surface of the Moon;
     Earth's Interior; Interior of Other Planets; Earth's Magnetic Field;
     Distant Magnetic Fields

VI   EARTH'S ATMOSPHERE   48
     Atmospheric Regions; Thermal Regions—Troposphere and Strato-
     sphere; Composition of the Lower Atmosphere; Solar Constant and
     Heat Balance; Rainfall and the Moon; Mesosphere and Airglow;
     Upper Atmospheric Regions; Ionosphere; Geomagnetic Cavity

VII INTERPLANETARY SPACE   59
     Space Regions; Electromagnetic Spectrum; Gases in Space; Magnetic
     Fields; Solar Wind; Solar Flares; Cosmic Rays; Zodiacal Light; Inter-
     planetary Dust

VIII ASTEROIDS, METEORS, AND COMETS   71
     Minor Planets; Meteors and Meteorites; Glassy Enigmas; Comets

IX   ATMOSPHERES OF THE MOON AND PLANETS   90
     Needle in the Lunar Haystack; Veils over Mercury; Models of Venus;
     Mixed Martian Clouds; Atmospheres of the Outer Planets

X   SURFACE OF THE MOON   105
     Lunar Topography; Surface Temperatures; Lunar Maps; Contacts
     with the Moon; Lunar Seas; Lunar Craters; Changes on the Moon?;
     Lunar Surface Composition

COUNTDOWN TO SPACE: A SPECIAL PICTORIAL SUPPLEMENT   121

XI   PLANETARY SURFACES   159
          Mercury's Dim Markings; Venus' Unseen Surface; Martian Kaleido-
          scope; Surfaces of the Outer Planets

XII   NATIVE LIFE ON OTHER PLANETS   174
          Evidence of Life in Meteorites; Origin and Conditions of Life; Lunar
          Life; Life on Mercury or Venus; Martian Life?; Life in the Universe

XIII ROCKETS AND SPACE VEHICLES   190
          Types of Rockets; Rocket Principles; Rocket Systems; Space Pay-
          loads; Nuclear and Electric Propulsion

XIV   ARTIFICIAL SATELLITES AND SPACE PROBES   200
          Earth-Satellite Orbits; Injection into Orbit; Reentry; Manned Orbital
          Flights; Missile Ranges and Tracking Networks; Astronomical Satel-
          lites; Lunar and Interplanetary Probes; Manned Space Flights; A
          Second Moon?

XV   MAN IN SPACE   215
          Noise and Vibration; Acceleration and Deceleration; Weightlessness;
          Life-Support Systems; Effective Human Action in Space; Meteors
          and Dust; Space Radiations; Solar Flares; Shielding

      TABLES   229

      BIBLIOGRAPHY   241

      INDEX   245

CHAPTER I

"For now we see through a glass, darkly; but then face to face; now I know in part; but then shall I know even as also I am known."

I Corinthians 13:12

For more than three centuries astronomers have scrutinized the dim images of the planets and satellites in the mirrors of their telescopes. The mirrors have grown larger and the images brighter decade by decade, yet these faraway bodies are still seen through a glass darkly, for their light flickers down faintly to us through the vapors and sludges of the earth's atmosphere.

A transformation like that of which Paul spoke is taking place in our view of the solar system—brought about by radio and radar astronomy, high-altitude balloons, artificial satellites, and unmanned and manned spacecraft. Astronauts have landed on the moon and viewed its farside from orbit. In time, Mars, too, will be seen "face to face," and this transformation will continue.

Seeing most of the planets is easy; no telescopes or rockets are required. By going out of doors a number of times on a clear evening, one can trace the stately march of the planets amid the stars. The first observers of the heavens did just this and, deceived by what they saw, thought that all these sparkling points of light move around us.

Over the centuries men finally penetrated to the realities behind these appearances. They distinguished planets and stars by their motions and then conceived the breathtaking panorama of the solar system, which no one has yet seen in its entirety. They discovered the omnipresent laws governing the movement of the bodies, large and small, within this system. Little by little, the actual position and movement of the whole solar system in the universe at large was worked out.

COUNTDOWN TO THE PLANETS

The third planet out from the sun, the earth, is the only planet we really know, of course. The earth sciences have recently taught us much about our spaceship earth. For example, by drilling deep down into the ocean floors it has been proven that the continents once formed a supercontinent, which broke into crustal plates that in time drifted apart.

Now orbiting earth satellites and space vehicles have added greatly to this knowledge. Study of the earth from space has revealed much about the size and shape of our planet and its gravitational and magnetic fields. The Van Allen belts have been discovered within the whole geomagnetic cavity. The earth's hydrogen corona has been photographed from the moon. And the earth's continents and oceans are being probed and analyzed with environmental satellites circling the poles.

The first spacecraft designed to leave the solar system, Pioneer 10, has been launched and is on its endless way. It carries a message about human beings on earth in case it should be intercepted thousands or millions of years from now by other intelligent beings. And the first experiments in the effects of weightlessness on astronauts for a month or two will be carried out soon on earth-orbiting Skylabs. It would take many months for manned spacecraft to travel to the closest of the other planets.

The rocks and soil brought back to earth by astronauts sampling a few locations on the moon have contained new minerals but no signs of native life. Water, so essential to life, has yet to be found on the moon. Its first pioneers may have to manufacture water from lunar rocks. But faint signs of water have been seen on planet Mars. The possibility of native life on Mars can be considered, although the odds for and against it seem even. The whole question of whether there is life elsewhere in the universe can be weighed with the help of the theory of probability.

With the advent of the space age a whole family of rockets and space vehicles has sprung into being. In less than 10 years unmanned and manned earth satellites have been successfully designed, built, and flown. Rockets are launched from the earth's surface, buffet their way up through the shearing winds of the dense atmosphere against the terrible tug of the earth's gravitation, and are injected into earth orbit, or plunge into interplanetary space. A totally new science of rocket propulsion has developed, together with advanced instrumentation and incredibly complex systems to track, predict, and control the orbits and trajectories of space vehicles.

To put man himself into space, the life sciences have been stretched to new limits. They now deal with the unprecedented strains of extreme acceleration and deceleration on the human system; the physiological effects of long-continued weightlessness and the hazards of the intense Van Allen radiation belts and the great space cyclones following solar flares; the utter detachment of space ships from humanity; the rapid oscillations between heat and cold; the range from vacuum conditions to impenetrable clouds; and the irradiations from space on the planets—all of these have made men question whether human beings can adjust to the strange new environments they seek to enter.

Space science is often called a distinct new field. The fact is that space exploration makes great demands on all the established scientific and engineering disciplines, requires them to work together, and infuses a spirit of creativity and adventure. Astronomy, once the only prober of the skies, is now but one among many sciences studying the planets and satellites. It will spearhead these studies with instruments stably mounted in satellites above the earth's atmosphere. And astronomy will continue its investigations of the nearby and distant stars, which must long remain beyond the physical approach of men—but not beyond feasible communication.

The full-fledged worlds of the planets and satellites are now being explored directly, not simply as points of light or tiny disks viewed from the surface of the earth. Not astronomy alone, but all of the sciences and their various branches will be required for successful manned exploration of the solar system beyond the narrow confines of the earth. Countless human efforts in scientific research, rocket technology, industrial development, and space flight must be integrated in our preparation to meet the planets face to face.

In the past we have marveled at the exploration of the West, the conquering of the poles, the climbing of the great mountain peaks, and the penetration of ocean deeps. The great new boundary that mankind is about to cross is space, leading to direct unmanned or manned contact with the moon, the other planets and satellites, in fact the whole solar system. Surely the unprecedented and hazardous conditions which must be overcome and the strange worlds which must be explored in the process will teach us more than we can possibly foresee about our own planet earth and about our own inner sources of weakness and of strength.

Surveyor 7 photographs from the moon prove that the earth turns.
—JET PROPULSION LABORATORY, CALIFORNIA INSTITUTE OF TECHNOLOGY

CHAPTER II

*"The world, the race, the soul—in space and
   time the universe,
All bound as befitting each—all surely going
   somewhere."*

WHITMAN

By day the dazzling sun moves through the
sky and the shadow it casts has been used for
thousands of years to tell the hours. At night
the moon floats overhead, measuring the
months with its phases, among thousands of
scintillating points of light—the planets and
stars. How far away are all these lights? Where
is the earth located in relation to them? Are
they moving, or is the earth moving, or are
both?

Almost all that is known about that sum
and substance of everything called the universe, and about our own position in it, has
come from observations of these lights in the
sky with instruments that have become ever
more precise and reliable over the years. But
this is not all—radical shifts have also been
required in the interpretation of these observations. "Things are not always what they seem
to be" has been a good maxim for astronomers. A clear distinction between appearance
and reality, what things seem to be as opposed
to what they really are, has proved essential
time after time in working out the position of
the earth in the universe and in explaining
many features of that one infinitesimal part
of the universe called the solar system.

Appearance and Reality

If you stop at a traffic light with another car
alongside, your car may suddenly appear to be
rolling ahead, even though your foot is down
hard on the brake. Only when you look at the
road between the cars or at the buildings in

OUR POSITION IN THE UNIVERSE

the background, do you realize it is the other
car, not your own, that is moving. Appearances have deceived you; the illusion was
caused by seeing the other car move without
an adequate reference system, like the road, to
which both cars could be related. Lacking such
a reference system, common to both the observer and the observed object, the observer
cannot know whether an observed body, he
himself, or both, are moving.

Until a few centuries ago the earth was quite
naturally taken as the stationary point in a
reference system to which all motion on the
earth, or above it in the skies, was referred.
This simple way of viewing things seems
naïve to us now, but an examination of the
realities behind certain appearances of the
moon and of the planet Mars will show how
easily they can be misinterpreted.

To any observer here on earth, the moon
rises, a huge sphere, over the eastern horizon,
swings upward across the sky, and sets in the
west. But the reality is entirely different.

The moon's apparent motion is caused by
the earth's rotation, which carries the observer
along with it. This is a well-known fact now,
but it is nevertheless striking to the observer
who must adjust his manually operated telescope every half-minute or so to keep the moon
in his eyepiece; he realizes how rapidly the
earth must be spinning with him. In actuality,
the moon revolves around the earth in exactly
the opposite direction, from west to east. If
it were possible—and it soon will be with
satellites and space stations—to look down on
the earth-moon system from high above the
North Pole, the direction of the moon's motion would be seen to be counterclockwise.
The sight of these two great spheres from
above should be magnificent. Our moon is so
much larger in proportion to the earth than
the other satellites in the solar system are in

proportion to their primaries that the earth and moon are thought by some to be a double planet—born and reared as twins.

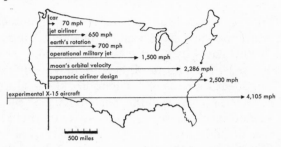

Comparison of some speeds on earth with the moon's speed in orbiting the earth, in miles traveled per hour.

The moon actually moves, then, from west to east. It goes slowly, astronomically speaking, circling the earth at an average rate of 2,286 miles an hour, or 0.635 mile a second. This is a dizzy pace, indeed, compared with the speed of a jetliner, which may travel at a mere 500 to 650 miles an hour, but some of the terrestrial speeds attained and planned can be compared with the velocity of the moon in its orbit. Already, speed records on earth have surpassed that of the moon. Artificial satellites and spacecraft, orbiting much closer to the earth, far exceed the moon's velocity, traveling at some 17,500 miles an hour.

The moon takes 27 days, 7 hours, 43 minutes, and 11.5 seconds to complete its 1.5-million-mile circumnavigation of the earth through space. This full circle through which it passes is divided by convention into 360 degrees, of which the moon only passes through a 13-degree arc each day. This daily inching of the moon from west to east explains why it rises about an hour later each night.

The moon's size is also misleading—it appears almost equal to the sun's. It looks this way because the moon's average apparent diameter in the sky makes an angle of 31 minutes (just over 0.5 degree of arc) for an observer on earth, while the sun's average apparent diameter covers an angle of 32 min-

utes. The ancients had great trouble trying to decide which was larger and which closer, the sun or the moon. In reality, the disk of the moon is a skimpy 2,160 miles across, about the straight-line distance from New York City to Salt Lake City. For comparison, the great ball of the sun is 870,000 miles in diameter, 110 times as large as the earth's diameter.

So the earth actually rotates from west to east, making the moon appear to move from east to west, even though it is actually revolving in the opposite direction. It is our own vision that makes the moon appear larger near the horizon. And almost every visual impression an observer has of the moon turns out to be an appearance which must be thoroughly examined to penetrate to the reality behind it.

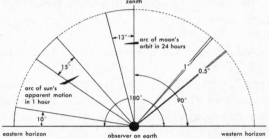

Apparent motion in the sky as measured by arcs and their corresponding angles for the observer. The sun's apparent motion through the sky is 15° in one hour, passing through an arc of 15° in the sky, or forming an angle of 15° for the observer from the beginning to the end of the hour. This is also called an "angular motion" or an "angular distance" of 15°. Each degree is divided by convention into 60 minutes ('), and each minute into 60 seconds ("). Half a degree (0.5° or 30') is about equal to the arc in the sky of the average apparent diameter of the sun and the moon.

Like the moon, the planet Mars appears to rise in the east and set in the west, because of the earth's rotation on its axis from west to east. Ancient observers, with the common-sense system of reference of a stationary earth, interpreted this to mean that the heavenly bodies revolved in nested celestial spheres or transparent orbs around the earth. On this

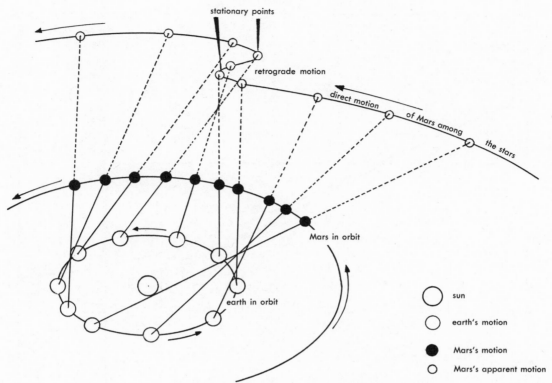

stationary points

retrograde motion

direct motion

of Mars among

the stars

Mars in orbit

earth in orbit

sun

earth's motion

Mars's motion

Mars's apparent motion

Apparent retrograde motion of the planet Mars among the stars as the earth passes it in a neighboring orbit.

basis, Ptolemy explained the motions of the sun and planets around the earth. Not until long after the death of Nicolaus Copernicus (1543), the Polish astronomer who refuted Ptolemy, was the movement of the earth in concert with the other planets around the sun acknowledged. A systematic doctrine of appearances, the Ptolemaic system was the view accepted by the early American colonists. It was taught at Harvard until 1656.

Mars provides an excellent example of how deceptive relative motion can be. Over a period of time, the reddish-white planet follows a path among the stars from west to east close to the ecliptic. (The ecliptic is the track that the sun appears to follow through the sky in the course of the year because of the revolution of the earth about it.) But at a certain point in its course each year, if it is watched for a week

or two, Mars appears to stop in relation to the stars, move backward for a time, and then return to its progression from west to east. When this apparent retrograde, or backward, action is plotted, it forms a loop in the sky. What is the explanation for this behavior?

Closer to the sun, and hurtling more rapidly around it than Mars, the earth completes its revolution in about 365 days, while Mars requires 687 earth-days. As the earth catches up to Mars in its neighboring orbit, Mars appears to hesitate in its course. As the earth passes it, Mars seems to turn back to trace the path shown in the diagram before continuing on its way along the ecliptic. The ancients had observed Mars's seeming retrograde motion, and Ptolemy was forced to add circles within circles to explain it in earth-stationary terms. Copernicus, who held the view that the planets

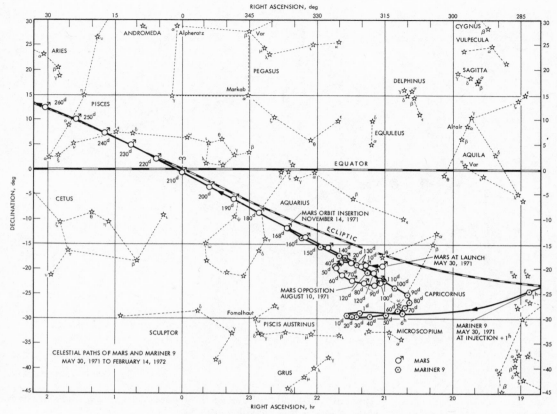

Paths in the sky of the planet Mars and the spacecraft Mariner 9 during its 168-day flight to Mars. Both the spacecraft and Mars went into apparent retrograde motion in the earth's sky.
—WILLIAM H. PICKERING, JET PROPULSION LABORATORY

Motion of the Earth

orbit the sun at different velocities in the same counterclockwise direction, provided the simpler explanation. Mars only appears to hover and rear backward among the stars because of the earth's changing relationship to it.

Such deceptive appearances abound in the solar system. One by one, the realities behind them are explained, but other appearances keep turning up to take their place. The looping path that the spacecraft Mariner 9 traced in the sky in 1971 as it flew toward Mars and then went into orbit of the planet, snarled with Mars' apparent motion as the earth caught up with it, would have made a thorny problem for Ptolemy.

Bit by bit the motion of the earth at any given time has been found to be the resolution or outcome of many different movements. Imagine the motion of a boy falling sideways toward the northwest from the bicycle he is riding southwest across the floor of a large elevator going up from deck to deck of an ocean liner plowing on an easterly course across the ocean, and pitching with a corkscrew motion in waves raised by wind from the southeast. Only during the twentieth century has the analogous complex motion of the earth been analyzed and our position in the universe fairly accurately pinpointed.

At the end of the handle of the Little Dipper, the North Star, Polaris, appears to be fixed in the night sky of the Northern Hemisphere. The other stars seem to sweep from east to west in arcs of circles, the centers of which are near the North Star. Photographic trails are made by the stars during an exposure of four hours, with the tiny arc of the North Star near the midpoint. The stars in the southern polar region make similar arcs, but no brilliant star like Polaris marks the pole.

Although the stars appear to move in this manner, they are called fixed because to the sight they are so nearly stationary in relation to each other. So infinitesimal is their actual motion in the patterns (like the constellations) they form in the sky that it can be verified only over years of time. But members of the solar system—the planets, sun, and moon—change their positions constantly in relation to one another and to the fixed stars.

The polar axis of the earth is tipped at an angle of 66.5° from the plane in which the earth orbits the sun, called the plane of the ecliptic. The sun and the moon exert unequal pulls on the slightly bulging equatorial region of the earth, and these forces make the earth wobble a little in its daily round on its axis. As a result, the north celestial pole of the earth, at the point near the North Star, has been found to shift over the centuries. This motion is called precession. The pole actually traces a circle in the sky around the pole of the ecliptic, completing a turn in a period of some 25,800 years. A wiggle in this traced circle is called nutation, and is caused by the unequal effect of the moon on the precession; this effect has a period of only 19 years.

The sweep of the stars in circles around Polaris brings home the fact of our own motion on the earth's surface. The speed of our motion varies with geographic location. At the equator, where the earth's circumference is about 25,000 miles, any point must move at about 1,040 miles an hour (0.3 mile a second) to complete the full circle in 24 hours. Since the circumference shortens toward the poles,

any point's speed with the rotation of the earth decreases correspondingly. A mere mile from either pole, the speed of rotation has diminished to a crawl of a quarter of a mile an hour, and the distance traveled to only 6 miles a day.

Again, the appearance of the earth's motion relative to the sun is deceptive. The sun seems to move with the rotation of the earth, rising in the east and setting in the west. But then, like the planets, it seems to move in relation to the stars day by day throughout the year from west to east along the ecliptic, the imaginary great circle formed on the celestial sphere by the sun because of the earth's actual revolution around the sun.

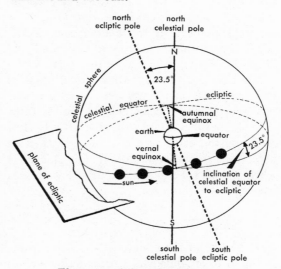

Elements of the celestial sphere.

The celestial sphere (an imaginary sphere of infinite radius with its center at the observer) on which the celestial bodies can be located, and the ecliptic, as well as other basic astronomical concepts, are diagrammed. The positions of the earth's north-south axis of rotation, extended indefinitely to the north and south celestial poles on the celestial sphere, determine the celestial equator midway between them—in the plane of the earth's equator, as shown in the diagram. The apparent path of the sun through the year defines the line of the ecliptic, and this determines the north and south ecliptic poles. The earth's axis of rotation is inclined at 23.5° to the poles of the ecliptic; the angle between the celestial equator and the ecliptic is also 23.5°. It is called the angle of obliquity or the obliquity of the ecliptic. The two points at which the celestial equator and the ecliptic intersect are called the vernal and autumnal equinoxes, and the line joining them the line of the equinoxes.

If an observer were looking down on the sun and the earth from above the planetary system (a space probe at an angle to the plane of the ecliptic is planned), he would see that the earth really moves along the plane of the ecliptic in a counterclockwise direction, making the sun appear to move along the ecliptic. Since the earth's actual motion around the sun is a little less than 1° of arc a day (360° in a year of 365 days), to the casual observer it is noticeable in the apparent motion of the sun only over a considerable period of time.

In dramatic contrast with its relatively slow spinning, the earth carries us along with it in its annual revolution around the sun at a whistling 66,000 miles an hour (18.5 miles a second).

Milky Way Galaxy

The stardust of the Milky Way Galaxy, which seems to have spilled in a meandering stream across the sky, forms the great community of thousands of millions of stars to which our solar system belongs. (See map of stars and constellations.) The discovery that this band arching the sky was a galaxy, and that our sun and its planets belonged to this system, was a great advance in itself. In the twentieth century astronomers have made intensive studies of our Galaxy, greatly aided in recent years by the radio telescope, which picks up many radio waves emitted by galactic sources. Gradually some light is being shed on our place in the Galaxy.

Through the analysis of the sun's motion in relation to the neighboring stars, astronomers first determined that our sun travels at about 12 miles a second (43,000 miles an hour) toward a point on the celestial sphere called the solar apex, not far from the bright star Vega in the constellation Lyra. The similar apparent motion of nearby stars away from this apex has proved the sun's actual motion toward it.

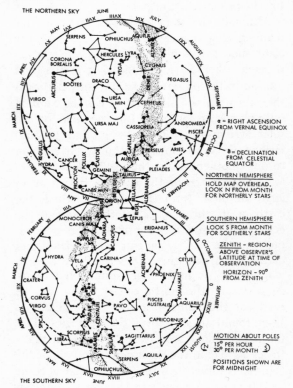

Arching path (gray) of the Milky Way across the stars and constellations in the skies of the Northern and Southern hemispheres. By following the directions given for each hemisphere, this map can be used to locate the major constellations in the skies during any month of the year.
—SPACE DATA, THOMPSON RAMO WOOLDRIDGE SPACE TECHNOLOGY LABORATORIES, INC.

The way in which astronomers locate positions on the celestial sphere is shown in the diagram. Declination, expressed in degrees, locates a body on the celestial sphere north (+) or south (—) of the celestial equator. Right

Ascension, expressed in hours, minutes, and seconds, starts from the vernal equinox on the celestial equator (see also diagram of the celestial sphere) and runs from I to XXIV hours eastward around the celestial equator. Declination is comparable to latitude and Right Ascension to longitude as used in locating positions on the spherical surface of the earth, although Right Ascension is continuous eastward, while longitude runs from 0° to ± 180°, east and west of the prime meridian at Greenwich, England.

In a whole series of relative motions that are a bit dizzying when put together, we go around with the earth, we revolve around the sun, and the whole planetary system moves with the sun as it speeds toward Lyra. The earth's orbit of the sun thus becomes helical,

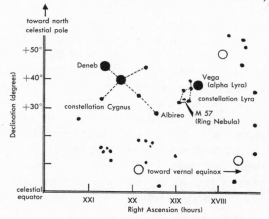

The constellations in the northern celestial sphere toward which the sun is moving. The solar apex is located near the fourth brightest star, Vega, in the constellation Lyra, while the other, much swifter, motion of the sun is toward the first magnitude star Deneb, in the constellation Cygnus. These constellations and stars can be located on the star map by looking upward from July on a line toward the center of the sky map for the Northern Hemisphere.

as if the earth were moving up a spiral staircase, the sun rising with it up the center of the stairwell.

But this is not all. Oddly enough, recent studies of the apparent motion of more dis-

tant stars, globular star clusters, interstellar dust, and gas clouds indicate that the sun is also speeding toward a point near the bright star Deneb in the constellation Cygnus, not far from Lyra. The velocity of this motion is almost literally "out of this world"—a whopping rate of about 140 miles a second (500,000 miles an hour). Escape velocity from the Milky Way Galaxy, the rate at which stars are freed from the Galaxy's gravitational system and can wander off into intergalactic space, is thought to be about 200 miles a second (720,000 miles an hour). A dozen and a half stars moving as fast as this have been discovered. Eventually they will escape from the Galaxy. The sun is not moving quite that fast.

Perhaps these two apparently contrary motions of the sun, in which the earth, of course, participates, can be reconciled in terms of the sun's actual position and movement within the Milky Way Galaxy. The Milky Way appears to be a spiral galaxy, shaped roughly like a disk with a somewhat thickened center and with a number of long, trailing, spiral arms. Much of it is hidden from view by massive clouds of galactic dust, so its shape, particularly that of its center, has only recently been determined. It took radio telescopes to do that mapping, since radio waves can penetrate dust clouds that light cannot. The Galaxy is estimated to measure about 90,000 light-years in diameter and perhaps a few hundred light-years in thickness. A light-year, one of astronomy's basic yardsticks, is simply the distance a ray of light will travel in one earth-year at its fantastic speed of about 186,300 miles a second.

The star nearest us, the triple system of Alpha Centauri, is about 4.31 light-years from the sun. Speeding approximately 6 trillion miles a year, the light from this system takes over four years to reach us, and the distances of most stars are much, much greater. Viewed in this way, the solar system diminishes to a tiny speck, almost alone in space. Only 25 stars are known within a sphere having a radius of 14 light-years from the sun. (Table 1 gives the names, distances, magnitudes, and motions of the 25 stars nearest the sun.) On the other hand, 10,000 stars flare within a radius of 100 light-years of the sun. On a large scale, then, the Galaxy's spiral arm is not underpopulated.

What explanation can be derived now of the apparently contrary motion of the sun within the Milky Way Galaxy? The sun, with the planetary system, is situated some 33,000 light-years from the Galaxy's center, and perhaps 13,000 light-years from its outer edge, although the boundary of the Milky Way is not sharply defined. The center is located in the direction of the star Alpha Sagittarii of the constellation Sagittarius on the southern celestial sphere. The stars of the Galaxy revolve around an axis at its center, much as the planets revolve around the sun. The swiftest motion of the sun, at 500,000 miles an hour, carries it in its elliptical, though nearly circular, orbit around the center of the Galaxy. Even at this speed, the cosmic year of the sun, the period in which it makes one complete revolution of the Galaxy, is judged to be about 250 million earth-years.

It is believed that the other distinguishable motion of the sun, at some 43,000 miles an hour, may reflect a wavering or vibration of the sun up and down, perpendicular to its orbital course in the Galaxy. The sun's motion may be like that of a needle on a warped phonograph record that waves slowly up and down with the warp, or like carousel horses moving up and down as they revolve with the merry-go-round. It has been suggested that the sun may go through two or three such perpendicular (transverse) vibrations, within a span of under 500 light-years from the galactic plane of revolution, as it completes one circuit of the Galaxy. On the other hand, the sun's minor motion may be only a small perturbation from a strictly circular orbit in the Galaxy,

caused by its relations with other stars or groups of stars in its vicinity. In addition, a much more rapid but smaller perturbation is caused by the gravitational pull of the planets orbiting the sun.

Beyond the Galaxy

Astronomers probing the depths of intergalactic space have found the Milky Way Galaxy to be one of a Local Group, a set of galaxies also called the Local Cluster. (Table 2 identifies these galaxies and gives some information about them.) It almost seems that the boxes enclosing boxes enclosing boxes continue in an infinite series. Nineteen galactic systems probably are members of this Local Group of ours. They vary in shape, from spiral galaxies like the Milky Way, to elliptical, flattened elliptical, irregular, and dwarf galaxies.

Farther out in space other galaxies and systems of galaxies appear literally by the million, each system forming a nucleus of denser matter in space. They appear most densely packed when the astronomer looks above or below the plane of the Milky Way system. In the plane itself, the stars and galactic dust are too concentrated to allow much to be seen beyond them, but there is no reason to assume that galaxies do not lie outward from the Milky Way in every direction.

On the surface of the earth, then, the actual motion is a constantly varying resolution of many motions: the rotation of the earth on its axis, its revolution around the sun, the sun's motion in the Milky Way Galaxy, and further, as yet undetermined, motions of the Galaxy in the Local Group of galaxies and the Local Group in relation to other clusters of galaxies. In the twentieth century, astronomers have achieved a fairly definite picture of the actual position of the earth and the solar system within the vast reaches of the universe.

CHAPTER III

". . . What if the Sun
Be Center to the World, and other Starrs
By his attractive Vertue and thir own
Incited, dance about him various rounds?"

MILTON

Unlike a celestial constellation, the solar system cannot be viewed on a clear night. It can only be seen as a unit in the mind's eye. From among the thousands of celestial lamps visible in the heavens, a very few have been identified as bodies that must travel endlessly around the sun under the thrall of its powerful gravitational attraction. The planets are the larger bodies in the system. They appear as disks when viewed through telescopes. All the stars but the sun appear only as points of light and cannot be resolved into disks by even the most powerful telescopes. Only five planets, Mercury, Venus, Mars, Jupiter, and Saturn are clearly visible to the unaided eye. The ancients called these brilliant bodies the wanderers, because they did not stay put in the regular wheeling pattern of the stars through the sky. Moving in orbit around some of the planets are natural satellites such as the earth's moon; the disks of a number of them can be resolved.

The Sun

While the sun is unique in the solar system, being the only one of its kind we possess, it is not an unusual star. It is a typical yellow-dwarf star, a member of the most numerous type of stars (spectral type dG2) and believed to be quite old. But it dominates the orbiting movements of the earth and the other planets, and they receive almost all of their light and heat from it. Life on earth would be impossible without it, but present forms of life would also be impossible if the atmosphere and magnetosphere did not shield us from some of

THE SOLAR SYSTEM

the sun's more intense radiations. While 95 per cent of its emissions are in the visible and near-visible range of wavelengths in the spectrum and are very constant, the sun also emits varying quantities of all other types of electromagnetic radiations, particularly ultraviolet, x-ray, and radio waves, as well as streams of high-energy particles with which no radiation equipment on earth can compete.

The sun is a great globular mass of hydrogen and helium atoms, with traces of all the other natural elements, many free electrons, and a scattering of molecules. Its diameter of 870,000 miles dwarfs that of the earth (7,927 miles) and even that of the largest planet, Jupiter (88,700 miles). Since it consists largely of lighter atoms, it is no wonder that the sun has a mean density only 1.41 times that of water, compared with the earth's density of 5.52. However, despite the sun's density it is so massive that what weighs 10 pounds at the earth's surface would weigh 280 pounds on the sun's surface.

Deep in the sun's interior, raging energies are created by the conversion of hydrogen to helium. The fireball of the hydrogen bomb on the earth mimics the sun on a small scale. The sun consumes over 4 tons of its mass per second in this thermonuclear reaction, which raises the calculated temperature of its center to some 25 million degrees F. This is the source of the dazzling incandescence of the sun's surface, which of course is the brightest thing we see.

Like the earth and the other planets, the sun's rotation is direct, but it spins flexibly like a mass of gases rather than a solid, with a period of about 25 days at its equator (moving from west to east some 14° a day) and 31 days in the higher latitudes toward its poles, so its average rotation period is called 27 days. It is perfectly spherical within a thousandth of a per cent, not ellipsoidal like the planets. Its equator is inclined at an angle of 7° 15′ to the plane of the ecliptic. Perhaps at one time it rotated much faster, but gave up most of its spinning momentum to its family of planets. The sun has a general dipole magnetic field, much stronger than that of the earth, and many other magnetic fields and disturbances, with one, two, or several poles scattered over its surface.

The sun has a complex atmosphere, at the base of which, comparable to a cloud layer in the earth's atmosphere, lies the photosphere or the visible disk, composed chiefly of hydrogen. It is perhaps 200 to 300 miles thick and at a temperature of some 10,000° F. Above this is the transparent, irregular chromosphere of tenuous gases, much hotter at 50,000° F. than the photosphere and visible up to about 6,000 miles above it. The chromosphere's upper regions are full of thousands of minute, luminous jets or spikes, called spicules, perhaps 250 miles across, which shoot up at high speeds, pause, and then fade out or retract, all within about 4 minutes.

The outermost feature of the sun's atmosphere, the corona, streams out above the chromosphere. It consists of a transparent, very low-density, high-temperature plasma, or fully ionized gas, perhaps as hot as 2 million degrees F. Normally it is invisible, since its luminosity is about half that of the full moon and the light from below overwhelms it. But during a total eclipse, the corona appears as a pearly-white, scalloped halo around the sun, visible out to several solar diameters. As it thins out at greater distances, it becomes the persistent solar wind, consisting largely of particles like protons and electrons, rushing out at velocities up to 700 miles a second at least as far as the earth, if not right out through the solar system.

Recent photographs of the sun from high-altitude balloons and rockets rising above most of the earth's atmosphere have revealed the details of solar granulation, looking similar to the surface of a hooked rug, visible in the photosphere over the whole surface of the sun. About 500 miles or more in diameter and lasting only 3 to 5 minutes, the granules are believed to be the tops of rising convection currents transmitting heat out of the interior. With the spicules, they give the sun its constantly changing appearance.

The "diamond-ring" effect in the total solar eclipse of July 20, 1963, photographed at Disraeli, Quebec. A sizable prominence appears at the upper right-hand side of the disk and the unevenness of the moon's limb is apparent above and below the "diamond-ring" effect.
—J. J. LABRECQUE, DOMINION OBSERVATORY, OTTAWA, CANADA

Solar prominences or filaments are spectacular knots or streamers of gas rising in and above the chromosphere, denser and cooler than the surrounding material. They shoot up for tens of thousands of miles, writhing and licking like flames, some looping back to the sun, others thrown up entirely clear of its surface. Their lifetime averages 50 to 75 days, so they are relatively permanent features of an always changing scene, though many of them suddenly disappear and then recur. Some of them seem to follow the lines of strong magnetic fields at the sun's surface.

The black etched sunspots are located in the photosphere. They may be round, elliptical, or irregular in shape, with a dark center or umbra edged by a grayish ring or penumbra. They appear dark only because their temperature (8000 to 9000° F.) is below that of the photosphere around them. Starting as tiny pores in activity centers on the sun, where very strong magnetic fields are usually to be found, they grow in a few days to full spots ranging in size from 500 to over 50,000 miles in diameter. Then they decay, lasting from a few hours to a week or longer, sometimes several months. Groups of many spots sometimes make irregular splotches on the sun's disk. Solar flares often occur in the activity centers also, rising at great velocities from the sun's surface. They may reach the earth with their showers of proton particles in from 1 to 5 hours. A day or two later, they may cause a sudden decrease in the intensity of cosmic-

The moon covered part of the blazing disk of the sun toward the end of the partial solar eclipse of March 18, 1969, photographed at Carnavon, Western Australia. Such dark filaments and sunspots, as well as bright flares, are all being investigated by solar scientists.
—GARY D. HECKMAN, U.S. NATIONAL OCEANIC AND ATMOSPHERIC ADMINISTRATION

ray radiation and severe storms in the earth's magnetic field.

Sunspot frequency runs in cycles, reflecting general periods of quiet and activity on the sun. This solar cycle averages 11.1 years, but has been known to range between 7 and 17 years, and the period is not yet predictable. At a period of "quiet sun," or sunspot minimum (1964–65), a few spots from the old cycle remain within 5° to 10° of the sun's equator and a few of the new cycle begin to appear at high latitudes of 25° to 35° north or south of the solar equator. With the increase toward a sunspot maximum, the spots spread throughout the sunspot zone between 35° north and south of the equator, usually passing the peak of activity about 5 years after the minimum, and then decreasing gradually over 6 years toward the sluggish activity of a quiet sun again. The magnetic polarities of sunspots are reversed in each 11-year period, so the full cycle of activity may be 22 years, though 88 years has also been suggested.

A solar probe, an unmanned spacecraft launched in toward the sun, has been proposed and is under study. It could carry out many experiments to increase our knowledge of the sun as it approached to within about 10 million miles of the surface, probably well down within the solar corona, and far closer than Mercury, which is about 36 million miles from the sun. Another probe launched up from the earth out of the ecliptic is feasible. Depending on the power of the launching rocket used, a position from 10° to 30° above the ecliptic in relation to the sun might be attained; there more could be observed of the high latitudes and the polar regions of the sun than is normally seen.

Senior Members of the System

New planets or satellites may, of course, be discovered tomorrow, but at this moment the

Jupiter VIII (left), and Jupiter IX (right), the outermost known satellites of the giant planet, both having retrograde motion. Jupiter VIII orbits Jupiter at a mean distance of 14.6 million miles in about 740 days. Jupiter IX orbits Jupiter at a mean distance of 14.7 million miles in about 760 days. Photographed with the U.S. Naval Observatory's 40-inch reflector in 1963, these plates of Jupiter VIII and IX had to be given 70- and 61-minute exposures, respectively, for the satellites have visual photographic magnitudes of 19.6 and 19.1 and are believed to be under 10 miles in diameter.

—ELIZABETH ROEMER, U.S. NAVAL OBSERVATORY, FLAGSTAFF, ARIZ.

known solar system consists principally of nine planets and a swarm of thirty-two satellites. Our own satellite, the moon, is included in this total, but not the artificial satellites and planetoids that man himself has placed in orbit around the earth or sun.

In the order of their distance from the sun, the first planet is little, silvery Mercury, flitting closest to the sun's fiery breath. Then comes cloudy Venus, nearly the earth's twin in terms of size, if in no other characteristic. Farther out from the sun than the double earth-moon system swings the ruddy planet, Mars, slightly more than half the size of the earth and two times the size of the moon. In the great, relatively empty band beyond Mars circle the minute planets called asteroids. Then the giant Jupiter looms white and belted, and beyond it the ringed, yellow Saturn, considered by many to be the most beautiful telescopic object in the heavens. The bulky greenish spheres of Uranus and Neptune follow, so distant that, except with

sizable telescopes, they look like faint stars. Finally, farthest out and last discovered, bobs tiny Pluto, which may well be a moon detached from Neptune in some long-ago dramatic shift in orbit.

No one knows how the satellites got into orbit around their planets, but there they are. Some have been discovered as recently as 1951. Enthusiastic reports of satellites around Mercury and Venus have never been confirmed. In fact, satellites do not thrive around the planets near the sun—perhaps these planets are too small to produce offspring. Our earth has its one moon, unusually large in proportion to our planet's size; the earth and moon are called a double planet. While other cloudlike satellites of the earth have been reported, preceding and following the moon in certain stable points in its orbit, these have yet to be confirmed.

Two pinpoint satellites moving swiftly in tight orbits around Mars were discovered at the U.S. Naval Observatory in Washington,

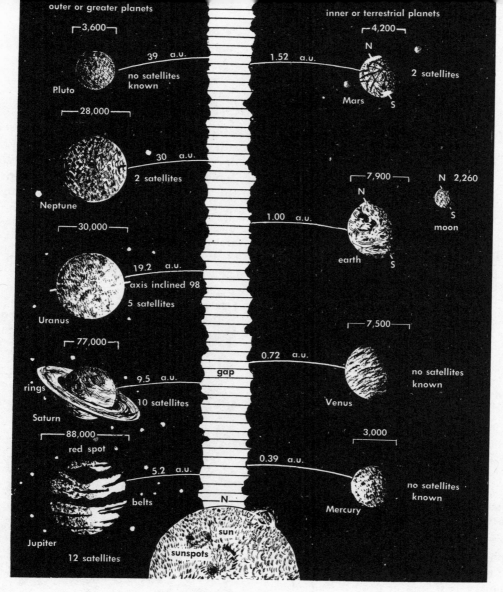

outer or greater planets

inner or terrestrial planets

┌─3,600─┐

┌─4,200─┐

N

39 a.u.

1.52 a.u.

no satellites known

2 satellites

Pluto

Mars

S

┌─28,000─┐

30 a.u.

N 2,260

2 satellites

N

Neptune

S

moon

┌─30,000─┐

19.2 a.u.

1.00 a.u.

axis inclined 98

N

5 satellites

Uranus

earth

S

┌─77,000─┐

┌─7,500─┐

9.5 a.u.

gap

0.72 a.u.

rings

10 satellites

no satellites known

Saturn

Venus

┌─88,000──┐

3,000

red spot

0.39 a.u.

5.2 a.u.

belts

no satellites known

N

Mercury

Jupiter

sun

12 satellites

sunspots

Known planets and satellites of the solar system. Approximate diameters of the planets are given above them (in miles). To the side are given their respective distances from the sun, in astronomical units (a.u.), 1 a.u. being about 93 million miles, the average distance of the earth from the sun. Most of the thousands of tiny bodies called asteroids fall in the gap between Mars and Jupiter.

D.C., in 1877, and they were named Deimos and Phobos after two attendants of the war god, Mars. Jupiter holds the record with its twelve known satellites, the first four discovered by Galileo in 1610, the twelfth, just barely visible, identified in 1951 in orbit at a great distance from the planet. Out beyond the marvelous geometrical disk of rings encircling its equator, Saturn has ten satellites. One of these, named Titan, is the only plane-

tary satellite that seems, on good evidence, to have an atmosphere. Uranus has five satellites and Neptune two. One of Uranus' satellites was discovered as recently as 1948, one of Neptune's in 1949. Like Mercury at the other extreme of the system, outermost Pluto appears to have no satellites of its own affecting its motion—which is natural enough if Pluto is indeed a stray from Neptune. Do satellites have satellites? To date nobody has seen one

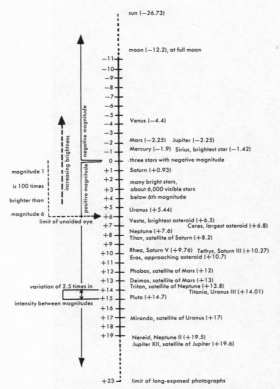

sun (−26.73)

moon (−12.2), at full moon

−11
−10
−9
−8
−7
−6
−5   Venus (−4.4)
−4
−3
−2   Mars (−2.25)   Jupiter (−2.25)
−1   Mercury (−1.9)   Sirius, brightest star (−1.42)
0   three stars with negative magnitude
+1   Saturn (+0.93)
+2   many bright stars,
+3   about 6,000 visible stars
+4   below 6th magnitude
+5   Uranus (+5.44)
+6   Vesta, brightest asteroid (+6.5)
+7   Ceres, largest asteroid (+6.8)
        Neptune (+7.6)
+8   Titan, satellite of Saturn (+8.2)
+9
+10   Rhea, Saturn V (+9.76)   Tethys, Saturn III (+10.27)
        Eros, approaching asteroid (+10.7)
+11
+12   Phobos, satellite of Mars (+12)
+13   Deimos, satellite of Mars (+13)
+14   Triton, satellite of Neptune (+13.8)
                            Titania, Uranus III (+14.01)
+15   Pluto (+14.7)
+16
+17   Miranda, satellite of Uranus (+17)
+18
+19   Nereid, Neptune II (+19.5)
        Jupiter XII, satellite of Jupiter (+19.6)

+23   limit of long-exposed photographs

increasing brightness

negative magnitude

positive magnitude

limit of unaided eye

magnitude 1 is 100 times brighter than magnitude 6

variation of 2.5 times in intensity between magnitudes

Placement of the sun, moon, planets, and some of the satellites and stars on the brightness scale of apparent visual magnitude.

except for the artificial lunar orbiters that have photographed most of the moon's surface.

The relative brightness of the planets, satellites, and stars is measured on a scale of visual magnitude, like a yardstick but with units in light intensity rather than inches. Between each magnitude is a brightness-intensity ratio of about 2.5; first magnitude is 2.5 times brighter than second magnitude, and so on. Magnitude developed in an odd manner, and as a result brightness decreases with decreasing negative magnitude (−5, −4) to 0 magnitude, and then decreases with increasing positive magnitude (+1, +2). Stars or planets down to about the first 6 positive magnitudes are visible to the sharp unaided eye. Bodies with magnitudes dimmer than +6 (those with higher positive magnitudes) can only be seen with telescopes.

The moon and the planets out as far as Saturn appear very bright in the sky, shining with the light they reflect from the sun. Under very favorable viewing conditions, Uranus, the next planet out beyond Saturn, with an average apparent visual magnitude of + 5.4, is so dim that to see it unaided one must know exactly where to look for it. Uranus was accidentally discovered on March 13, 1781, by the English astronomer Sir William Herschel. Thought for a year to be a comet, it was finally correctly identified.

Neptune, with a mean apparent visual mag-

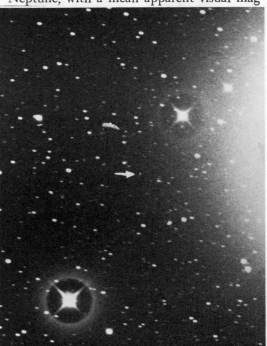

Phoebe (Saturn IX), Saturn's smallest satellite and the one most distant from the planet, photographed on June 19, 1960, with the 40-inch reflector of the Flagstaff Station, U.S. Naval Observatory. Phoebe orbits its primary with retrograde motion at a mean distance of 8 million miles in a period of 550 days. With a diameter of perhaps 100 miles and a photographic magnitude of +15.5, Phoebe can barely be picked out in Saturn's glare (right), caused by the 33-minute exposure required to record its satellite.
—ELIZABETH ROEMER, U.S. NAVAL OBSERVATORY, FLAGSTAFF, ARIZ.

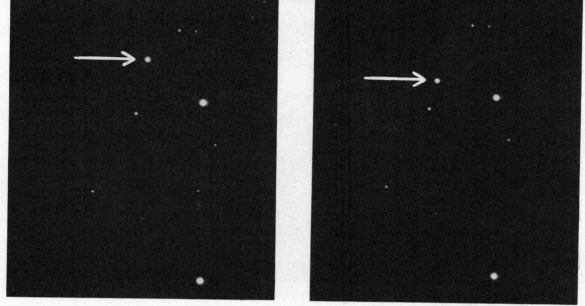

Motion of the planet Pluto in relation to the "fixed" stars. The photograph on the right was taken with the 200-inch Palomar reflector 24 hours after that on the left.

—Mount Wilson and Palomar Observatories

nitude of + 7.6, cannot be seen without a telescope. Its discovery was an international affair. Independently, at about the same time, John C. Adams of Cambridge, England, and Urbain J. J. Leverrier of Paris, predicted its approximate position on the basis of small variations, called perturbations, in the orbit of Uranus. It was seen and identified in this position on September 23, 1846, by the German astronomer Johann G. Galle. For 84 years thereafter it was generally believed that the planetary count was in. But in 1930, Clyde W. Tombaugh at the Lowell Observatory at Flagstaff, Arizona, discovered Pluto. A sizable telescope is required to view this small planet, which has a magnitude of +14.7.

Search for New Members

The discovery of Pluto reawakened interest in searching the skies for new planets and satellites. On two or three successive nights, photographs are taken of the same tiny overlapping regions of sky, producing a night-to-night series of negative plates. The areas studied are either along the path of the ecliptic, where planets might be situated, or in the space around a planet governed by its gravitational attraction rather than that of the sun, where

satellites might be found. The negative plates for two evenings are then viewed side by side in a blink microscope, dot by dot, to see if any of the celestial objects photographed have moved from one night to the next. If one has, its image appears on the viewing of the first plate, disappears on the viewing of the second, reappears with the first plate, and so on, producing the blinking effect for which the instrument is named. Pluto was discovered with a device of this kind.

In the combing of the heavens many nibbles or planetary suspects have to be tracked down. These usually turn out to be small asteroids or dim variable stars, whose light has diminished below the threshold of visibility on one of the plates. On the basis of changes in the motions of Halley's and two other comets, a scientist has predicted a tenth planet, "X," three times as massive as Saturn and nearly six billion miles from the earth, far out beyond Pluto. Astronomers at the Royal Greenwich Observatory, England, have photographed and studied the sky area where Planet X was predicted and found no planet there.

In recent surveys, three new and faint satellites orbiting Jupiter at a great distance were discovered by Seth B. Nicholson, on photographs made with the 100-inch telescope of

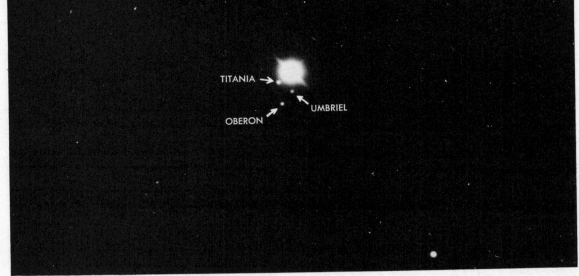

Three of the satellites of the planet Uranus—Titania, Oberon, and Umbriel—photographed on May 15, 1961, with an exposure of 3 minutes with the 120-inch Lick Observatory reflector. Miranda is not shown, and Ariel was invisible in front of the planet, whose image had to be overexposed to bring in the satellites clearly.

—LICK OBSERVATORY, MOUNT HAMILTON, CALIF.

the Mount Wilson Observatory in California. In 1948, Gerard P. Kuiper discovered Miranda, a fifth satellite of Uranus, and in 1949, Nereid, orbiting the planet Neptune. Both of these satellites are small, not more than 200 miles in diameter. Then in 1966, Audouin Dollfus picked out Saturn's tenth satellite, named Janus, so tiny that it was only glimpsed when Saturn's rings were tipped edgewise to the earth. Table 3 presents in summary fashion some basic facts about all the known planetary satellites.

Planetary Orbits

With the development of Greek geometry, and for centuries thereafter, celestial bodies and their movements were analyzed only in terms of the sphere and the circle. These forms were "perfect," and it was thought that

The two satellites of Neptune—Triton and Nereid. An exposure of 10 minutes was made on June 25, 1962, with the 120-inch Lick Observatory reflector. Triton's orbit is so nearly circular that its eccentricity has not yet been determined. Nereid orbits Neptune at a mean distance of 3.5 million miles with direct motion, in a period of 360 days.

—LICK OBSERVATORY, MOUNT HAMILTON, CALIF.

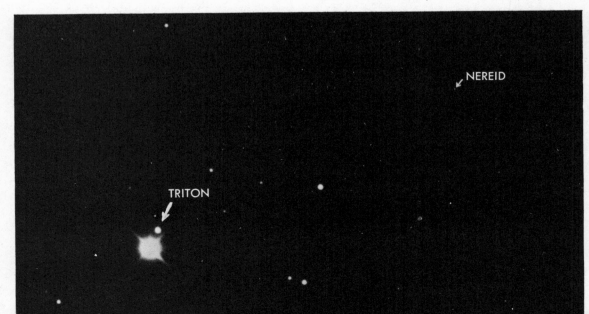

whatever lay in the heavens, associated with the gods, must manifest perfection. But any observations of the natural world soon demonstrate that it is always much more complex than man's simple, abstract concepts. Man's own creations, like roads and buildings and furniture, do accord with the straight lines, circles, and spheres of man-made geometry. Such forms are rarely approached in the natural world.

What struck early observers as straight lines, perfect squares, or other geometric figures on the moon's surface have been resolved into natural components—the wasted rims of old craters, abrupt shifts in contour, or rills slashed deep into the surface. The moon and planets themselves are not perfect spheres, but oblate spheroids, or ellipsoids, bulging out at their equators, but still beautiful to see. And as Johannes Kepler, the great German

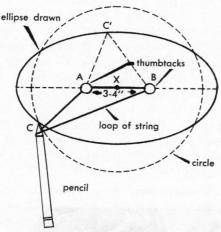

One way to make an ellipse. The ellipse is drawn around two thumbtacks, A and B, pushed into a piece of heavy cardboard or soft wood. The ends of a piece of string 8 to 10 inches long are tied togther to form a closed loop. The string is wound a few times around a pencil or pen, C, and held near its point with a finger. The remainder of the loop is placed around the two tacks and the string tightened by pulling the pencil away from the tacks. The pencil is held upright and a full revolution traced around the tacks with the pencil, the string being kept taut at all times.

astronomer, finally and reluctantly concluded, the planets move in elliptical, not circular orbits.

These orbits are like squashed or flattened circles. A true ellipse is not easy to draw, although the shape abounds in nature. Ellipses cannot be constructed with a compass, and it is almost impossible to draw one correctly freehand. A practical way to draw ellipses is illustrated in the drawing, which at the same time demonstrates some of their characteristics.

Points A and B in the drawing are called the focuses of the ellipse, which is contrasted with a circle whose center, X, is also the center of the long (horizontal) and short (vertical) axes of the ellipse. The drawing shows why ellipses are often called flattened circles. The greater the distance between A and B, keeping the length of the loop of string constant, the greater the amount of flattening, or eccentricity, of the ellipse.

As the pencil traces the ellipse, the distance ACB always remains the same, as at AC'B, or at any other point on the ellipse. The distances AC and BC do change as the pencil moves, but the length of the whole loop of string, ACBA, does not change, nor does the distance AB. So the length of string left for ACB or AC'B is always the same, and this constitutes one mathematical definition of an ellipse.

Think of the earth as being at C, with the sun at B, which is called the principal or primary focus by astronomers. Then there is nothing at A, called the unoccupied or secondary focus. This is the elliptical path that the earth and the other planets follow in orbiting the sun and the moon and other satellites follow as they orbit their primaries.

Another method of making an ellipse reveals how it is related to conic sections, fundamental to celestial mechanics, the branch of astronomy that analyzes the motions of celestial bodies. An ice-cream cone (without the ice cream), or some other fairly stiff cone-shaped object, can be cut in a number of ways. The shape of the edge of each of the slices cut will be a conic section: either a circle, an ellipse, a parabola, or a hyperbola, depending on the angle of the cut.

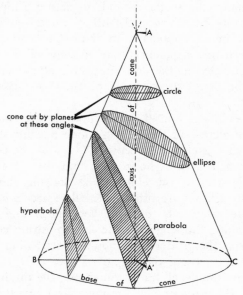

Various sections through a right circular cone, forming a circle, an ellipse, a parabola, and a hyperbola. Line AA' is the axis of the cone, formed by the rotation of the right triangle AA'C around the axis. The sides of the cone thus formed are AB and AC, representing its surface. A slice across the cone at right angles to its vertical axis AA' makes a circular section. Every point on its circumference is equally distant from its center on the vertical axis. An oblique slice across the cone produces an ellipse; the greater the angle from a right angle, the more eccentric the ellipse. A slice exactly parallel to one side of the cone, AB or AC, makes a parabola, an open curve, not closed as are the circle and ellipse. To round out the family of conic sections, a hyperbola is formed when the cut is at a greater angle to the base of the cone than the side of the cone, making another open curve.

The eccentricity of a great proportion of the orbits of bodies in the solar system is so small that they are very nearly circles, although certain planets and satellites ride in more elongated orbits. Many of the asteroids have very flattened orbits, and so do some of

the meteoroids. The most unusual orbits are those of some of the comets. A very few (perhaps 1 out of 100, it has been estimated) may move in hyperbolic orbits, although this is a highly controversial point. If they do, these comets come plunging in to barely round the sun and then go zooming off into cosmic space, never to return.

Elements of Orbits

Each member of the great astronomical triumvirate, Copernicus, Kepler, and Newton, contributed to the explanation for the swift and relentless coursing of the planets through the vast reaches of the solar system. Their principles reveal the foundations of celestial mechanics, which makes it possible to calculate the orbits of distant planets or meteorites plunging through the earth's atmosphere, to predict such events as eclipses and occultations in the planetary system, and to plan the orbits that space vehicles follow on their forays through interplanetary space.

The Polish astronomer Nicolaus Copernicus blew the earth-centered scheme of thought sky-high in his work, *On the Revolutions of Celestial Bodies* (1543). His insistence that the sun is the center of the entire system, with the planets and their satellites moving around it, was the beginning of the modern concept of the heavens, creating a ferment in men's minds.

Tycho Brahe, the famed Danish astronomer, designed many of his own instruments and was a stickler for precise observations, which were passed on after his death to Johannes Kepler, the German astronomer who had assisted him. Brahe thought, with Copernicus, that the five known planets spun around the sun, but held to the conviction that the sun and moon revolved around the earth, leaving it fixed as the center of the system.

In 1577, however, Brahe followed the path of a comet in the sky in his methodical manner. His notes show that he wondered whether this comet might have moved in an elongated, egg-shaped, rather than circular, course around the sun. In using these terms, Brahe very nearly described an ellipse. Unfortunately, it was simply a passing fancy on his part.

So it remained for Johannes Kepler to discover laws strictly governing the orbits of the planets and to announce them in his book, *New Astronomy, or Celestial Physics* (1609). For year after frustrating year, he had tried to fit the accurate data left him by Brahe into circular planetary orbits. All at once, while working on the orbit of Mars, Kepler hit upon the solution. The planets, he conceived, move in elliptical, not perfectly circular, orbits. Then he formulated the three laws of ellipses which explained their movements.

Kepler's first principle, sometimes called the law of ellipses, states simply that orbits like those of the planets are some form of ellipse, with a central attracting body, like the sun, at the principal or primary focus, and nothing occupying the other focus.

Kepler's second generalization, called the law of equal areas, displays his remarkable insight. He stated that the line joining the sun and the planet, known as the radius vector (*BD* or *BC* in the diagram), sweeps over equal areas of the plane of the ellipse in equal times as the planet moves around in its orbit. In the diagram, the radius vector is shown sweeping over the shaded areas, each covered by the planet in its orbit in, say, 10 days, and each having an area equal to the others, though very different in shape. The orbiting body, then, must speed up as it goes through perihelion, the point nearest the sun, having a much longer arc of the ellipse to cover within 10 days. But the orbiting body slows down as it passes aphelion, the point farthest from the sun, covering a much shorter arc of the ellipse in a 10-day period as it sweeps an equal area.

The artificial satellites orbiting the earth speed up in the same way as they near it in

their ellipses and slow down as they move out farther from the earth, although in some cases their orbits are so nearly circular that little change in velocity is noted. The comets in highly elliptical orbit vary greatly in orbital velocity. They speed very quickly, in a few days, through that portion of their orbits in the vicinity of the sun including their perihelion, and then take years for the remainder of their orbits including aphelion, so far out from the sun that they are not observable by the largest telescopes.

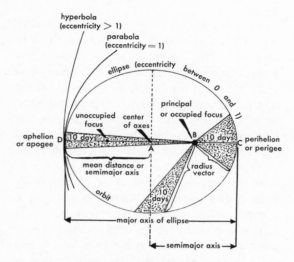

Elements of an elliptical orbit. An ellipse is a closed curve in a given plane, such that the sums of the distances of each point on the curve from two points (the focuses) are equal. When occupied by bodies, the principal focus is that containing the central attracting body (*B*), like the sun, with a planet (*C*) or some other body like a space probe in orbit, while the unoccupied focus is the same distance from the center of the axis (*A*) as the principal focus.

Kepler's third generalization is known as the harmonic law or law of squares. As formulated later by Sir Isaac Newton, it states that the squares of the time periods for the complete revolution of the orbiting planets around the sun are proportional to the cubes of their average distances from the central attracting body, the sun. The average distance from the earth to the sun, about 93 million miles, is called the astronomical unit (a.u.).

A little later, Newton integrated the ideas of Copernicus and Kepler, as well as those of Galileo and some of his own, into the three sweeping laws of motion and the law of gravity, which he stated in 1687 in his *Mathematical Principles of Natural Philosophy* (*Principia Mathematica*). With superlative insight, he grasped the universal law that governs the fall of all bodies here on earth and the continuous fall of the planets as they orbit the sun.

Newton's first law of motion states that every body, anywhere in the universe, here on earth or in the heavens above, remains at rest, or in a state of uniform motion, unless it is acted on by some outside force. In space, friction is negligible, and the principal outer force is gravity. According to his second law, any body acted on by a force is accelerated in the direction of the force's action, in direct proportion to the force's strength and in inverse proportion to the mass of the body. Mass is the amount of matter or material in a body, as opposed to its weight, which in turn is the result of the gravitational pull of one mass on another. In his third law of motion, Newton stated that for every action of a force there is an equal and opposite reaction. This third law is vital to the movement of missiles and spacecraft driven by rocket engines.

According to Newton's final or fourth law, that of gravitational attraction, every body in the universe attracts every other body with a force that varies directly with the product of their masses (such as the masses of the earth and the sun) and varies inversely with the square of the distance between them (such as the square of the astronomical unit in the case of the earth and the sun). The orbits of our artificial satellites and spacecraft are governed just as surely by this attraction as is the wheeling of the planets around the sun, and it determines the actual routes taken by interplanetary probes as they plunge up through the earth's atmosphere and out into space.

The laws of Newton and Kepler are all that are necessary for a grasp of the nature of orbits. The revolving motion of the earth in its orbit follows all of these laws. The gravitational attractions of the sun, and to a lesser extent, the other planets and satellites, act on the earth, varying directly with the product of their masses and that of the earth, and varying inversely with the square of the distances between them. Against the earth's centrifugal motion, acting at right angles to the direction of the sun, is the centripetal force of the attraction of the sun, tending to make the earth fall directly toward it. These forces are just balanced at the earth's distance from the sun, in the elliptical orbit actually followed by the earth and representing the resolution of all the forces acting on it.

The unit of gravity on the earth, commonly called a "g," is defined in terms of its effects —the acceleration it produces in a body falling freely toward the earth under its influence. This acceleration for a body in free fall is approximately 32 feet per second. For a body falling from rest under the influence of 1 g, the velocity at the end of the first second of fall is 32 feet per second, at the end of the second second it is 64 feet per second, and so on. At the end of 10 seconds, the body's velocity will be 320 feet per second and it will have fallen 1,600 feet. The American astronauts endure a force of from 6 to 8 g's when they are accelerated into orbits in their capsules and again when the retrorockets are fired. These decelerate the capsules enough so they will not burn up as they reenter the dense atmosphere and drop down to earth.

CHAPTER IV

". . . And measure every wand'ring planet's course,
Climbing after knowledge infinite."

MARLOWE

When Mariner 2 caught up with Venus in mid-December, 1962, and shaved past the planet, crossing from the nighttime to the daytime side, the seething, straw-colored cloud masses rolling past just 20,000 miles below would have made a magnificent scene for any passengers. Live scientists using many instruments could have unveiled more of Venus' mysteries than the radiometers and magnetometers that Mariner 2 carried.

But the few instruments did quite a job. This first space vehicle to pass in the vicinity of another planet telemetered reels of coded information about Venus back to earth before it veered away into its own 348-day orbit of the sun. For example, as it approached Venus, the planet's gravitational field gave Mariner a sharp tug, which accelerated it by about 3,500 miles an hour. By this effect on the craft, the most accurate estimate yet was obtained of Venus' mass—0.81485 ± 0.015 per cent of the earth's mass. The mass which had been calculated in 1943 on the basis of Venus' perturbations of the orbit of Mercury (0.813 ± 0.34 per cent) was remarkably close, and that of 1954 from the perturbations of the asteroid Eros (0.8148 ± 0.05 per cent) was even more accurate.

Many other proofs have come in that our knowledge of the solar system rests on stalwart foundations. Three different methods have shown, for instance, that the mean distance from the earth to the moon is 238,866 miles, with a probable error of only 1 mile more or less. Over the decades, measurements of the distances of the planets from the sun, the details of their orbits, and their sizes,

THE PLANETS AND THE MOON

masses, and periods of rotation have been made more and more accurate. The structure of the solar system and the bodies composing it has been outlined in broad strokes. But its origin is a conundrum to which a dozen or more answers have been given, and no one yet knows which one is right.

Birth of the Planets

Remarkably steady and unvarying as the planets are in their rounds, no one believes that the present scheme of things has always existed. What came before? Explanations of the origin and development of the solar system and the universe, sometimes called cosmogonies, have been offered from time immemorial, in myths and legends, religious dogmas, and scientific hypotheses.

Galileo, who in 1609 first used a telescope to observe the moon and then-known planets, discussed a theory of the birth of the solar system in his *Dialogues* (1632 and 1638), which he credited to the Greek philosopher Plato. In his *Timaeus*, Plato had described how the world-soul or creator made the original bodies (which he identified as the sun, moon, Mercury, Venus, Mars, Saturn, and Jupiter) and brought time into being by setting them in motion. "Now all of the stars which were necessary to the creation of time had attained a motion suitable to them . . .," Plato wrote, "they revolved, some in a larger and some in a lesser orbit; those which had the lesser orbit revolving faster, and those which had the larger more slowly."

While this contains the germ of Galileo's theory, Galileo went beyond Plato with his idea that the planets were originally all gathered in one place, out beyond Saturn, the most distant from the sun, and that they then began to fall toward the sun with their "natural acceleration," later known as the force of gravity. When each reached its present velocity, the creator halted its straight fall toward the sun and changed it into the familiar uniform circular motion.

Plato and Galileo were correct in asserting greater planetary orbital velocities with lesser orbits but with his law of gravitation Newton easily refuted Galileo's notion that the natural acceleration of gravity on the earth was the same throughout the solar system.

After Newton had hit the nail on the head with his universal laws of motion and gravitation, there was a rash of more or less scientific theories of the origin of the universe and of the solar system. Immanuel Kant, the German philosopher, said that the universe originated out of a scattering of cosmic dust. This theory appeared in his highly ambitious *General History of Nature and Theory of the Heavens* (1755). The concept later came to be known as the nebular hypothesis, after being more scientifically formulated in 1796 by Pierre Simon, Marquis de Laplace, the French mathematician and astronomer.

At the very end of his *Explanation of the System of the World*, Laplace described the primitive atmosphere of the sun as extending far out beyond the orbits of the later planets and resembling a nebula: a thin, hot, slowly rotating cloud of gases and particles.

Eventually the cloud cooled and condensed, with first one and then other equatorial, doughnut-shaped rings forming around the sun. These unstable whirling rings were then clotted and compressed into planets—the outer planets first, then the inner ones—all revolving in the same direction around the sun at the center of the system. Laplace's theory won general acceptance for about a century, then increasing knowledge of the system led to a rapid succession of new theories in the twentieth century.

Laplace's nebular theory had refuted the notion of the French naturalist, Comte de Buffon (1707–88) that a comet had approached very close to the sun and torn out a stream of matter from it that was gathered at a distance into smaller and larger globes. Buffon thought that these globes then became opaque and solid by cooling, thus forming the planets and their satellites.

In the so-called planetesimal hypothesis formulated early in the twentieth century, a star or stars was substituted for Buffon's much too insubstantial comet. Passing close to the sun, this star pulled out bulging masses or filaments of hot materials from the sun, leaving clumps of debris revolving around it. The planets were formed from these clumps by gradual accretion. In a later variation of this same idea, Sir James H. Jeans's tidal hypothesis (1928), a passing or colliding star drew out from the sun a long, cigar-shaped prong, conceived as thicker in the middle than at either end, explaining the formation of the giant, gaseous mid-planets like Jupiter and Saturn. But none of these hypotheses satisfactorily explained the solar system's great circling movement, or angular momentum, with the sun rotating and the planets revolving about it.

Collision theories like Jeans's, moreover, made the birth of the solar system seem a very fortuitous and unusual occurrence, because it is most unlikely that stars will approach close to one another in the vast voids of interstellar space. During the 1940's and 1950's such theories began to be replaced by ideas harking back to the old nebular hypothesis of Laplace. In the contemporary view, the solar system condensed in a cool state from some original source of cosmic dust, the sun in the center being or becoming a radiant star because of its increasing mass and density, which initiated the fierce torch of nuclear fusion reactions within the sun.

Some modern theories assume that the sun was surrounded by a large contracting cloud of gas and dust like a nebula; others, that the sun, as a member of the primitive Milky Way Galaxy, passed through a dense cloud of gas

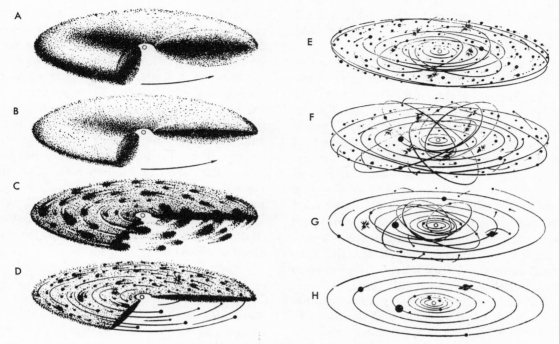

One form of the nebular theory of the formation of the solar system, developed from that of the Soviet mathematician, O. J. Schmidt. Left: The formation of asteroidal bodies from a gas and dust cloud around the sun. Dust collecting into a disk, A and B; disintegration of the disk into agglomerations which turn into asteroidal bodies, C and D. Right: the formation of planets from a swarm of asteroidal bodies. Increase in eccentricity and inclination of orbits as a result of mutual gravitational perturbations, E; accretion of asteroidal bodies and their fragments to form planets, F and G; the planetary system today, H.
—B. J. Levin, O. Schmidt Institute of Physics of the Earth, Moscow, U.S.S.R.

and dust in its path, drilled out a globular mass, and bore it away. Such a bulbous cloud, containing the natural elements found in the solar system, might have originated in one or more of the titanic explosions of earlier stars, called supernovas.

The primitive cloud became disk-shaped, some think, and the planets and satellites are assumed to have formed from great whirlpools in it. Others believe that the planets agglomerated from denser nuclei of gas and dust which gradually formed into larger bodies by attracting the clouds of debris along their courses—meteorites, comets, larger asteroids, or even infant planets. In the inner portion of the disk-shaped cloud, the denser planets formed from the heavier, less volatile elements, while the huge outer planets formed from lighter, gaseous elements driven farther out from the sun.

These present-day theories contend that the planets were formed in a cold state. They were conceived and incubated in a refrigerator, not in an oven. Later they were heated by the disintegration of radioactive elements within them, the radiation of the sun, and the explosive shocks of smaller bodies plunging into them. Gigantic infalls of debris plastering the planets' surfaces, cataclysmic explosions caused by collision with larger bodies, and clouds of dust and hurtling bodies blasted loose by their impact may sketch the lurid early history of the planets.

The pocked, jagged, and blasted surface of the moon is thought to show the scars of the earlier history of the solar system, unhealed by the effects of oceans and an atmosphere. In one view, the moon may be a small, primitive protoplanet which fell under the control of the earth late in the formation of the system. Others believe that the earth and moon formed separately and at about the same time as a double planet. The old notion that the moon was torn from the earth while both were still in a plastic state, leaving a vast hole, the remnants of which are the deep trenches in the Pacific Ocean, is now given little credence, though it still exists.

Another hypothesis shows promise because it integrates the development of the sun with that of the planets. A contracting nebula of material rotated very fast, as do most stars. Soon the inner core reached the mass and density necessary for nuclear fusion reactions. In fusion, helium is created from hydrogen through a multistage cycle, losing mass and releasing excessive energy. This formed the radiant sun. In this process the sun became unstable when it had shrunk to a diameter about equal to the present orbit of Mercury; from it a disk of gas sloughed off and expanded, although still coupled by magnetic fields to the sun. From this disk the protoplanets condensed, acquiring most of the angular momentum of the sun and solar system. As the planets coalesced around the sun, they moved farther away, shaping up into the present solar system.

Meteors that plunge into the earth's atmosphere and are large enough to survive a fiery journey to its surface are beginning to furnish some clues as to the history of the solar system. Particles of a stony meteorite that fell near Bruderheim in Alberta, Canada, in 1960, revealed, when pulverized, glassy, pea-sized spheres called chondrules or spherules, composed largely of silica. Samples of the minute amounts of xenon gas found in these chondrules and in other portions of the meteorite were analyzed in a very sensitive mass spectrometer. Greater proportions of an early isotope of xenon (xenon 129) were found in the chondrules than in the surrounding materials. It may be that such glassy beads were among the first tiny particles to separate and solidify from the primitive material in the solar system about 4.6 billion years ago. The elements composing the meteorites are estimated to be about 4.7 billion years old, perhaps dating from the time the solar system itself began to form from an original gaseous cloud of elements.

Scale of the Planetary System

Some idea of the relative sizes and distances of the planets can be obtained by reducing them all to a very small scale. Suppose earth is scaled down from its near 8,000-mile diameter to a sphere 12 inches in diameter, about the size of a basketball. What happens to the other planets on this same scale? The planet Mercury becomes a ball 4.5 inches in diameter. Venus is almost the size of the earth —11.5 inches across. The moon, surprisingly, is 3.25 inches in diameter, and Mars is about 6.4 inches across, slightly more than half the diameter of the earth and about twice that of the moon.

Coming to the blown-up gas giants, on this same scale Jupiter is about the width of the average room, 11.25 feet in diameter; Saturn is nearly as large at 9.5 feet. Then the size decreases sharply to Uranus and Neptune, large beach balls 3.8 and 3.5 feet in diameter, down to Pluto, with a diameter of only about 5.5 inches. The sun itself is an enormous balloon 59 feet in diameter.

Using exactly the same scale, and expressing the distances of the planets from the sun in miles, Mercury is placed about 0.9 of a mile from the sun. The earth is about 2.25

miles and Mars over 3 miles away. Across the asteroid belt, Jupiter is 11.6 miles distant from the sun. Pluto wanders almost lost in the vastnesses of space at about 67 miles out from the 59-foot-wide sun!

Revolution of the Planets

Since they are the immediate objectives of early space exploration, the names of the inner planets and the moon, our near neighbors in space, are appearing more frequently and familiarly in the news headlines and stories of the 1970's. Basic facts about their orbits, sizes, and rates of rotation are given in Table 4. With radio telescopes analyzing the faint radar echoes returning from their surfaces, and with probes heading off toward the moon, Venus, and Mars, the figures in this table can change overnight, but in most cases the changes will be small. For convenient comparison, the same kinds of facts about the outer planets have been collected in the same table. A number of doubtful gray areas remain, as the question marks in the table demonstrate, particularly concerning the physical features of Mercury, Venus, and the distant Pluto, all of which are difficult to observe and measure accurately.

The orbits of the planets are elliptical and follow the principles of orbital motion and gravitation. They are most stable in their rounds and are spaced out quite neatly from the sun. Bode's law, named for the German astronomer Johann E. Bode (1747–1826), who popularized it, is a geometric progression that fits their distances quite well, though it is irregular for Mercury at the start and breaks down for the outer planets, Neptune and Pluto, undiscovered when the law was formulated.

Frequently the distances of the planets from the sun and other distances in the solar system are given in terms of astronomical units (a.u.), the "yard" of the solar system. For the planets, these represent the mean distances from the center of the sun to the centers of

the planets, in comparison with that of the earth, which is by definition 1 a.u. from the sun. In Table 4, 1 a.u. has been taken as 92,-956,000 miles, based on recent radar measurements of the distance of Venus.

Motion of the Moon

Unless it has a surface too thickly heaped with dust, the bulky, malformed globe of our moon should make a fine space station, one that could never be duplicated in many respects by an artificial structure. In comparison with the velocities of other bodies in the solar system, the moon revolves very slowly around the earth, or, more correctly, around the center of mass (barycenter) of the joint earth-moon system, which lies about 1,000 miles under the earth's surface. The moon jogs along its orbit at a leisurely 0.64 mile a second, some 2,300 miles an hour. In revolving around the moving earth, the moon swerves back and forth from one side of the earth to the other in relation to the sun. Basic facts about the moon's orbit, size, and rotation are given in Table 4 for direct comparison with those of the earth and the other planets.

At its average distance of 238,866 miles from the earth, the moon takes 27 days, 7 hours, and 43 minutes on its terrestrial round and rotates on its axis in the same time. Thus the moon always keeps the same face toward the earth, as it moves around it. But the moon's farside, away from the earth, is not always dark and cold. The moon has a hemisphere of sunlight sweeping around it from east to west as does the earth. Since the moon passes from one conjunction to the next (new moon to new moon) in about 29.5 days, any point on its surface endures a 14-day day and a 14-day night, alternately frying and freezing. Since the moon is not inclined on its axis to the ecliptic nearly as much as the earth, certain points on high ground near its poles may be continuously lighted by the sun, a feature

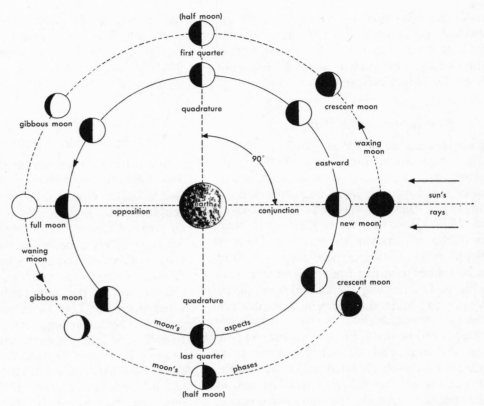

Phases and aspects of the moon. The phase is that portion of the sunlit hemisphere of the moon seen from the earth in a given configuration or aspect of the sun, earth, and moon.

which may prove useful in obtaining constant power from solar radiation.

Although the moon keeps the same face toward the earth, about 59 per cent of its surface can be seen over a period of time because of certain "librations," or variations in the whirling earth-moon system. Those additional sectors at the edge of the moon's face which are rarely seen have been painstakingly studied and mapped. Forty-one per cent of its surface is observable whenever lighted by the sun, and up to another 18 per cent can be seen on occasion because of the variations. But none of the remaining 41 per cent was ever viewed until the Russians photographed a part of it from a space vehicle. Launched on October 4, 1959, the "automatic interplanetary station" Lunik 3 telemetered

back to earth the first crude photographs of about 70 per cent of the moon's far side.

The moon and the planets, most notably Mercury and Venus, pass through the familiar phases. Their visible shapes, known as their apparent figures, change as increasing or decreasing proportions of their hemispheres lit by the sun are seen from the earth, depending on the relative positions of the sun, the body, and the earth. The dark portion of the moon with the terminator, or line dividing day from night, is not the shadow of the earth across the moon. Rather, the terminator is the edge of the sunlit hemisphere of the moon, and the dark portion beyond it is the part of the hemisphere turned toward the earth and unlit by the sun. This dark portion is sometimes lit very dimly by "earthshine," the sunlight reflected onto the moon from the earth and reflected back again.

Other Earth Satellites

In 1961 the Polish astronomer K. Kordylewski claimed that without the aid of a telescope he had seen and photographed some faint, cloudlike spots or blurs in the sky in the same orbit as the moon. His observations have not been confirmed, but there is a possibility that he may have caught sight of several earth satellites other than the moon, so tiny and scattered that they are very difficult to see.

Kordylewski reported that one such group of satellites was 60° ahead of the moon in its path and the other group 60° behind the moon, in positions that could result, astronomers believe, in quite stable orbits. The dim objects or luminosities, separated by a few degrees in the sky, might be clouds of dust, tiny meteoroids, or even fair-sized chunks of solid material; they could derive from either the earth or the moon, or have been captured in the moon's orbit from interplanetary material.

The positions (L_4 and L_5) in the moon's orbit of the earth in which these dust clouds or satellites were indicated are 60° ahead of and following the moon as seen from the earth. These points, as well as L_1, L_2, and L_3, are known as the Lagrangian libration (or equilibrium) positions, after the French mathematician and astronomer Joseph Louis Lagrange. He proved (1772) that bodies could be in stable orbits at these points within a system, like those of the earth and the moon or the sun and the planets, in which two bodies rotate about a common center of gravity. The so-called Trojan asteroids or planets were found in the L_4 and L_5 positions in Jupiter's orbit of the sun, forming 60° angles with Jupiter from the sun and confirming Lagrange's calculations.

The stability of a hypothetical body in position L_4, 60° ahead of the moon and traveling

with its velocity, has been calculated, and its perturbations would be relatively small. The gravitational fields of the sun, moon, and earth affect the body slightly, but the other planets do not disturb it appreciably over a period of several years. After 120 days it would have returned to within about 50 feet of its original position, and at the end of 180 days would have moved no more than about 200 feet from the L_4 point, with deviations in its movement from the earth-moon plane of less than 1 per cent. Another study has shown that the body would remain on a trajectory about the libration point for at least 700 days, but then the sun's gravitation and radiation pressure would be likely to detach it from the position, unless its movement were corrected. This may provide some clue as to why Kordylewski's claims have not been confirmed.

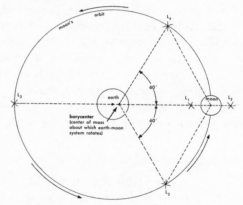

Lagrangian libration points in the earth-moon system. All in the plane of the moon's orbit, these points are indicated by the x's at L_1, L_2, and L_3 on the line passing through the centers of the earth and the moon, and L_4 and L_5 in the moon's orbit, preceding and following it by 60°. Both L_4 and L_5 form equilateral triangles with the earth and the moon at the other two apexes of the triangle.

It has been suggested that space observatories or scientific laboratories might be placed in the L_4 or L_5 position, where they could be maintained with only small adjustments in their orbits. The L_1 and L_2 points have been proposed as sites for communication-relay satellites above the moon. These might be necessary since the moon's small size makes its curvature so sharp that line-of-sight (straight) communication signals, even from some height on the moon's surface, would not carry very far on the surface, and there appears to be no lunar ionosphere as there is in the earth's atmosphere to bounce signals back to the surface. Calculations show, however, that the L_1, L_2, and L_3 points (called the collinear points, since all are on the same line intersecting the earth and the moon) would be quite unstable in comparison with the L_4 and L_5 positions.

These points are being investigated extensively from the earth; their study by unmanned spacecraft could be fitted into the over-all study of cislunar space and would divulge valuable scientific information.

One of the minor planets, Toro, has been called an earth satellite because Toro's orbit of the sun crosses that of the earth and is influenced by it. Toro is not another moon, however, because it swings in its orbit far in close to Venus and far out beyond the path of Mars. Toro's orbit is being analyzed, however, because it may be that our moon was in such an orbit before it was captured by the earth.

Size and Shape of the Planets

The earth is by all odds the largest of the huge, nearly spherical bodies nearest the sun. The earth outranks all of the other inner planets in diameter, volume, sheer mass or amount of substance, and the density of its composition. However, compared with the outer giants, it is small, even though it does have the highest average density of all the planets, great and small.

Smallest among the inner group of planets and satellites are Deimos and Phobos, Mars's two streaking satellites, with insignificant diameters of perhaps 8 to 16 miles, respectively. On the basis of searches made, Venus, within the orbit of the earth, cannot have a distant satellite larger than about 3 miles in diameter,

or a very close satellite larger than perhaps 7.5 miles.

The peanut-sized satellites of Mars race around it only 3,700 and 12,500 miles away from the surface, closer than the orbits of some of the earth's artificial satellites. Phobos revolves around Mars in less than a third of Mars's own rotation period; from the planet,

Satellites of Mars—Deimos (left) and Phobos (right), photographed with the 24-inch refractor of the Lowell Observatory. The image of the planet itself had to be overexposed to bring in its satellites, the visual magnitude of which is about +13.
—LOWELL OBSERVATORY, FLAGSTAFF, ARIZ.

it would appear to rise in the west and swing across the sky to set in the east three times a day. The period of revolution of Deimos is also relatively fast, taking only about 30 hours compared to Mars's 24.5 hours, so that it rises in the planet's east and sets in its west, as does our moon. Appearing to move very slowly westward as seen from Mars's surface, Deimos goes through the entire cycle of phases twice between each rising and setting, in a period of nearly 3 earth days.

Views of Mars rolling huge beneath one of its satellites have been a favorite subject for space artists, who sketch in imaginary details of Mars's surface. Long before these actual Martian vistas have been seen, breathtaking photographs and television views of the varied surface of Mars and of the whole earth have been taken from space.

Mars' satellite Phobos in a computer-enhanced image from Mariner 9 spacecraft in its 34th orbit of the planet. An asteroid may have chipped off the upper right corner of rocky Phobos and others made the craters.
—JET PROPULSION LABORATORY, CALIFORNIA INSTITUTE OF TECHNOLOGY

The next largest body in the inner group is the moon, 2,160 miles across, roughly two-thirds of the breadth of the United States; it would fit easily within the North Atlantic Ocean. Then comes Mercury, with a volume only 6 per cent that of the earth and a diameter of roughly 3,030 miles. At 4,220 miles in diameter, Mars is less than halfway between the moon and the earth in size, and it has only 15 per cent of the earth's volume. Venus is nearly the size of the earth (it once was believed to be larger), with an estimated diameter of 7,550 miles and roughly 86 per cent of the volume of the earth.

Because they are nearer the sun than the earth, Mercury and Venus can occasionally be viewed as small, dark, round spots gliding from east to west and making solar transits across the brilliant face of the sun's disk. These solar transits, when the orbits of the planets bring them directly between the earth and the sun, have provided a chance to correct the calculated orbits and diameters of Venus and

Mercury and to study their atmospheres, or, in Mercury's case, apparent lack of an atmosphere.

The diameters of Mercury and Venus are still uncertain, but very significant, because the estimated volumes depend on them, and densities in turn depend on volumes. Mercury's most recent transits have resulted in the figure of 3,030 miles for its diameter, plus or minus a few per cent, and this probably is fairly reliable. Mercury's latest solar transit, shown in the photograph, came on May 9, 1970. This gives an idea of how small the planets are in relation to the great disk of the sun.

In 1959 the planet Venus occulted the first-magnitude star Regulus; Venus passed directly in front of the star as viewed from the earth. Here was a chance to check on the diameter of Venus under the dense layer of clouds enshrouding it. When such a bright star moves behind a planet with an atmosphere, its light does not snap out, but takes some seconds to disappear, still visible but fading as its light passes through the increasingly thicker atmosphere near the planet's surface.

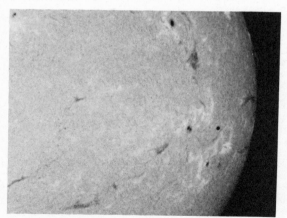

Planet Mercury is the perfectly round black spot and the irregular patches are sunspots in this hydrogen-alpha photograph of Mercury's transit of the sun on May 9, 1970.
—GODDARD SPACE FLIGHT CENTER, NATIONAL AERONAUTICS AND SPACE ADMINISTRATION

The measurements of this occultation reported by two dozen observatories were analyzed. With a deduction of 114 miles for the thickness of Venus' atmosphere derived from temperatures (a roughly estimated element entering the determination), it was concluded that Venus' diameter was about 7,550 miles, knocked down from 7,700 miles. With such careful, multiple observations, the limits of accuracy for micrometer measurements of Venus' diameter have just about been reached. An uncertainty of plus or minus 7.5 miles was set on this measurement.

Venus does not treat us to the spectacle of a solar transit as often as Mercury does. Observations of Venus' last transit on December 6, 1882, were used to revise its calculated orbit. The planet's next two solar transits will come in a pair on June 8, 2004 and June 6, 2012. They will not be awaited as anxiously as they were in 1874 and 1882, for long before then the descendents of the Mariner spacecraft, now named Voyagers, will have ventured often enough and close enough to Venus to supply far more exact reports than mere transits or occultations can offer. Soviet space capsules, parachuting down through Venus' thick atmosphere, have already taught us much more about Venus.

When Mars was closest to the earth in its oppositions of 1954, 1956, and 1958, the French astronomer Audouin Dollfus and his colleagues used a double-image micrometer to check its diameter. Working with the 24-inch refractor telescope at Pic-du-Midi Observatory and the micrometer, they figured out that the Martian equatorial diameter was probably 4,219 miles and the polar diameter 4,169 miles, allowing 19 miles for the apparent increase in radius due to the Martian atmosphere. The older figure for the equatorial diameter had been 4,200 miles. The new measurements gave an average density (water equals 1) of 4.09 for Mars, compared with the earth's 5.52. Though they used five different color filters, ranging from red to blue, in all cases the mea-

sured diameter was the same. Nothing was changed markedly by all this work, but this correction was basic to a great many determinations about Mars.

Amazingly enough, with the Mariner spacecraft flybys of Mars in 1965 and 1969, the best estimate of Mars' equatorial diameter was 4,218 miles, within 1 mile of that of Dollfus. The final returns from Mariner 9, which went into orbit of Mars in late 1971 to become the first artificial satellite of a planet other than the earth, should change this figure very little.

Seen in the telescope, the disks of the larger planets actually seem to be slightly flattened at the poles and bulging at the equators, as is Mars. The motion of the axial rotation of the planets is faster at their equators, and the centrifugal forces are stronger there, so they naturally tend to slump in this direction. Flattening, or the index of oblateness, is arrived at by subtracting the polar diameter from the equatorial diameter and dividing the result by the equatorial diameter. The index is indicative of the inner constitution and structure of the planets (Table 4).

The oblateness indices of Mercury and Venus have not been figured out, since their diameters cannnot be measured accurately enough. The radius of the moon pointing toward the earth is estimated to be about 0.7 mile longer than its polar radius, but the radius on the far side of the moon and its resultant shape are not yet accurately known. Probably the moon is very nearly a perfect sphere, with equatorial diameter no more than a mile or two greater than its polar diameter.

The gas giants, Jupiter, Saturn, and Uranus, show particularly high oblateness; that is why they look a bit mashed down in a telescope. Although the edge, or limb, of Jupiter's disk is not too sharp in the photograph of the planet on page 103, you can note Jupiter's oblateness if you measure its equatorial diameter (across) and compare this with its polar diameter (up and down).

The density of an object, that is, its mass per unit volume, tells a great deal about its nature. The moon's density (Table 4), 0.6 relative to an earth density taken as 1.0, is by all odds the lowest of any body in the inner solar system, although Mars comes close with its relative density of 0.76. Scientists were long puzzled over what materials could compose the moon, so light in comparison with the earth. Since lunar rocks have been brought back to earth by Apollo astronauts, study has revealed that the rocks on or near the surface are mostly of lighter kinds, such as basalts or anorthosites. Also, the moon may entirely lack or have only a very small nickel-iron or iron core, such as the earth is supposed to have.

Planetary Rotation

Since the moon's large surface features can be seen (sharp and crisp under good viewing conditions), as can the surface of Mars (though quite indistinctly), the tilt of their equators to their solar orbits, or of their polar axes to these orbits, and their periods of rotation are known (Table 4).

As if some giant playing with tops had given them exactly the same kind of spin, the inclination of the axes of the earth and Mars coincide within 2°. Early Mariner 9 reports from Mars showed that its axis inclination determined from earth had been accurate within half a degree. Consequently, Mars and the earth have many other features in common. The path of the sun moves north and south of their equators annually, the polar regions of both are frigid and never entirely thaw, and their seasons change in similar cycles. The Martian days and nights are very like ours—the planet takes only 37 minutes longer to complete a full rotation. In certain respects, men would feel right at home on Mars; in others, the red planet would be a completely alien world. If they were unable to create a terrestrial environment around them, with a plastic dome, or living quarters under the surface,

human beings would be promptly asphyxiated in the Martian air, which consists largely of carbon dioxide, and pass into a deep freeze at night, even in Mars' most tropical areas.

The periods of rotations of Mercury and Venus and the inclinations of their equators to their orbital planes have long been the subject of controversy, and only their rotation periods are now well known.

For many years, astronomers thought that Mercury rotated on its axis once in 24 hours. They observed dim features apparently coming into nearly the same position where they had seen them the day before. Later observers believed that Mercury turned only once around on its axis in 88 days, with each revolution of the sun. With such rotation, like that of the moon in its orbit of the earth, Mercury would always keep the same face toward the sun, with one half blowtorch hot and the other half always dark and cold.

Finally, it was suggested by radar data, from radio signals bounced back to earth from Mercury, that its period of rotation is 59 days, just two-thirds of its swift revolution of the sun. This was then proved by careful scrutiny of previous photographs of Mercury's surface features. But Mercury still poses so many questions that a proposed flyby of a space probe would be a major scientific event.

Dozens of early observers concluded that Venus' period of rotation was about 24 hours, like that of the earth. Then late in the nineteenth century, the Italian observer, Giovanni V. Schiaparelli, of Martian "canali" fame, attributed to Venus a synchronous or captured rotation of about 225 days, coinciding with its year; others set the rate at 30 days, and some estimates ranged down from this to 24 hours.

During the 1960's, however, radio-echo or radar analysis finally came into its own. Venus' rotation proved to be retrograde, or backward, from east to west, the only planet with such a spin. And its rotation had a period of about 245 days, longer by far than the planet's year!

CHAPTER V

*". . . This Earth, a spot, a graine,
An Atom, with the Firmament compar'd . . ."*

MILTON

The age of the Milky Way Galaxy, in which the solar system and our earth were born and nurtured, has recently been set at from 10 to 15 billion years on the basis of two entirely separate lines of evidence, one galactic, the other subterrestrial.

Tens of thousands of light-years away, near the center of the Galaxy, a large number of old dwarf stars have been discovered that are deficient in metals as compared with the sun. Since the heavier elements such as the metals are thought to have appeared late in stellar history, produced by star explosions (novas and supernovas), this implies that the dwarfs were formed at an earlier period than the sun. From this it has been deduced that the Galaxy itself must be at least 10 billion years old.

The other line of evidence derives from the extent of the radioactive decay of rhenium 187 to osmium 187 as measured in samples from ores found deep within the crust of the earth itself. According to this slow-ticking clock, the heavy chemical elements of the earth may have formed from 5 to 10 billion years before the creation of the solar system, consistent with a galaxy from 10 to 15 billion years old.

The texture of the earth's surface has been compared with that of a dried prune or wrinkled apple, full of spots, pits, and fissures, and obviously very old, with a long and rugged history. Fossil organisms discovered in a flinty-iron deposit near Lake Superior, Ontario, probably are 2 billion years old, and it is surmised that life may have existed on earth for 4 billion years. On the basis of radioactive decay, the earth itself has been roughly estimated to be between 4.5 and 5.0 billion years of age, young

PLANET EARTH

in relation to its parent galaxy, old in relation to the fleeting 2-million-year period during which man has evolved.

In the decade of the 1970's, we are really beginning to see the earth as a planet. A great array of new vehicles and instruments—high-altitude balloons, rockets sounding the atmosphere, artificial satellites, and unmanned and manned spacecraft—have made this possible. We have seen the nearside and the farside of the moon. We have looked down on the wonders of Mars. And most important, we have viewed the whole great earth from space.

At the same time, vessels and instruments have been designed that can withstand the gigantic pressures of the ocean depths and probe their secrets. With ultrasensitive recording devices, more and more has been discovered about what lies far beneath our planet's thin crust. The broad outlines of knowledge of the planet earth from its very center to the outer limits of its atmosphere have been blocked out over a couple of centuries. But such is the acceleration of scientific advance that in the last decade a great proportion of these facts have had to be revised.

All this knowledge of our own planet forms the backdrop against which the other planets will be investigated. In turn, whatever new information is obtained by space exploration about the moon and the other planets, under the extremely different conditions of each, will reshape our explanations of the earth itself, its origin, and its history.

The Great Globe Itself

The size and shape of the planet earth, the exact location of places on it, and the variations in terrestrial gravitation and magnetism are the province of geodesy. The earth's basic shape or figure has long been known to be spherical. With more accurate measurements, a little flattening at the poles was found: the polar diameter was calculated as about 26.5 miles shorter than the equatorial diameter. This flattening or oblateness comes from a "slumping" toward the equator caused by the rotation of the earth on its axis. It is very small, 0.34 per cent or 1/297 of the 7,927 miles of the earth's equatorial diameter. The earth's figure is like that of an ellipse rotated on its minor axis, or diameter of least length, to form an ellipsoid. The greater diameter of the earth at the equator is called the equatorial bulge. The differential pulling and hauling of the gravitation of the sun and moon on this bulge produces polar precession, the slight wobbling of the earth as it spins on its axis.

The variations of the earth's surface are familiar: the flatness of its plains, plateaus, and oceans, the roughness of its mountains, valleys, and deep-cut gorges. But these variations in elevation above or below sea level are not extreme, viewed in relation to the full sweep of the earth's surface. They are probably smaller in proportion to the earth's diameter than the variations on Mars' surface, and still smaller than those standing out so clearly on the surface of the moon. The moon, however, has turned out to be smoother than its stark shadows make it look. And Venus, judging by the radar echoes, has a generally smoother surface than the moon, although the same echoes have shown great mountains or mountain ranges on Venus, too. Apparent gradations on Mercury's surface have been glimpsed with telescopes, and recent radar contacts with the planet have indicated that it may have approximately the same surface characteristics as the moon, if not somewhat rougher.

Mercury, Venus, Mars, and the moon then, are all rocky bodies like the earth, battered by impacts from space and smoothed more or less by the various forces acting on their surfaces. Hidden under their thick atmospheres, the surface contours of the giant planets have yet to be determined. Both the Russians and

Americans have reported the bouncing of radar signals from Jupiter, but little interpretation has been possible.

To map the continually varying surface of the earth accurately, obtaining the exact locations, elevations, and distances of points all over the globe, the basic shape of this surface must be known precisely. To provide a uniform and standard international reference system, geodicists have adopted the so-called international spheroid. This spheroid, a theoretical construct, is a perfectly regular ellipsoid, the minor axis of which is parallel to the axis of rotation of the earth, with values assigned to its major axis and its oblateness or eccentricity that fairly closely approximate those of the earth itself.

To make the center and the axes of this standard ellipsoid coincide with the center and the axes of the earth, another theoretical construct, the geoid, has been developed. The geoid is the surface coinciding with the mean sea level in oceans, everywhere normal to the gravitational field. While the geoid is warped and twisted by varying gravitational forces at different places on the earth and affected by the heavy land masses of the continents, it holds fairly closely to the basic reference spheroid, not varying from it normally by more than a few hundred feet.

Artificial satellites in orbit around the earth already have extended geodetic knowledge. The gravitational field of the earth's equatorial bulge, over which the satellite Vanguard 1 flies twice in each orbit, pulls the satellite observably out of line each time it crosses the equator, causing the plane of its orbit to move slowly clockwise along the equator at a rate which depends on the size of the bulge. Measurements of this orbital perturbation indicate that the flattening ratio of the earth's geoid is $1/298.2 \pm 0.2$, rather than the adopted value of 1/297.00. The earth is slightly less flattened than had been thought. It has been calculated that the flattening ratio would be 1/299.8 if the earth were relatively fluid in its interior,

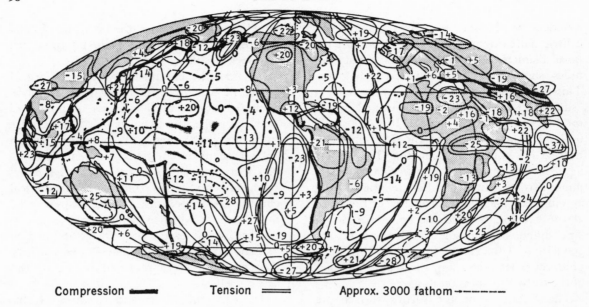

Compression ▬▬▬   Tension ═══   Approx. 3000 fathom ▬ ▬ ▬

The gravity field of the earth as measured by its effects on the orbits of artificial earth satellites. Numbers on contour lines are (+) high gravity and (−) low gravity in milligals, a gal being a small gravitational unit. Heavy lines indicate compression of the crust, double lines indicate crustal tension, and dashed lines show ocean basins at the 3000-foot depth.

—WILLIAM M. KAULA, UNIVERSITY OF CALIFORNIA, LOS ANGELES

because of the high temperatures and pressures deep down inside it. So the observations of Vanguard 1, and many later satellites, far out in orbit, have shown that the interior of the earth must have less plasticity and more mechanical strength than had been predicted.

On the basis of many thousands of satellite tracking photographs by powerful telescopic cameras situated around the world, revealing the swervings of the satellites in their orbits, a map of the earth has been made showing very precise computer-derived contours of the geoid, or geopotential surface, of the earth. The gravitational bulges and indentations around the earth make it appear quite misshapen.

The bumps and hollows stand for areas in which the density of the earth is large or small, resulting in a greater or smaller pull of gravity toward the center of the earth. These areas were later found to be related to the action of the plates of crust on which the continents

ride. Along the edges of these plates a rift system was found, running along the deep bottoms of the earth's oceans. Along the tensional rises in the rift system, where new crust is rising from below and adding to the edges of the plates, the gravitation is high. Along the compressional trenches, where old crust of the plates is sinking deeper down into the earth, the gravitation is low. So discoveries of rises and trenches along the ocean bottoms were related to variations in the earth gravitation revealed by the swerving of satellites orbiting the earth.

The Anna series of "winking star" satellites carried xenon-flash tubes somewhat like the electronic-flash lamps used by photographers. As these flashed on command from earth, they provided nearby reference points for triangulation, permitting places on earth on different continents to be located for the first time with an accuracy of a few feet.

Views of the Earth

We all know now what our own planet looks like from space. But the earliest such photographs showed only a part of the great curve of the earth, beneath which swept views of large areas of it never before seen entire.

The first photographs from great heights were made by cameras carried up in high-altitude balloons to heights of 15 to 20 miles and later lofted in sounding rockets as high as 300

Sequence of 5 pinpoint light flashes (circled) from the geodetic satellite Anna, photographed by a telescopic satellite-tracking camera at Aberdeen, Maryland, against the background of the stars in the region of the Big Dipper. (The star arcs are caused by the earth's rotation during the exposure.)
—Coast and Geodetic Survey,
U.S. Department of Commerce

miles. Made in black and white by V-2 and U.S. Navy Aerobee rockets, they revealed how details begin to merge at such heights, how the view changes with the season, and the reflec-

tivity of various surface materials. The actual colors seen from such heights would be pastels, ranging from red-brown to blue-green.

The astronauts have told us what the earth looks like from 100 to 160 miles up and what kinds of details can be distinguished. As John Glenn passed above the nightside of the earth (February 20, 1962), he commented that "before the moon comes up, looking down is just like looking into the Black Hole of Calcutta." He could see the horizon silhouetted against the light of the stars, but the earth below was dark. After the full moon came up, the clouds were visible below, so that the earth was whiter than the dark background of space. The citizens of Perth, on the west coast of Australia, turned on their lights when Glenn was scheduled to be over their city. He observed the lights as like the dim glow of a small town seen at night from an aircraft. "Knowing where Perth was," Glenn reported, "I traced a very slight demarcation between the land and the sea, but that's the only time I observed a coastline on the nightside."

The earth was clearly visible to Glenn as he passed over the dayside. "A lot of things you can identify, just as from a high-flying airplane," Glenn commented. "You see by color variations the deep green woods and the lighter green fields and the cloud area. I could see Cape Canaveral clearly, and I took a picture which shows the whole Florida Peninsula; you see across the interior of the Gulf." Glenn could identify such cities as New Orleans, Charleston, Savannah, and El Paso, Texas. The different types of clouds below his craft and the cloud fronts and weather patterns could be clearly seen. He could pick out the larger irrigated areas by the patterns around El Paso and El Centro, Texas, and watched dust storms swirl in western Africa. He identified the Gulf Stream in the Atlantic Ocean by its difference in color from surrounding waters. Photographs of the earth taken by Glenn and other American astronauts almost transfer us to their space capsule.

Smaller objects than these were not actually seen, Glenn believed, but identified by their surroundings. "For instance, you see the outline of a valley where there are farms, and the pattern of the valley and its rivers and perhaps a town. You can see something that crosses a river, and you just assume that it's a bridge. As far as being able to look down and see it and say 'That is a bridge,' I think you are only assuming that it's a bridge more than really observing it."

Before the advent of earth-orbiting artificial satellites from which photographs could be returned to earth, the appearance of earth from space had to be determined indirectly. Ancient observers noticed that occasionally the dark portion of the moon (not at the moment illuminated by the sun) could be dimly seen, particularly at the new-moon and crescent-moon phases. The dark areas seemed to glow, so they thought the moon was phosphorescent like the sea, or perhaps translucent, allowing a little light to filter through it. Kepler finally gave the correct explanation for this moon glow: It was light from the sun, he said, reflected on the moon by the earth, and then back to earth. Analysis of this doubly reflected light, called earthlight, earthshine, or ashen light, indicated how the earth must look from a distance.

Astronomers found that the earthlight was predominantly bluish light, rather than bluish-green or olive-green as had been believed. They correctly inferred that the earth would look mainly blue or bluish-white from a great distance, as opposed to Mars' predominantly reddish light and Venus' yellowish-white light. The photometer, by which light from unknown sources is compared with light from standard sources, showed the earthlight to consist largely of the blue wavelengths. Its color appears to vary, both at random and with the seasons, becoming bluer as its intensity increases.

Now we know that, as astronomers predicted, the bluish earth appears more bluish-white or gray when its cloud cover is heavy. Where the clouds are scattered, the dim outlines of bluish

to blue-black oceans, and brownish to red-brown continents and islands, are traceable. The closer an observer is to the earth, the more variation is noted in its colors, like that seen by John Glenn, because cloud cover and seasonal effects are then more noticeable.

The reflectivity of the earth, called its albedo, has also been studied by means of the earthlight. The albedo is the proportion of light a body reflects of the total light striking it. The earth's mean albedo is about 0.36 on a scale in which a perfect reflector of light would be 1.0 and a perfect absorber of light (black body) would be 0.0. The mean-visual albedos of the planets are compared with that of the earth in Table 5, along with their solar constants, the amount of radiation per unit area arriving at the planets from the sun. Substantial confirmation of the figure of 0.36 for the earth's mean albedo has been found in direct measurements of the average proportion of light from the sun reflected by the earth's surface and its atmosphere together, with instruments on earth-orbiting satellites. The earthlight has been correctly interpreted.

The magnitudes of the earth, moon, and the other planets are also compared in Table 5. The type of magnitude described in Chapter 3 is the measure of the apparent visual brightness of celestial bodies viewed from the earth. The magnitudes given in the table are the unit magnitude, the planets' brightness at the same unit distance from the earth and the sun, and the mean opposition magnitude, their brightness when the earth stands on the straight line extended from the sun to the other planet.

Weather and environmental earth satellites now carry a host of different instruments, in addition to their cameras. They send back information and photographs of the weather, ocean currents, and many features of the earth's land masses that are used in many ways by scientists. The study of cloud formations has helped weathermen to locate large cyclonic storms and hurricanes and to make their weather predictions more accurate, believe it or not.

The Himalayas in the India-Nepal-Tibet border area photographed by astronaut Gordon Cooper from a height of 100 miles.
—NATIONAL AERONAUTICS AND SPACE ADMINISTRATION

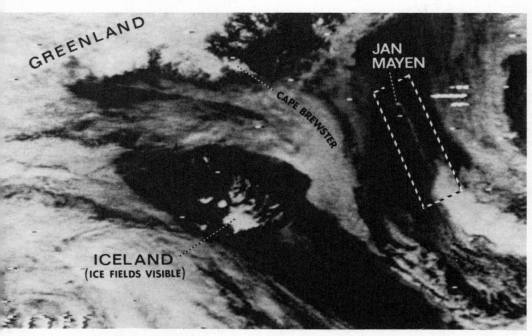

The Nimbus IV experimental weather satellite was 700 miles high when it transmitted this photograph of Iceland and Greenland to Greenbelt, Maryland. A plume of ash (box) trails from Beerenberg Volcano on Jan Mayen Island. No such active volcanoes have been spotted on Mars from nearly the same height by Mariner 9.
—GODDARD SPACE FLIGHT CENTER, NATIONAL AERONAUTICS AND SPACE ADMINISTRATION

The continents as they had started to drift apart about 135 million years ago as the ridge system opened and the oceans expanded. Arrows indicate direction of drift and arrow length relative speed of drift. Note India drifting rapidly toward Asia and Australia breaking from Antarctica.
—R. S. DIETZ AND J. C. HOLDEN, *Journal of Geophysical Research*

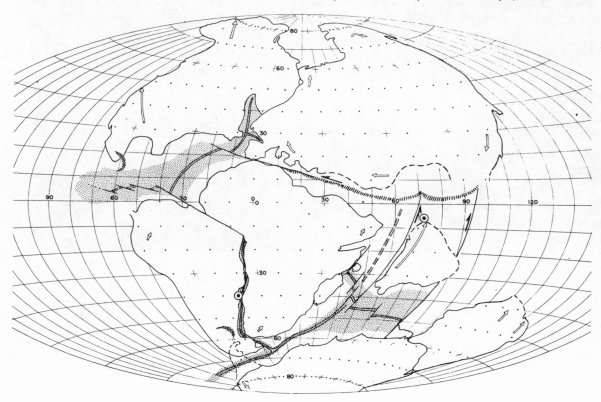

Earth's Surface

On globes of the earth, the possibility that South America may have fitted snugly against Africa is at once apparent. That these two continents had once been united with the others in one supercontinent, called Pangaea, was proved during the 1960's.

The ridge or rift system running through the oceans of the earth revealed the prior existence of Pangaea, its breakup, and the drifting apart of the continents, gradually spreading around the earth. First, it was discovered from the magnetization of rocks that the magnetic field of the earth had gone through many reversals in the past over millions of years, with the north pole becoming the south pole, and vice versa. Then the magnetic reversals turned up in long, narrow bands running along both sides of the newly mapped ridges in the ocean bottoms. Also, volcanoes and earthquakes were most frequently located along the lines of the ridge system.

These discoveries implied that melted rock was pouring up along these cracks in the earth's crust, pushing apart the plates of crust along the ridges, and forming wider and wider oceans, like the North and South Atlantic. This spreading of the oceans made the continents on the plates drift apart. This theory was confirmed when deep-sea drilling into the bottom showed that the rocks grew older and older the farther they were from the ridge.

At other places, the ocean bottom revealed deep trenches. Along these trenches and along certain volcanic island arcs like those in the South Pacific and Japan, it is believed that the earth's crust is being gobbled up. Here the crust is plunging down into the earth's interior to make up for the crust being added along ridges elsewhere. So the continents and oceans of the earth have not been permanent features. Instead they have changed with these events going on in the crust and under it, moving continents one or two inches a year.

Earth's Interior

The mass, or quantity of matter, of the earth totals the almost incomprehensible figure of about 6×10^{21} tons. This has been determined in the laboratory by comparing the gravitational attraction of a large sphere for a tiny sphere with the gravitational attraction of the earth for one of these spheres.

The mean or average density of the earth (its mass per unit volume) has been calculated to be 5.52 grams per cubic centimeter. A cubic inch of average earth stuff equals about 0.2 or $\frac{1}{5}$ of a pound. By comparison, a cubic inch of average moon stuff weighs about $\frac{1}{8}$ of a pound (0.12). The same amount of Martian material weighs about $\frac{1}{6}$ (0.15) of a pound and that of Saturn only $\frac{1}{40}$ (0.025) of a pound. The comparative volumes, masses, and densities of the planets are given in Table 4.

The mean density of the earth seems very great compared with familiar substances. It is about 5.5 times that of water, 2 times that of aluminum, $\frac{2}{3}$ that of iron, $\frac{1}{2}$ that of lead, and $\frac{1}{3}$ that of gold. It is almost twice as great as the mean density of the rocks underlying the continents and about 70 per cent greater than that of the heaviest of the common igneous rocks at the surface. What can lie beneath the surface of our globe to make its mean density so high?

Something about the nature of the earth beneath its crust has been inferred from the action of seismic waves as they pass under the surface and through the earth. The back-and-forth oscillation of some earthquake waves in long periods of from 1 to 6 minutes has been interpreted as indicating that the whole earth rings, somewhat like a struck bell, with heavy earthquakes.

Some evidence about the earth's interior comes, too, from observations of the strength and variation of the earth's magnetic field, variation in the gravitational field at the surface, and heat-flow measurements. The study

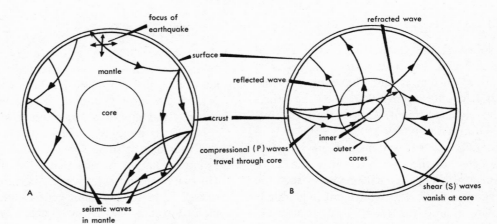

(A) Representative paths of seismic waves through the mantle of the earth; (B) waves reflected or refracted by the earth's core or absorbed in it. The compressional waves are longitudinal and the shear waves are transverse; they react differently at the earth's core, as shown in (B).

of both natural earthquake waves and vibrations caused by man-made explosions, including nuclear detonations, indicate that the earth is not completely uniform inside, but consists of a series of concentric spherical shells or layers.

The crinkled and jumbled outer layer of the earth, appropriately called the crust, runs down to an average depth of 10 to 15 miles, ranging between 3 and 30 miles. The mantle of the earth, the next layer beneath the crust, is roughly 1,800 miles thick. Some theories distinguish two shells in the mantle: the sima, extending from 600 to 725 miles below the crust, below which the focuses of earthquakes do not occur, and consisting of a *silicon-mag-nesium* material, and the transition layer from the sima to the core, which may consist of silicates interspersed with iron. The sima is more often called the athenosphere, a name for the rather plastic shell in which convection may be bringing melted rock to add to the crust at the ridges and into which the crust may be swallowed at trenches.

Beneath the mantle is the core, that part of the earth running from about 1,800 miles down to the center, a huge sphere with a diameter of about 4,320 miles, a little larger than the planet Mars and almost exactly twice the size of the moon. The outer core probably is in a liquid or plastic state and the inner core solid. The great density of the earth is explained by assuming that its core consists largely of iron or nickel-iron, with traces of other elements.

The crust of the earth, sometimes called the lithosphere to distinguish it from the hydrosphere (oceans) and the atmosphere, averages about 15 miles in thickness, though continental crusts are thought to be much thicker (up to 30 miles) than the crust under the oceans. A skim of sedimentary rocks such as sandstone, and light silicic metamorphic rocks like granite, composes the upper surface of the crust, down to a depth of perhaps 4 to 6 miles. Beneath this, the material may be largely basalt, a dark-colored, fine-grained volcanic rock, or a heavier, dark "mafic" rock containing magnesium and iron. This material runs down for 5 to 10 or more miles, depending on the thickness of the crust, and the temperature increases with depth, perhaps up to 2000° F. at 25 miles down. In many places under the continents what is called the Conrad discontinuity defi-

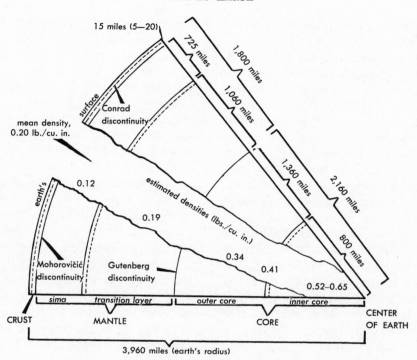

15 miles (5—20)

725 miles

1,800 miles

1,060 miles

1,360 miles

2,160 miles

800 miles

surface

Conrad discontinuity

mean density, 0.20 lb./cu. in.

earth's

0.12

estimated densities (lbs./cu. in.)

0.19

0.34

0.41

0.52—0.65

Mohorovičić discontinuity

Gutenberg discontinuity

sima   transition layer   outer core   inner core

CRUST   MANTLE   CORE   CENTER OF EARTH

3,960 miles (earth's radius)

Cross section of a segment of the earth's interior, with the approximate depths and estimated densities of the distinguishable layers or shells and the location of the inferred discontinuities.

nitely separates the crust into these two levels, while in other places they may not be as distinct.

Between the earth's crust and the mantle lies the Mohorovičić discontinuity, named for the Yugoslavian seismologist, Andrija Mohorovičić, who inferred its presence in 1909 from observations of an earthquake's seismic waves. This discontinuity is a boundary below which earthquake and other seismic waves begin to behave very differently than in the crust, indicating a change in the materials composing the interior, or perhaps a change in crystal structure or phase. (Water changes phase, for example, when it becomes ice or vapor under different temperature-pressure conditions.) The seismic waves accelerate from an average rate of 4.3 miles a second above this boundary to about 4.9 miles a second below it.

A similar discontinuity farther down, between the mantle and the core of the earth,

is called the Wiechert-Gutenberg discontinuity. The compressional or primary waves travel more slowly below this point, while the shear or secondary waves vanish at this discontinuity. Compressional waves oscillate or compress back and forth in the direction of the wave motion, while the shear waves oscillate or vibrate at right angles (transversely) to the direction of the wave motion and are sometimes called the shaking waves. This probably indicates the beginning of some different phase of the material, either a liquid or plastic core. The position of the Gutenberg discontinuity has been measured very accurately at 1,800 miles beneath the surface. The mantle therefore extends slightly less than halfway to the center of the earth, but it contains roughly ⅚ of the earth's total volume.

With the crust, the upper mantle is the site of earthquakes and is thought to be in a state of slow-motion turbulence. The mantle may consist primarily of olivine, a magnesium-iron silicate, or it may be peridotite, classed as an igneous rock. Its density and the velocity at which sound waves travel through it are similar to those of the mantle. But to determine the exact composition of the mantle is, because of its remoteness, like trying to determine the nature of the moon's surface. In fact, less is known about the mantle than about the surface of the moon, since the latter can at least be seen. What is really needed in both instances is to have sample chunks to examine in a laboratory.

Scientists believe small samples of the lower crust or upper mantle have been found near or on the surface. In Italy, for example, the thrusting in the forming of Southern Alps may have shaved up a chip of the mantle, which curled up to the surface. Here the basalt of the lower crust exposed over a distance of 3 to 6 miles changes suddenly with a discontinuity to the peridotite that may well be a chip from the mantle.

Recent analysis of seismic-wave travel and other data implies that the Mohorovičić discontinuity may not be nearly as significant as a plastic-behaving layer found in the earth's mantle at a depth of from 35 to 155 miles below the surface. Seismic waves appear to travel more slowly through this layer of the upper mantle, which may reflect the effects of higher temperature, with rocks near the melting point and thus undergoing some kind of phase change.

The older view of the earth's interior was that it consisted of an iron core beneath a basaltic mantle; the core was believed to be liquefied, containing small amounts of nickel or cobalt mixed with the iron (the three most abundant heavy metals in the stars), or possibly silicon. This theory, it was believed, best explained the earth's high mean density and the formation of its magnetic field. In addition, it related the earth's interior to the composition of the iron type of meteorites, which consist largely of such nickel-iron and may be remnants of a small disintegrated primitive planet or planets.

The liquid core of the earth may be moving or swirling slightly, for the north magnetic pole is shifting northward in a fairly regular manner. True, it moves about as fast as cold molasses pours, at a rate of only 5 miles a year. In 1950 the magnetic pole was at the northern end of Prince of Wales Island in the Canadian Arctic Islands. When checked in 1962, it had moved north of Prince of Wales Island to a point near the southwest corner of Bathurst Island in the Queen Elizabeth group. The south magnetic pole also moves, faster if anything than the northern one.

According to one view, all the materials of the earth began to condense together in a cold state from the original gassy nebula from which the solar system was formed about 4.5 billion years ago. After the earth had taken shape, its radioactive elements heated it to the melting point of iron. This liquid then settled slowly by gravity toward the center to form the core.

The first earth materials to condense, in another theory, were oxides and silicates with additional radioactive elements. Then the iron would condense around this nucleus as the temperature decreased, followed by the mantle materials. The radioactive nucleus would melt itself and the surrounding iron, which would settle within to form the inner core. The lighter radioactive nucleus would rise to make the present liquid outer core. The heat from this outer core would be a source for motions in the mantle and continental drift. But more must be learned before final theories are reached.

Interior of Other Planets

The observations, models, and theories of the interior of our own planet have been applied by analogy to the other terrestrial planets—Mercury, Venus, and Mars. These are presumed to have compositions and structures somewhat similar to that of the earth, since their mean densities (running from 0.15 to

0.19+ pound per cubic inch) are close to that of the earth (0.20 pound per cubic inch or 346 pounds per cubic foot), and their sizes do not vary tremendously from that of the earth.

The density of Pluto is not known definitely. Jupiter, Saturn, Uranus, and Neptune, the Jovian planets, or gas giants, have very low mean densities, running from 0.03 to 0.09 pound per cubic inch. A density of 0.10 pound per cubic inch (173 pounds per cubic foot) is just about the point which divides the terrestrial from the Jovian planets. This places the moon, with a density of 0.12 pound per cubic inch, almost on the borderline, though slightly closer to the rocky bodies like the earth. The plethora of varied theories about the moon's inner composition reflects this borderline density position.

On the basis of density, the heaviest satellites of Jupiter, Io (Jupiter I) and Europa (Jupiter II), can also be grouped with the terrestrial planets. On the other hand, the inner satellites of Saturn, Mimas (Saturn I), Enceladus (Saturn II), and Tethys (Saturn III), are light enough to fall into the Jovian planet class, while Saturn's large satellite, Titan (Saturn VI), which probably has an atmosphere, may fall, like the moon, on either side of this classification, as may a number of other satellites.

Recent work on models of the interior composition of Jupiter and Saturn indicates that they may be composed largely of highly compressed hydrogen, with some other elements mixed in, particularly helium. Another model suggests that while Jupiter and Saturn may have cores of heavier materials, like rock and iron, that are relatively small in proportion to the planets' size, their mantles consist of fluid or solidified hydrogen and helium, largely hydrogen, or several varieties of water-ice and other ices. Now preliminary studies of Jupiter's radio emissions, however, indicate the possibility of sources of heat under a solid surface. The outward heat flux from Jupiter may be

60 per cent greater than the solar radiation it is receiving, while that from Saturn is about twice as great as the incoming solar radiation (Table 5). As a result, the models of Jupiter's and Saturn's interiors are under revision.

Uranus and Neptune are slightly more dense than Jupiter and Saturn, too dense to be simple solid balls of hydrogen. It has been suggested that they may be composed mainly of helium with admixtures of ice and solid methane (CH_4) and ammonia (NH_3) gases.

If contemporary theories are correct, these Jovian planets, as well as their gaseous satellites, do not provide a fit environment for man. On the other hand, the terrestrial-type satellites, Io (Jupiter I) and Europa (II), and possibly Titan (Saturn VI), which may have compositions somewhat like the moon, are worlds on which man might fashion for himself a livable, though closed, environment.

Earth's Magnetic Field

A compass needle swinging toward the north magnetic pole demonstrates the existence of the earth's magnetic field positively. But the magnetic phenomena of the earth are, in fact, extremely complex, and their large-scale investigation is only beginning.

The measured intensity of the general terrestrial dipole (two-poled) magnetic field is very slight compared with that of natural or artificially created magnets. The gauss is a measure of the magnetic-flux density, named after the German scientist Karl Friedrich Gauss, a pioneer investigator of the earth's magnetism. A gamma is $\frac{1}{100,000}$ of a gauss. At the geomagnetic north and south poles, the intensity of the magnetic field is about 0.63 gauss or 63,000 gammas. At the geomagnetic equator, which does not correspond to the rotational equator, the intensity is more than halved to about 0.31 gauss. The surface magnetic field ranges from a high of about 0.725 to a low of about 0.245 gauss, averaging around 0.5 gauss or 50,000 gammas.

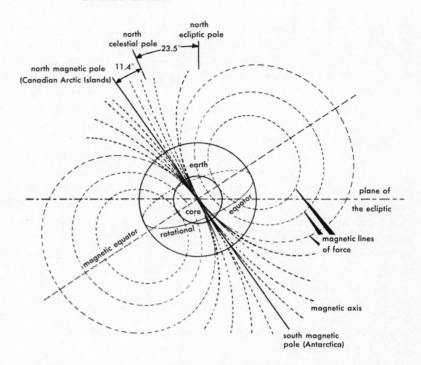

Cross section of the lines of force of the earth's dipole magnetic field. The north and south magnetic, celestial, and ecliptic poles are distinct from each other. (See also diagram of the geomagnetic cavity in Chapter 6, within which the magnetic field is probably largely contained.)

The form of the earth's magnetic field is about like that of any uniformly magnetized sphere. Within this doughnut-shaped (toroidal) form, however, the earth's field shows large short-term fluctuations. The great variations probably result from excess magnetic materials, such as magnetite or other mineral deposits in the crust of the earth. One of these spots is at Kursk, in the U.S.S.R. near Moscow. It has an intensity of 1 gauss, much larger than that of the main field and even reversing the direction of the lines of force on occasion. The earth may have other minor magnetic poles in addition to its two major ones, forming a multipolar field. The shape of the outer portions of the lines of force of the earth's magnetic field may vary extremely from the doughnut shape, dependent on conditions in the geomagnetic cavity and space, becoming lopsided, with the side opposite the sun tapering off into a tail.

The strength of the earth's magnetic field has been found to wax and wane from studies of rock magnetization, which have also indicated the reversals in the field. About three such reversals per million years have taken place in recent geologic times. It is possible that the movement of the earth's magnetic poles are related to this field reversal. It has been theorized that when the magnetic poles move close enough to the earth's poles of rotation, or geographic poles, within about half a degree, the currents in the core that create the field are reduced, and this reduces its strength to nearly zero and causes its reversal. In this connection, it has been determined that the sun's magnetic poles reverse in 11-year periods. Worldwide studies of these magnetic shifts are going forward under the International Upper Mantle Project.

The most widely accepted explanation of the earth's magnetism is the theory that the field is generated by the movement of material conducting electric currents in the fluid or plastic core of the earth, causing the core to operate as a self-exciting dynamo and create the magnetic field. The field could hardly be generated in the crust, because magnetic changes do not appear to be correlated with the crust's geology, or in the mantle, which, although turbulent, does not seem mobile enough either to generate the field or cause the magnetic variations discovered. A fluid or liquid core, with electric currents generating the field proper and swinging or surging enough in relation to the earth's surface to carry along with it the lines of magnetic force and flow of current, appears to be the most plausible explanation at the moment.

A working laboratory model of the type of self-maintaining dynamo which may originate the earth's magnetic field was recently built by British scientists. Two iron-alloy cylinders, with their axes at right angles to each other, about 3 inches apart, rotate in cavities in a block of the same material, somewhat as portions of the earth's core may rotate. Electric contact between the cylinders and the block is provided by a thin layer of mercury. A small magnetic field applied with the rotation of the cylinders induces an electric current and magnetic field which is larger than the applied field and is self-maintaining. Magnetic fields of 1,000 gauss can easily be produced in the block by the rotation of the cylinders. The success of such a model tends to substantiate the self-maintaining dynamo theory of the earth's magnetic field.

Distant Magnetic Fields

Evidence that the moon once had a magnetic field has been found in remnant magnetization of lunar rocks returned to earth and by a satellite orbiting the moon. Whether such a lunar field came from a small core or from an external field is not settled. No magnetic field has been observed on Mercury.

No final conclusions are possible yet about Venus' magnetic field, although the probability is that it is either small or nonexistent. Venus had been believed to have a magnetic field stronger than that of the earth. At periods of inferior conjunction, when Venus lay nearly on a line between the earth and the sun, the streams of electrified solar particles (solar plasma or wind) reaching the earth appeared to be affected. It was thought that an intense magnetic field on Venus might explain these effects. But in 1962, when Mariner 2 carried a magnetometer past Venus at a distance of 21,650 miles, there was no indication of a magnetic field, though the instrument was sensitive to a 0.5-gamma variation or a field strength 100,000 times weaker than that of the earth. Venus' field was tentatively concluded to be at most no more than $\frac{1}{10}$ to $\frac{1}{20}$ of that of the earth, or so weak that it was pressed in by the solar wind to a very limited region close to the planet. On the other hand, Venus may have a different magnetic structure than the earth; it may consist of a series of poles rather than two poles. Mariner 2's magnetometer has created another Cytherean mystery.

Jupiter's peculiar radio emissions have indicated that it probably has a very strong magnetic field, in addition to sizable belts of trapped radiation similar to the Van Allen belts around the earth. From polarized radio bursts from Jupiter, its surface magnetic field has been estimated to reach an intensity of up to 5 gauss, ten times stronger than the earth's 0.5 gauss. A large magnetic field around Saturn is doubtful, however, since its radio emissions do not show Jupiter's characteristics. Although the magnetometers on Mariner flybys of Mars were working, they did not report a magnetic field on the planet. If it has a weak field, it has not as yet been detected.

CHAPTER VI

EARTH'S ATMOSPHERE

"This most excellent canopy, the air."

SHAKESPEARE

A teardrop-shaped mass of gases that weighs less than one-millionth of the earth's mass, the atmosphere comprises all that lies between the earth and interplanetary space. A necessary condition for nearly all living things, and the conveyer of climate and of the weather, the atmosphere is one of the most significant features of the human environment. What goes on in the chaos of these great billowing and streaming masses of gases, of particles mixing in them, of plasmas, of gravitational and magnetic fields, and of turbulent showers of radiation and massive shock waves composing the atmosphere?

In the past 10 years, a great deal has been learned about the atmosphere, largely because of the drive into space. Balloons have soared into the atmosphere's lower levels, embryonic rockets have penetrated higher, and earth satellites have collected reel upon reel of data as they orbited through its upper levels. These developments have forced complete revisions of what was solid textbook doctrine only a few years ago and have given birth to a new discipline, aeronomy, the science of the physics and chemistry of the atmosphere's upper levels.

Atmospheric Regions

The atmosphere has been mapped out vertically in a number of concentric spherical regions or shells, somewhat like the strata in rocks. Such regions are only roughly demarcated, as much to indicate the type of investigation as actual significant atmospheric variations, but they have proved valuable. Many of these atmospheric layers or spheres are recent discoveries. The suffix "sphere" denotes each of these regions, as in the term troposphere, the lowest level of the atmosphere, which runs up to 5 to 10 miles (25,000 to 50,000 feet) above the surface of the earth. The roughly defined upper boundary or borderline of each region is denoted by the suffix "pause." The tropopause is the shifting upper limit of the troposphere, at an elevation of 5 to 10 miles. In the past the customary atmospheric scale was in terms of feet of elevation, now it is in terms of miles, and our knowledge of what goes on up there has increased by about the same ratio.

The main regions of the atmosphere and the principal elements composing it are shown in the diagram. The basic thermal classification of the atmosphere distinguishes the troposphere, stratosphere, mesosphere, thermosphere, and exosphere—distinctions based on variations in temperature with height. Data on the atmospheric temperature zones are given in Table 6. Classified according to composition, the main regions of the atmosphere are called the homosphere, the lower region in which the composition remains constant and the gases are fairly well mixed, and the heterosphere, in which the composition begins to change markedly through molecular dissociation and recombination, diffusion, and photoionization. Classified according to dominant processes, several regions of the atmosphere are significant, such as the ionosphere, characterized by changes in intensity of ionization, and the chemosphere, where chemical reactions are dominant. In the ozonosphere, ozone (O_3), vital for most life on earth, plays a major role; in the magnetosphere, the earth's magnetic field is the dominant force.

Thermal Regions—Troposphere and Stratosphere

Making up the cellophane-thin skin called the

earth's lower atmosphere, the troposphere and stratosphere are the regions most directly affecting our lives. Most weather is made in the troposphere, but the clouds that abound in it are few and far between in the stratosphere. The dividing line between them, the tropopause, varies in elevation from about 10 to 12 miles at the equator, decreasing to 4 to 5 miles at high latitudes, and varying from between 4 to 9 miles in the north and south temperate zones.

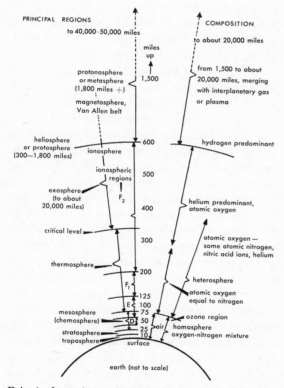

Principal regions distinguished in the earth's atmosphere and its major constituents.

During the 1930's, the stratosphere was considered a windless, cold place of dead calm and quiet. The discovery of jet streams at the tropopause and in the lower stratosphere came as a surprise. These ribbons of wind 6 to 8 miles up may be as much as 300 miles wide, but only 1 or 2 miles thick; they have normal velocities of 35 miles an hour in the summer

and about 90 miles an hour in the winter, but they occasionally whip along at 100 to 200 miles an hour. Aircraft often fly in the stratosphere because of the slighter resistance of its thinner air, and can take advantage of these jet streams to speed them even more quickly on their way.

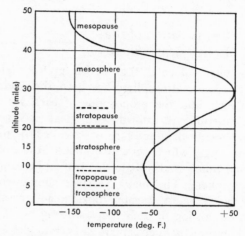

Temperature variation in the lower part of the earth's atmosphere up through the thermal regions to the mesopause.

The temperature decreases with height up through the troposphere, plummeting down steadily from a mean of about 50° F. at the earth's surface to a low of between —50° and —75° F. The cold remains quite constant up through the stratosphere, which reaches an altitude of 20 to 25 miles, although the temperature begins to rise in its upper levels. With greater altitude, the west winds that are found in the lower stratosphere diminish, and easterlies take over at very high levels.

The bulk of the earth's atmosphere is concentrated near its surface in the troposphere and stratosphere. At an altitude of 3.4 miles (18,000 feet) about half of the total mass of the atmosphere is below. About 95 per cent of the atmosphere's mass is below the 15-mile (80,000-foot) level. Conditions at the top of the dents which the earth's crinkled surface makes in the atmosphere attest to this fact. Atop Mount Everest, for example, at 29,028 feet, the air is already getting very thin, and even well-acclimated climbers require extra oxygen to avoid altitude sickness. Even at elevations of 5,000 to 7,500 feet, distance runners at track meets are badly slowed by the lack of a normal concentration of oxygen in the air.

Composition of the Lower Atmosphere

The blend of gases that makes up the "air" of the lower atmosphere has been thoroughly analyzed, and the proportions of each gas in the mixture are known with great accuracy. Air consists of a mixture of so-called permanent gases, which remain at nearly the same proportions up through the troposphere and stratosphere. The major variable components are water vapor, carbon dioxide, and ozone; the proportion of each keeps shifting.

The principal permanent gases, nitrogen and oxygen, together constitute about 99 per cent of the atmosphere near the earth; argon accounts for almost another 1 per cent. Other gases, such as neon, helium, methane, krypton, nitrous oxide, hydrogen, and xenon, are present in extremely limited quantities. Oxygen becomes dissociated into atomic oxygen, which with greater elevation makes up a greater and greater proportion of the total atmosphere until at a height of about 300 miles above the surface the atmosphere is composed almost entirely of oxygen atoms along with helium and hydrogen.

Many other gases and particles can be found in the air in very small and varying proportions, either taken up from the earth's surface, or, occasionally, produced by chemical processes in the air itself. These include carbon monoxide (CO), sulfur dioxide (SO_2), ammonia (NH_3), and nitrogen dioxide (NO_2) among the gases; compounds, mainly from the oceans, such as sodium, magnesium, calcium, and potassium salts; very fine soil particles such as quartz, mica, calcium carbonate, and feldspar; radioactive isotopes such as iodine 131 and strontium 90, the products of nuclear explosions; and some fine particles of dust that have sifted down from cosmic space or from meteors consumed in the atmosphere.

Once in a while great quantities of fine dust or ash are belched out with large volcanic eruptions, like the blasting of Krakatoa on Java in 1883 or the eruption of the Agung volcano on the island of Bali in 1963. The Agung eruption was thought to have produced a thin layer or sheet of particles 15 to 20 miles above the earth's surface, which spread as far as the Northern Hemisphere and gave rise to brilliant golden sunrises and sunsets.

Oxygen—one-fifth of the atmosphere—is what sustains the lives of animals and men. While there are anaerobic bacteria that do not require oxygen, and organisms of this kind may have been the predominant form of life under primitive terrestrial conditions, the metabolism of more complex life is based on oxygen. Plants, on the other hand, use the carbon dioxide in the atmosphere for their metabolism and produce most, if not all, of the atmospheric oxygen, which was probably not part of the primitive earth atmosphere.

The carbon dioxide in the atmosphere comes from several sources: from natural processes, like the gases emitted by volcanoes, and as a product of animal metabolism and decomposition and of such oxidizing processes as fires and internal-combustion engines. The proportion of carbon dioxide probably remains quite constant, although some have surmised that it may be increasing because of greater industrialization. A large-scale increase would wreak havoc at the earth's surface by creating a "greenhouse" effect—absorbing and then trapping solar radiation and radically warming the terrestrial climate. But the earth's variable water vapor absorbs nearly six times as much of the radiant energy from the sun as all the other gases combined, as well as accounting for nearly all the absorption by gases of the radiation emerging from within the earth, so it has the major role in determining the weather. It is debatable whether changes in the concentration of carbon dioxide in the atmosphere have caused any permanent temperature changes on earth.

The ozone in the atmosphere is largely concentrated in the region from 10 to 30 miles above the earth. This is fortunate, for one thing, because of ozone's penetrating and rather nasty odor, which can sometimes be whiffed near a short circuit, after lightning has struck nearby, or around electrical machinery in operation. While some ozone can be identified at the earth's surface, it becomes more concentrated with increasing height and it is most dense at 14 to 15 miles above the equator and at 6 to 8 miles up in the high latitudes. Stratospheric winds move it from equatorial regions toward the north and south temperate zones, where it moves into the troposphere through gaps in the tropopause related to jet streams. Recently instruments have been designed to be carried aloft by balloons, to observe the circulation and vertical distribution of ozone; these indicate the mixing and flow of large masses of air, so ozone is proving very valuable in the study of the atmosphere.

Solar Constant and Heat Balance

High-altitude balloon and sounding-rocket observations have made possible a more precise determination of the total radiation from the sun on a unit area of the earth. Called the solar constant, this unit measure of the heat available to the earth has an average intensity of about 2 gram calories per square centimeter per minute when the solar radiation is at normal incidence (that is, at right angles) to the atmosphere and directly above it, at the earth's average distance of one astronomical unit from the sun. While recent determinations of the solar constant have varied from 1.96 to 2.02 gram calories, the unit figure, like that of the a.u. itself, is rapidly becoming more exact. The constant represents a tremendous flood of

radiant heat being received over the entire hemisphere of the earth exposed to the sun. This solar energy generates the monumental forces involved in the earth's weather—in comparison with these forces thermonuclear explosions are popguns.

Insolation is that portion of the solar radiation not reflected, scattered, or absorbed in its passage through the atmosphere—it actually reaches the earth's surface. Under average conditions, about 35 per cent of the solar radiation is reflected and back-scattered in the atmosphere, and about 17 per cent is absorbed by water vapor, by other gases, and by dust, so that only 48 per cent finally warms the surface. These average proportions vary greatly with time of day and season, latitude, and type of surface and cloud condition. With overcast skies, perhaps only 35 per cent of the solar radiation finally reaches the ground; with average cloudiness (52 per cent) over the earth's disk, the insolation is about 50 per cent.

The solar constant itself, however, appears to be remarkably stable. Changing conditions in the atmosphere or on the earth are believed to have brought about long-range climatic changes. For a time it was thought that conditions on the sun, such as sunspot variations, which reach periods of maximum incidence in 11-year cycles, might affect the earth's solar constant, or that large sun flares or prominences might make the constant vary somewhat. The sun's brightness did seem to increase by about 2 per cent with the increase in sunspot activity from 1954 to a maximum in 1959. However, this finding was based on indirect measurements and has not been definitely established. But such solar events as flares and sunspots do produce significant variations in x-ray radiation from the sun, with increases running from ten to several hundred times the normal intensity. It has not yet been proved that such variations affect the solar constant. The very steadiness of the constant helps to provide the stable environment in which life on earth thrives.

Over a period of a few years, at least, a very steady balance also is maintained between the thermal radiation received by the earth from the sun and the earth's own radiation back into space. If this equilibrium did not exist, the earth would be heating up or cooling down slowly—which does not seem to be the case. The complex movements of thermal radiation into and out of space, the earth's atmosphere, and to and from the earth's surface somehow maintain this balance.

The sun's radiation filtering down through the stratosphere and the troposphere, warming large masses of air and ocean and land surfaces, creates the earth's weather. The rotating earth and its atmosphere act like a huge heat engine, transforming radiant (or heat) energy into kinetic energy (the energy of motion) by heating its working fluid, the atmosphere, at a high pressure and cooling it at a low pressure. In this process the heat is transported from the hot source (that part of the atmosphere most heated by the sun's radiation) to the coolest atmosphere—the "cold sink," as it is called. For the earth, the hot source consists of the equatorial regions and the cold sinks are the polar regions; the heat is carried by winds (kinetic energy) from the equator toward the poles and the cold from the poles toward the equator. With the rotation of the earth, the lower atmosphere breaks into a series of great planetary waves, 2,000 to 3,000 miles long; on the average, about six of these are active at a given time. These wave masses, moving in huge circumpolar vortices, create the high- and low-pressure regions in the atmosphere, the stock-in-trade of the weatherman.

Now for the first time large-scale atmospheric conditions, visualized in mathematical models and satellite photographs of cloud cover, are helping meteorologists to forecast the weather. The Nimbus series of American weather satellites has carried cameras above the greatest proportion of the atmosphere, to altitudes of 400 or more miles, to photograph and transmit pictures of the major cloud formations and their motions. Global coverage, particularly over oceans and sparsely inhabited mountainous regions and deserts, had never been achieved by ground-based weather stations, so the satellite data have meant a great immediate step forward in meteorology. Since Nimbus photographs show cloud waves and masses which could only be dimly inferred from the ground, they have made even detailed short-range forecasts more reliable.

In addition, the weather satellites have carried sensitive infrared (heat) detectors to measure the intensity of the radiation transmitted or reflected up through the atmosphere from the earth. This varies greatly with changes in temperature, water-vapor density, and the extent and height of the cloud coverage. As the techniques are developed, the maps based on these measurements will become most significant in the long-range forecasting of the weather and the understanding of its causes, for they will reveal the complex, over-all heat-transfer processes taking place in the earth's atmosphere, which can now be only roughly diagrammed.

Rainfall and the Moon

Farmers have always known somehow that a relationship exists between the weather and the moon, and farmers' almanacs have supported this tradition. One of the old, and now shattered, myths of scientists has been that there was no such relationship. Like many categorical statements, this one has eventually had to be retracted. The data on the relationship between the earth's weather and the moon has been available all along to anyone who cared to examine routine U.S. Weather Bureau reports on the dates and places of maximum 24-hour precipitation per calendar month.

In 1962 United States scientists reported the analysis of a total of 16,057 of these maximum precipitation records, representing 6,710 dates, recorded at 1,544 American weather stations

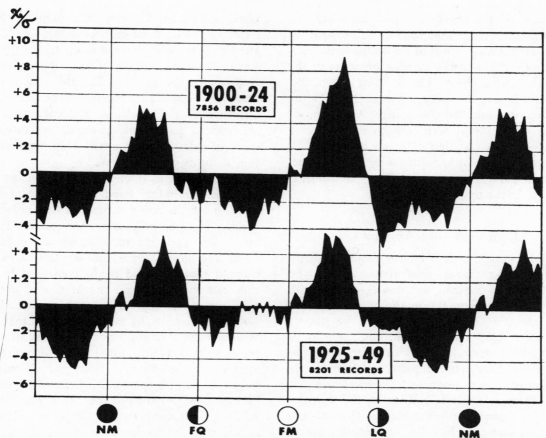

Changes in maximum precipitation in the United States with phases of the moon, divided into two separate 25-year series for comparison. The changes are treated in terms of variations (+ or —) from standard measure of ten-unit moving totals of synodic decimals for 16,057 record dates of maximum precipitation in the calendar month of 1,544 U.S. weather stations, for the years 1900 to 1949.

—DONALD A. BRADLEY, MAX A. WOODBURY, AND GLENN W. BRIER, *Science*

over the 50-year period from 1900 to 1949. They divided these records into two groups, 1900–24 and 1925–49, and tabulated the days on which the maximum monthly precipitation occurred against the phases of the moon. The tabulations showed very clearly, with high statistical significance, that in North America there is a strong tendency for the extreme precipitation to fall near the middle of the first and third weeks after new moon, especially on the third to fifth days after new moon and full moon. In the same way, the weeks after

first quarter and third quarter were lacking in such heavy precipitation, the low point falling about 3 days prior to full moon and new moon.

While day-to-day rainfall cannot be reliably predicted from the moon's phases, the discovery of this relationship does open up new vistas for atmospheric research. What kind of underlying mechanism could produce such a relationship? It might be a common factor related to both the heavy precipitation and the lunar phases. In 1963 scientists reported that similar links seem to exist between lunar phase and the number of meteorites burning up in the earth's atmosphere, disturbances in the earth's magnetic field disrupting radio communications, and the amount of ozone in the earth's atmosphere. The mystery, then, becomes deeper, involving the magnetic field, the meteorites, and the atmosphere. A solution of this mystery may well lead to a more satisfactory understanding of how our over-all atmosphere operates, particularly in its outer reaches, and have profound effects on the long-range forecasting of the weather.

Mesosphere and Airglow

Oddly enough, little is known about the mesosphere, the region immediately above the stratosphere. Ranging up from the stratopause at 20 to 25 miles, the mesosphere extends to an elevation of 50 to 75 miles. The heavier meteors reach down into it, and their trains have been studied there with radar. The temperature rises quite rapidly up through its lower portion, where many chemical reactions take place. At about 30 miles up, the temperature is roughly comparable to that at the earth's surface. Then the temperature about-faces and drops rapidly again, down to around —150° F., far lower than stratospheric temperatures.

The mesosphere is a borderline region, hard to investigate because its constituents are, on the one hand, too rarefied to support high-altitude balloons and, on the other, too dense for orbiting satellites to remain any length of time without their orbits decaying. The mesosphere's radiation is almost too feeble to be studied from the earth, as it is drowned out by other radiation originating lower down in the atmosphere. Most of our knowledge of the mesosphere is coming from the sounding rockets, which can flash swiftly through it, and from the increasingly significant spectroscopic studies of the airglow.

The chemosphere falls within the mesosphere and here many chemical reactions take place, stirred in the very thin soup of this region. These chemical processes involve atomic (O) and molecular (O_2) oxygen, ozone, hydroxyl radicals (OH), and sodium compounds; during the day they mop up masses of penetrating solar energy such as ultraviolet rays. Then these chemical reactions are reversed, particularly at night, and various forms of radiation are emitted. One manifestation is the so-called airglow.

On his Mercury flight, John Glenn saw a luminous tan-to-buff band at a height of 6° to 8° above the red-blue band of the horizon, perhaps 55 to 65 miles above the surface. The tan ribbon appeared more luminous to him when the moon came up. At first it was thought that the band might consist of multiple reflections between the inner and outer windows of the Mercury capsule (which did not slant at quite the same angle), or that it resulted from auroral phenomena in the atmosphere, since there are auroral lines which might appear on the horizon from the 150-mile height of Glenn's spacecraft. Some of Glenn's photographs also showed the band. The later Mercury astronaut, Scott Carpenter, determined that the luminous band was the well-known airglow in the mesosphere.

The ultraviolet light or x-ray radiation from the sun breaks down the ozone (O_3) into its atomic oxygen and molecular oxygen components; these recombine at night, and the excited electrons involved emit the airglow, forming ozone again by recombination. The ozone may also combine with molecular hydrogen (H_2) to form molecular oxygen and hydroxyl, which emits an infrared glow at night when the process occurs; a good proportion of the night airglow in the infrared region of the spectrum may derive from the hydroxyl reaction. A photochemical cycle of nitrogen oxides is believed to give the visible emissions in the green region of the spectrum.

At a height of about 50 miles, near the top of the mesosphere and in the lower levels of the ionosphere, the strange and beautiful streamers or waves of the noctilucent clouds appear. Prevalent during the summer in the northern latitudes, they have been investigated in joint Swedish-American experiments using sounding rockets to trap samples in the region of the clouds. Such samples were obtained from rockets sent up from Kronogard, in northern Sweden near the Arctic Circle.

The noctilucent clouds drift mostly toward the west, and sometimes toward the southwest. Their form continually changes as if the motions of the atmosphere at this level were very irregular. They travel with a mean velocity of about 30 miles an hour, with occasional velocities of up to 60 miles an hour. It had been theorized that the noctilucent clouds consist of ice crystals (it is very cold in this region) or perhaps of dust sifting down from meteorites that have burned up in the atmosphere at slightly higher levels.

A rocket penetrating a noctilucent cloud collected about 10 million particles per square inch—many times more than a rocket that did not. About 20 per cent of the particles showed a halo effect on the film on which they were gathered, indicating that they had had a coating of ice. Analysis of the particles themselves indicated some nickel, which meteors contain, and the particles were so large that their origin was more probably from space than from the earth. So the clouds evidently consist of ice frozen on minute particles, both theories thus having an element of truth. But how could water vapor have risen through the low-temperature regions below the clouds without freezing and being precipitated? It is possible that the water vapor is formed by chemical reactions at the same level. More rocket-trapping flights are planned to attempt to reach a full explanation of the noctilucent clouds.

Upper Atmospheric Regions

Above the mesosphere lies the thermosphere, a region of the atmosphere ranging from about 50 miles up to between 250 and 300

miles elevation, where the exosphere sets in. The thermosphere is appropriately named, since it is a mighty hot place, in which the temperature rises with increasing elevation until it levels off at an average of about 2250° F. Over the solar cycle the temperature of the upper thermosphere may vary between a minimum of about 700° F. and a steady maximum of 2250° F., with occasional brief excursions to about 2500° F. This high-level heating of the atmosphere has not yet been satisfactorily explained. It may have something to do with extreme ultraviolet solar radiation.

Atomic oxygen is the principal constituent of the thermosphere, with small and diminishing proportions of the heavier atomic nitrogen and nitric-oxide ions, and increasing proportions of helium. Gas diffusion, dissociation, ionization, and recombination are the principal processes taking place. The thermopause is close to the critical level where atmospheric evaporation may occur. The mean free path of travel without a collision of neutral (un-ionized) particles of gas is long enough so that some of the gases may begin to escape en-

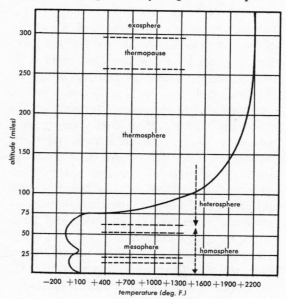

Temperature variation in the atmosphere up through to the exosphere.

tirely into interplanetary space; they may be replaced by gases flooding in from space, principally hydrogen.

A region of helium appears in the thermosphere and exosphere at an elevation of about 200 miles, running up to about 600 miles. This then gives way to a region of tenuous hydrogen, the lightest gas and the simplest element. It is believed that this atomic hydrogen is generated by the dissociation of water vapor and methane at heights down to 50 miles in the mesosphere and diffuses upward from there, though the upper levels may be supplied in part from space. The hydrogen gas runs far out, getting thinner and thinner, for at least 15,000 to 20,000 miles and perhaps more, then tapers off, merging with the interplanetary gas. The helium region is sometimes referred to as the heliosphere or protosphere, and the hydrogen region as the protonosphere or metasphere; this whole tenuous mist of gas around the earth is appropriately called the geocorona.

Ionosphere

The ionosphere is a deep region of the atmossphere distinguished by the fact that the particles in it are largely ionized, or charged. It overlaps many of the thermal regions so far described, and may, in fact, be conceived as encompassing the ionization features of the whole atmosphere, rather than existing as a distinct region. It contains ions, positively charged atoms which have lost outer electrons, and free electrons of negative charge. This ionization is produced by the powerful solar and galactic or cosmic ionizing radiation entering the atmosphere from space and penetrating to great depths in it.

The ionosphere begins with what are termed the C and D regions, which are transient; with the C below and the D above, it ranges between about 30 and 50 miles up in the mesosphere. Then come the E and F regions, and the helium- and hydrogen-ion regions above.

The concentration of ions increases with altitude, although the particle density decreases.

Many features of the ionosphere have been studied with earth satellites, particularly with the American Vanguard and Explorer series, and probably by Soviet satellites as well. Much information on the ionization regions and the density of the atmosphere at various levels has also been obtained by the use of radio sounders, such as the huge, 22-acre antenna system of the United States Bureau of Standards, jointly run by the United States and Peru on a site at Jicamarca, near Lima, Peru. At this location near the equator the horizontal character of the earth's magnetic field overhead keeps the electrified components of the atmosphere also horizontal, so that they reflect a great enough proportion of the radio pulses from the antenna to reveal the density of ionized fields from a low level right on up into the magnetosphere.

Geomagnetic Cavity

The discovery of the Van Allen belt region, related to the earth's magnetic field, was probably the most significant discovery yet made about the earth's immediate environment by means of artificial satellites. It was predicted by the American physicist S. Fred Singer in 1956, and was discovered in 1958 by James A. Van Allen of the State University of Iowa by means of an analysis of the Geiger-counter data transmitted from the instruments of the Explorer 1 satellite correlated with data from sounding rockets. Since then, Explorer 12 and many other satellites have furnished mountains of information about these belts and the magnetosphere that encloses them.

As shown in cross section in the diagram, the earth's magnetic field consists of a huge doughnut-shaped ring of magnetic lines of force fanning out from the magnetic poles, its effective strength running out at least some 10 to 14 earth radii (40,000 to 60,000 miles) from

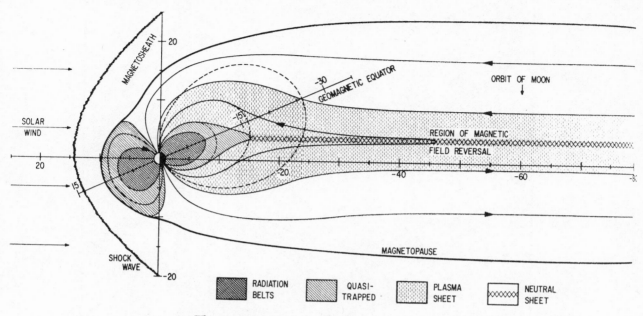

The geomagnetic cavity and earth's magnetosphere are contained within the magnetopause in this summary cross section of the earth's immediate environment in space inferred from reports from many spacecraft. The scale is indicated in earth radii. The earth may lose some of its hydrogen and helium from its upper atmosphere through its geomagnetic tail, which extends far out beyond the moon, millions of miles into space.

—NORMAN F. NESS, LABORATORY FOR SPACE SCIENCES, NASA, AND *Review of Geophysics*

the earth into space. The magnetosphere is the name of this region where the atmospheric plasma is strongly affected by the earth's magnetic field; it encompasses both the ionosphere and the exosphere. At its outer limits this "geomagnetic cavity" is battered into weird shapes by the solar wind. Most questions about the nature and operation of the magnetosphere remain unanswered.

The Interplanetary Monitoring Platform or "Imp" satellite, launched on November 26, 1963, into a highly elongated orbit reaching 122,800 miles from the earth at apogee, began to sketch in the picture of the shock wave and turbulent zone at the boundary of the earth's geomagnetic cavity. In its fifth 4-day orbit, the Imp also touched what appeared to be a

"wake" or cavity streaming out on the nightside of the moon away from the sun. Imp's instruments measured the turbulent regions of the earth's cavity where the solar wind strikes the magnetosphere some 40,000 miles from the surface, and showed that the earth too has an extended turbulent wake and a tail which may reach out from the nightside of the earth as far as the planet Mars or beyond.

The Van Allen region consists of somewhat artificially distinguished belts of positively charged protons and negatively charged electrons in the inner portion of the magnetosphere. These protons and electrons have been trapped by the earth's magnetic lines of force and spiral back and forth between mirror points around the magnetic poles, with ener-

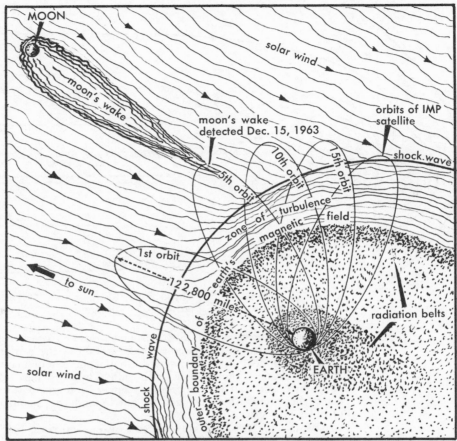

Orbits of the Imp satellite through the geomagnetic cavity, showing how the fifth orbit may have carried it through the moon's wake. The earth's zone of turbulence may have somewhat the same shape as that of the moon shown, reaching out great distances on the far side of the earth, perhaps to 50 a.u.

gies ranging from what is called "soft" to very "hard" or intense radiation. The Van Allen belts extend from a height of about 700 miles above the earth at the magnetic equator out to perhaps 20,000 miles. At their horns, where they curve in closer to the earth's surface at its magnetic poles, they may approach within heights of 200 to 300 miles, and other areas have been found where they may dip down as close if not closer to the surface.

While the study of the Van Allen belts is still preliminary, it is believed that these belts constitute plasmas, involved in the transfer of energy from the sun down through the iono-sphere to the earth, particularly when the great solar flares cause such effects on earth as magnetic storms, auroras, and blacking-out of radio communications. The relation of the belts to the ionosphere is not understood. It seems possible that the protons and electrons trapped in the belts may sometimes be shaken out at the tips near the poles and cause the auroras, which occur most frequently in what are called the auroral zones around the earth's North and South poles. There is speculation also that the belts may have other still obscure effects on the earth and its atmosphere, even on the weather.

CHAPTER VII

"The wreck of matter and the crush of worlds."

ADDISON

Not so long ago the space beyond the earth and its atmosphere was thought to be black, cold, inactive, and empty, an all-pervasive void, characterized by three dimensions and little else. Space was conceived in a mathematical manner, as a kind of receptacle having length, breadth, and depth, but no contents; it was that which surrounded or existed between material things. Mathematical concepts are fine for measuring, relating, and explaining, but they do not exist in nature as such any more than do straight lines or points. This concept of space as a complete void mirrored human imagination and ignorance more than it expressed any knowledge.

In the twentieth century, and particularly in the last decade, scientists have contrived to turn space inside out. It has proved to be fairly bursting with activity—a jungle rampant with all kinds of gases, particles, dust, large and small bodies, and a multitude of fields and waves. Through it all only the first exploratory trails have been hacked.

Space Regions

When confronted with a blooming, blinking chaos, about which little is known, usually the first steps are to sort out similar things, make preliminary classes, and begin to describe and analyze. Something like this has happened as information about space has come rolling in and its complexity been grasped.

A number of regions have been distinguished within space, so that each can be studied and characterized in its own right. Near space extends from the earth's atmosphere out around its immediate vicinity, and

INTERPLANETARY SPACE

more is known about it than any other sector of space. Instrumented rockets, earth satellites, and space vehicles have probed it and checked it out. The point at which near space is assumed to begin above the earth depends on the feature being investigated. Even 20 miles above the surface, 99 per cent of the gases in the material atmosphere lie below. At 100 miles up in orbit, though still within the ionosphere, the astronaut sees the stars brilliantly steady against a dark, velvety background. If he opened a port in his capsule, he would experience most of the untoward effects of space. At about 15,000 to 20,000 miles up—another cutoff point for near space—the hydrogen and helium gases peter out to about the consistency of interplanetary gas, perhaps 160,000 particles per cubic inch. Near space may also be conceived as ending at the boundary of the geomagnetic cavity where the solar wind meets the earth's magnetic field and a region of turbulence has been discovered, 50,000 to 60,000 miles out from the earth. Within this huge sphere many fascinating materials and effects have already been investigated.

In relation to the moon, the earth's gravitational field is dominant out to about 220,000 miles, or to within 20,000 miles of the lunar surface, and this may also be thought of as the limit of near space. At this "null point," the gravitational fields of the earth and the moon are balanced in strength, though both fields actually extend beyond this to infinity, growing weaker and weaker as the inverse square of the distance. A body at exactly this point, which varies with the motions of earth and moon, would not fall toward either body until moved in one direction or the other by the shifting forces. A sphere around the whole earth-moon system, the center of gravity of which (the barycenter) is under the surface of the earth itself, is called cislunar space. The gravitational forces of the sun and the earth-moon system are balanced at about 4 million miles from the earth, and this sphere of 4-million-mile radius around the earth is often called "terrestrial space."

A number of spacecraft have been given an escape velocity of more than 7.1 miles a second (in addition to the earth's own velocity of 18.5 miles a second) that is sufficient to drive them out of the earth's effective gravitational field into that of the sun, which becomes more powerful at the 4-million-mile limit of terrestrial space. These space probes either bypassed the moon at sufficient distance and speed not to be drawn down to it or were aimed directly toward Venus or Mars. On their courses toward these planets, they moved under the sun's gravitational influence in interplanetary space, becoming satellites in solar orbit. Later, if they passed fairly close, they came under the perturbing influence of the target planet's gravitational field but stayed in solar orbit.

The first such artificial planet was the Russian Lunik 1, launched toward the moon on January 2, 1959; the next were the American Pioneer 4 space probe and its fourth stage, launched as a lunar probe on March 3, 1959. These tiny, silent bodies in solar orbit, sometimes called planetoids, became man's first additions to the architecture of the solar system. The Mariner 2 space probe, launched toward Venus on August 27, 1962, reached the limit of terrestrial space on September 18, when the earth's gravitational field had slowed it to less than 2 miles a second from its initial escape velocity of 7.1 miles a second. Then it began to pick up speed under the influence of the sun's gravitational tug in its new orbit. Sweeping into passage by the planet Venus, it then went on in a nearly eternal orbit of the sun.

Interplanetary space is the space between the major bodies in the solar system, the planets and the sun. The radiations from the sun and their reflections from the planets must cross this space in reaching the earth. And it is this jungle that unmanned and then manned vehicles must penetrate to approach the inner planets, protected somehow from severe radiations and penetrating particles. Interplanetary space is very cluttered, and as yet we have barely an inkling of what dangers may lurk there.

On March 2, 1972, the first interstellar space probe, Pioneer 10, was launched from Cape Kennedy. Its orbit will carry it to many firsts if all goes well. Pioneer 10 was the first to fly beyond Mars, at which Mariner 9 had stopped in orbit, to reveal more Martian wonders. If Pioneer 10 can successfully cross the dangerous zone in which the minor planets, or asteroids are thick, then the space probe will swing past the largest planet, Jupiter, at the end of 1973 and report on its vast radiation belts and magnetic and gravitational fields. Jupiter's strong pull will, in fact, speed up the space probe enough as it passes to toss it right out of the solar system into interstellar space. There it will speed through the stars interminably.

Pioneer 10 carries an aluminum plate etched with a message for any intelligent beings who might intercept it. The message depicts our solar system, the path of the space probe, male and female human beings, and the location of our system in relation to 14 pulsars and the center of the Milky Way galaxy, so that any intelligent and scientific being could find us.

The outer reaches of interplanetary space have often been thought to be at the farthest known planet, Pluto, some 39 a.u. (3.6 billion miles) from the sun. Actually, the sun's gravitational influence extends much farther, out to perhaps 150,000 to 200,000 a.u. from the sun, where myriads of refrigerated comets may roam, and where, on occasion, the gravitational influence of nearby stars may become greater than that of the sun. Sometimes these stars may so perturb the courses of comets that they plunge in a highly elliptical orbit toward the sun, becoming visible from the earth as they dart in to round the sun.

Beyond interplanetary or solar space lies interstellar space, which stretches out between our star, the sun, and the other stars. The nearest nine known stars are 4 to 10 light-years (25 to 60 trillion miles) away (Table 1). The more distant stars of our Galaxy are up to 100,000 light-years away. Yet this is all part of galactic space within the Milky Way. Out beyond the Galaxy lies intergalactic space; it, too, may contain much more than the vast emptiness previously assigned to it.

Electromagnetic Spectrum

The most evident signals from beyond the earth arrive in the form of light and heat, so naturally light and heat waves, or radiations, were studied in the first investigations of what lies in space. As physics and astronomy amassed more and more information about events on earth, in the solar system, and beyond, it became apparent (about 1870) that light and heat were only tiny segments of a broad range of radiating waves of various kinds. With the gradual analysis and integration of this array of waves, called the electromagnetic spectrum, came many of the historic discoveries of science. As it took shape, gaps in the spectrum were filled in gradually (although some still exist), and the spectrum was greatly extended at both ends.

The principal known types of waves found in the electromagnetic spectrum are the gamma- and x-rays of very short wavelengths, the longer ultraviolet and visible light, infrared or heat waves, and radar and radio waves, down to very long pulsations.

Since the range in wavelengths of the whole electromagnetic spectrum is so immense, and since knowledge of it has been gained in fits and starts, scientists have come to use different units in measuring various parts of it, or different units for the same parts, which is sometimes confusing. Angstrom or micron units are often used for waves of shorter length, while centimeters or meters may be used for longer waves. An angstrom is one hundred-millionth of a centimeter and a micron is one ten-thousandth of a centimeter. For the larger units, conversion charts that aid in translating metric units into the customary United States system appear in Table 7. Length, area, volume, capacity, weight, temperature, pressure, density, power, heat, and other measurements can be easily converted with these charts from the metric units used by scientists into the more familiar units of everyday life. Thus, one meter is 39.37 inches, one kilometer is about 0.6 of a mile, one foot is slightly over 0.3 of a meter, a yard is about 0.9 of a meter, and one mile is a little more than 1.6 kilometers. Scientists use the metric system exclusively throughout the world.

The earth's atmosphere absorbs or reflects a great many of the waves and particles rolling in toward it from space. On occasion the atmosphere may absorb as much as 60 per cent of the visible and near-visible radiations, but some wavelengths manage to penetrate to the surface. Visible light and heat (infrared radiation) were the earliest known radiations penetrating the earth's atmosphere, permitting a view of the universe beyond. They cover extremely narrow segments of the whole electromagnetic spectrum. In the late nineteenth and early twentieth centuries, man's vision was broadened with the recognition of other kinds of radiation: ultraviolet light, x-rays, and radio waves, which were first called Hertzian waves after their discoverer, the German physicist Heinrich Hertz. Shorter-wavelength radio waves (the radar frequencies) come through another very revealing window into space that was discovered and exploited only in the last quarter century.

As explored by radio and radar astronomy, as well as by ultraviolet and infrared astronomy, the view of the universe has changed in many respects from that known by means of visible light alone. As instruments on earth or on balloons above the atmosphere survey the radiations in the radio, far-infrared, and

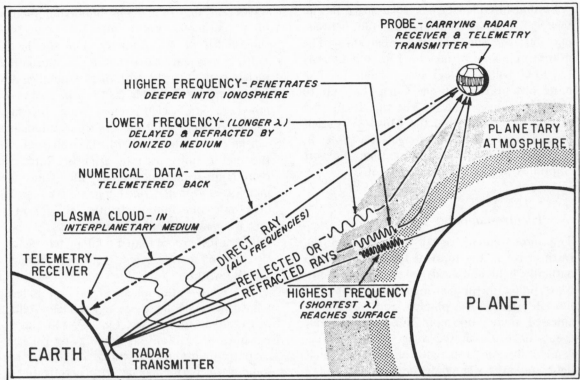

Bistatic radar astronomy technique for study of features of interplanetary space and of the atmospheres and surfaces of planets: retransmitting rays from earth which have passed through or been reflected by various media from a transmitter on a space probe back to a telemetry receiver on earth for analysis of the effects of their travel (wavelength = λ).
—INTERNATIONAL GEOPHYSICS BULLETIN

ultraviolet bands of the spectrum, a whole new range of phenomena are being observed in space. These include cooler, nearly dark celestial objects of which there have been preliminary hints, and the remarkable quasi-stellar radio sources (quasars) that may be near the far boundaries of the known universe.

A new system known as bistatic radar may prove most valuable in the more intensive study of the solar system. Bistatic radar involves a powerful radar transmitter as well as a large receiving antenna on the earth and a smaller receiver and retransmitter on a space vehicle or probe. The earth transmitter would send out radar pulses which would go directly to the space probe as it flew by the moon or another planet and would also be reflected

from the other body and back to the space probe. The probe would then directly retransmit both signals and they would be picked up by the receiving antenna on earth. This process should show a clear difference between the signals which had only gone directly to the probe and those which had echoed from the other body, and so make it possible to cancel out the effect of interplanetary space on the signals. It is expected that bistatic radar systems can be used in detailed studies of planetary ionospheres, atmospheres, and surfaces, without landing on them or even approaching very close to them. Such systems may well be used in analyses of the solar corona, comets, and the interplanetary medium itself, to say nothing of enabling the more precise measurement of interplanetary distances.

Gases In Space

The tenuous outermost sphere of the earth's atmosphere, largely composed of hydrogen gas, begins to shade off into the consistency of interplanetary space at a distance of about 20,000 miles from the earth's surface. While there are undoubtedly a great variety of the atoms of other elements in interplanetary and interstellar space, spectroscopic studies have shown that the basic constituent is hydrogen —ionized hydrogen atoms, stripped of their single outer electrons. Among the other atoms and molecules identified in space are calcium, sodium, potassium, iodine, titanium, and cyanogen (CN) and hydrocarbon (CH) ions.

The hydrogen atoms are probably very thinly scattered through space, forming what is called a low-density plasma. It is estimated that there are probably only 1,000 to 2,000 particles (largely hydrogen) per cubic centimeter in space at 1,850 miles elevation, from 10 to 1,000 hydrogen protons and electrons (separated by ionization) per cubic centimeter in interplanetary space, and perhaps fewer than 1 per cubic centimeter in interstellar space beyond. How empty is this? By comparison, a few years ago the most highly exhausted vacuums that had been produced here on earth still contained about a million particles per cubic centimeter.

In 1963, the hydroxyl radical (OH) was discovered in interstellar space with an American radio telescope, and since then intragalactic chemistry has thrived. Hydroxyl has been found in two other galaxies as well. Some 20 molecules found in the Milky Way include water, ammonia, formaldehyde, methyl alcohol, methyl acetylene, and formamide. The more complex molecules are typical of chemicals probably involved in the beginning of life. They should form in planetary atmospheres more readily, as a basis for the complicated chemicals of organisms. Since the distribution of the hydroxyl radical appears to be different from that of hydrogen in space, the former may provide a good tool for mapping the Milky Way Galaxy. No strict distinction now seems possible between organic and inorganic chemistry.

Magnetic Fields

The relationship between the relatively stable magnetic field (magnetosphere) around the earth, extending out to perhaps 40,000 miles at the earth's equator, and the activities going on in interplanetary space beyond are shown in the diagram of the geomagnetic cavity and tail in Chapter 6.

Astronomers have proposed that a similar structure may form around the sun with its whole solar system as it moves with the solar corona and wind through the interstellar magnetic field and its hydrogen atoms. The solar shock front may be 85 a.u. ahead of the sun, with a highly disturbed magnetic field just within the front. Behind the system a shadow cone or tail of streaming solar wind in magnetic lines may form, similar to the earth's smaller geomagnetic tail through which hydrogen escapes from its atmosphere. Evidence for such solar architecture has not yet come in.

Interplanetary space is rarely free of magnetism, with fields of at least a few gammas present. During periods of low magnetic activity, the field is about 5 gammas, rising to about 20 gammas or more during magnetic storms, and falling to about 2 gammas during very quiet times. Occasionally the field will show almost no change for an hour or two, but usually it varies irregularly, in periods of from 40 seconds to several hours.

Solar Wind

A rather steady solar wind was clearly identified by Mariner 2 as it traveled toward the planet Venus. This is a stream of charged or ionized particles, largely protons, and the electrons which had been associated with them, moving out from the surface of the sun into interplanetary space and transporting some solar energy to the earth. Also called the solar plasma, this wind was found to be fairly continuous by the instruments aboard Mariner, though it occasionally rose to "hurricane" force in solar-flare activities which were accompanied by sharp changes in the magnetic field around the space probe. It was measured by an electrostatic spectrometer kept pointed toward the sun, and there was always a flow of plasma from its direction.

The velocity of the motion of protons (hydrogen nuclei) and alpha particles (helium nuclei) away from the sun was usually between 250 to 435 miles a second, ranging infrequently as low as 200 or as high as 775 miles a second. Eight geomagnetic storms were recorded in a 2-month period (August 29 to October 31), with some increases in plasma flux and energy. The velocity of the wind was higher than had been predicted from the changes in direction of the tails of comets, and was in agreement with that discovered by Explorer 10 and the Lunik satellites. The magnetic fields in interplanetary space were found to be carried along by the plasma, since the energy of the plasma was much greater than the energy density of the magnetic fields.

Solar Flares

Perhaps the most dramatic events in interplanetary space are caused by solar flares, invisible in ordinary light, but discernible in red light. These flares are most frequent during the periods of intense sunspot activity, which follow a mean 11-year cycle, ranging from 7 to 17 years. But solar flares or subflares can occur at any time. Right now they are being studied carefully because their intense radiations pose a threat to astronauts on space flights.

Solar flares, or bursts of solar cosmic-radiation particles, are known to be mostly protons, with energies ranging widely, from less than 10 Mev (million electron volts) up to almost 50 Bev (billion electron volts), an electron volt being the energy acquired by an electron in falling through a potential difference of 1 volt. Low-energy solar-flare events, with particles up to 400 Mev, and medium-energy events, with particles up to several Bev, occurred from five to thirteen times a year over the period from 1935 to 1959. High-energy events were relatively rare; their bursts of solar cosmic radiation up to 20–50 Bev occurred only once or twice every 4 or 5 years. The flares appear to originate in active, disturbed, and turbulent regions of the sun.

The sun's own complex system of magnetic fields must play a role in these solar flares which have such spectacular effects on the earth. Special instruments called magnetoscopes have been developed for observing the magnetic sun. These are basically spectroscopes fed the sun's light by large telescopes, such as the special tower telescope at Mount Wilson Observatory in California and the 60-inch solar telescope at the Kitt Peak National Observatory, Arizona. Spectral lines are split by magnetic fields. In typical solar magnetograms, the distances of the wavy lines from the horizontal lines represent the strengths of magnetic fields on the sun. Wavy lines above the horizontal lines represent positive polarity, while the wavy lines below indicate negative polarity.

It is known that extremely strong (up to 1,000 gauss) magnetic fields exist in the vicinity of the active areas of the sun where the solar flares occur, and may have something to do with the origin of the solar flares. According to one theory, magnetic fields of great strength collapse and produce a turbulence on the solar surface which results in the flare. Another theory states that changes in the magnetic fields on the sun cancel each other out and cause the release of enough heat to start a thermonuclear reaction at the surface, involving the fusion of hydrogen atoms and

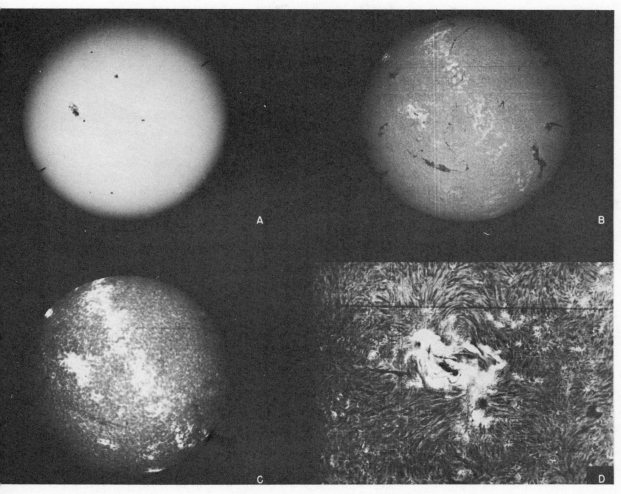

Views of the sun: (A) the whole disk with a number of large sunspots in white light; (B) the whole disk in the red light of the hydrogen alpha line; (C) the whole disk in the violet light of the calcium K line; and (D) an enlarged section around a sunspot taken in the red light of the hydrogen alpha line. (September 15, 1949)

—Mount Wilson and Palomar Observatories

helium atoms and resulting in tremendous explosions that eject the materials composing the flare and create the radiations which accompany it. Theories of this kind cannot be confirmed with the scanty evidence at hand. However, solar flares appear to form in bright patches or plages in the sun's outer envelope (the chromosphere), above the photosphere, in which the sunspots originate.

With a solar flare, which may last from several minutes to a few hours, a mass of charged particles is flung out from the sun's surface and spurts into interplanetary space. The photograph made in calcium light shows a solar flare arising on the sun. The flare material, consisting of a variety of particles and radiations, is believed to drag a strong magnetic field with it, drawing out the magnetic lines of force like loops of taffy and creating a magnetic field enveloping the earth but still rooted in the site of the flare on the sun's surface. With these flares a magnetic storm occurs on the earth, but the cosmic-ray intensities from interstellar space decrease; they must be deflected by the magnetic field accompanying the flare. The earth also experiences auroras and radio blackouts. The cosmic-ray intensity, chiefly of the lower-intensity rays, decreases by about 5 per cent.

Many statistical studies of the correlation between solar flares and various features of the sun are done for purposes of their reliable prediction. Typical flare indicators tentatively tried in such predictions are large and complex sunspot groups, complex magnetic fields in such groups with many poles of opposite sign, very bright plages, frequent radio bursts from the sun, hot spots in the corona, the presence of many small flares in active regions, and certain types of prominences.

Reports from a solar observatory orbiting the earth have shown that the sun has polar caps cooler and less active than its other regions. The sun was blotted out by a disk just as the moon covers the earth during a solar eclipse. In this way the sun's hot atmosphere, or solar corona, can be studied. This corona appears as a beautiful halo around the sun during total solar eclipses. The corona's temperature over the caps is about 1.8 million degrees, as opposed to the average coronal temperature of 3.6 million. Temperatures above active regions of the sun may rise to 8 million degrees and to 70 million degrees above flares. During the active periods at the solar-cycle maxima, the sun's corona runs even higher temperatures.

Everything vaporizes in a split second at such temperatures. Shooting rockets carrying radioactive wastes directly into the sun has been proposed as one way to get rid of them for good, but it would be all too costly!

Cosmic Rays

With its solar flares the sun produces some of the low-energy cosmic rays which bombard the earth, but it is believed that the sun's contribution is only a small part of the whole. Cosmic rays consist of very high-energy protons (hydrogen nuclei), alpha particles (helium nuclei), and the nuclei of heavier elements (up to iron, at least, with an atomic weight of 55, and possibly to heavier elements), with some high-energy electrons. The energy of the particles ranges from 100 Mev up to stupendous billions of electron volts (Bev). Solar flares, producing particles with energies ranging up to 20 to 30 Bev, do not begin to approach the energies of some cosmic rays which arrive from other sources.

The particles in the cosmic radiation from the sun seem to consist almost entirely of protons, with only a very minute proportion of alpha particles. Typical cosmic rays, however, contain about 13 per cent alpha particles, so the sun does not appear to be their primary source. Cosmic radiation from the sun decreases during periods of the quiet sun, while the cosmic-ray bombardment from interstellar space rises, perhaps because the interplanetary solar magnetic field is then weaker, and more of the cosmic rays can penetrate to the earth's vicinity.

Cosmic rays may be accelerated in the vast magnetic fields of galaxies, produced in the cosmic explosions of whole stars (called supernovas) or of a number of stars, or may originate in some other manner not fully understood. Fred Hoyle, an astronomer at Cambridge University, has developed a theory to explain the production of the intense radio sources discovered by radio telescopes, as well as the origin of cosmic rays. He has suggested that huge masses of stellar material in the center of galaxies begin to contract. When they have been reduced to about the size of the whole solar system, these great masses "implode," or contract catastrophically, in about 100 seconds; much more energy is produced by the force of gravitation acting in this manner than could ever be produced by thermonuclear fusion reactions. These tremendous energies might then shoot out a mass of material ten million times greater than that of the sun at speeds up to 62,000 miles a second, producing energy in the ranges required by the strong radio sources. This energy might be responsible, Hoyle suspects, for the main injection of cosmic rays into space. These rays might then be given even higher energies by the operation of natural acceleration processes, probably in extragalactic space. Whether

This jet of a galactic explosion appears to have shot out of the nucleus of Galaxy Messier 87 (NGC 4486), perhaps with the release of cosmic rays. The photograph is a short exposure made with the 120-inch reflector of the Lick Observatory.
—LICK OBSERVATORY, MOUNT HAMILTON, CALIF.

such implosions by masses of stars or galaxies actually can explain cosmic rays or the recently identified quasi-stellar sources is an open question.

Zodiacal Light

Tons upon tons of dust in fine, microscopic particles like those that glint in sunlight sift down gradually to the surface of the earth through its atmosphere. A small proportion of this comes from the earth itself, swept into the upper atmosphere by winds and rising convection currents, having been blown up into the air by volcanic eruptions or blasted into the stratosphere by large nuclear explosions. But most of this dust comes from outside the earth's atmosphere. Earth-satellite data have shown that the dust immediately around the earth, evidently trapped within the earth's gravitational and magnetic fields, has a considerable nickel-iron content similar to that of meteorites, from which it may have been ground in innumerable collisions.

Before the space around the earth and the moon had been thoroughly explored by earth- and moon-orbiting satellites, it was feared that there might be a giant dust ball around the earth, cluttered with larger particles and chunks, which would be very dangerous for spacecraft rapidly plunging through them. The unknown is always feared. And reports from the first probes into near space tended to strengthen this fear, with estimates of a dust-particle flux in near space 100 times greater than that in interplanetary space.

More orbiting and probing spacecraft soon settled this fear, however, with more comprehensive samples. Spacecraft have not been damaged in near space, nor have spaceships bearing astronauts been holed by flying natural missiles, although accidents have happened from internal breakdowns.

The dust and small particles are there, however. The existence of the interplanetary dust had been suspected for almost 300 years, ever since Jean-Dominique Cassini, director of the Paris Observatory, had studied the phenomenon known as zodiacal light in 1683 and attributed it to fine particles of dust in space.

The zodiacal light is a dim cone or pyramid of light, usually barely visible over the normal nightglow in the sky. It appears in the west about an hour after the sun sets and can be seen in the east an hour before the sun rises, probably often contributing to the lightening effect in the sky known as the false dawn. Seen under favorable conditions the zodiacal light is about as bright as the band of the Milky Way. The cone's base is on the horizon, and its apex is almost along the path of the ecliptic, normally fading out toward the observer's zenith, though some visual photographic and photoelectric observations have traced it right across the sky along the band of the zodiac, from which it derives its name.

On the side of the earth opposite the position of the sun occurs an even dimmer phenomenon, called the counterglow, or Gegenschein. This is a dim patch, 8° to 10° deep and 5° to 7° wide, occasionally visible in the night sky. While it has not been studied to the same extent as the zodiacal light, there are at least two theories as to its cause: a phase-angle effect on sunlight that has been reflected from interplanetary dust particles, or a kind of tail of the earth, pushed out away from the sun by its radiation, consisting of such atmospheric atoms as hydrogen or other particles, and reflecting light from the sun.

The zodiacal light does not show a parallax (differing angles of observation varying with the location of the observer on the earth), so there has been no direct evidence as to where it occurs. Over the centuries, and even in recent years, there has been much speculation as to the nature of this cone of light. Some have viewed it as an extension of the solar corona, the bright glare of light visible around the sun during total solar eclipses, resulting from the scattering of sunlight by electrons from the sun. Others have believed it was a phenomenon in the upper atmosphere of the earth. Still others have agreed with Cassini that it represented the sunlight scattered by a lens- or discus-shaped cloud of interplanetary dust more or less centered about the sun and extending out at least past the earth. The basic question concerns what scatters the sunlight.

Free electrons in the ionized solar atmosphere are believed to cause the great solar corona by the scattering of sunlight; this view is supported by the fact that the light of the corona is polarized. But the corona, even as seen and photographed from high in the stratosphere during solar eclipses, cannot be seen to extend out beyond 10° to 13° from the limb of the sun. This may represent in part the limitations of the photographic plates or the effect of light scattering in the earth's atmosphere. On the other hand, the zodiacal light seems to start at about 18° to 20° from the sun, and is not visible until the sun is that far below the horizon. It is difficult, therefore, to see how it can be solely an outward extension of the solar corona, the intensity of which diminishes rather sharply toward its outer edge. Furthermore, the symmetrical disk of the zodiacal light is not exactly in line with the plane of the ecliptic, but more in line with that of the plane of Jupiter's orbit, which has an inclination to the ecliptic of 1.3°. So the zodiacal light seems more likely to be reflections of some sort from interplanetary space, where the gravitational effects of Jupiter's sphere appear to affect it, than reflections in the atmosphere of the earth, though these may play some minor role in its creation.

The spectra of zodiacal light more closely resemble the absorption spectrum of the sun, with its dark Fraunhofer lines, than the continuous spectrum of the solar corona. Electron scattering of the light probably makes only a minor contribution to it, and tiny dust particles in inner planetary space are probably the major cause of it, reflecting and scattering it into the faint, cone-shaped blur in the sky. This conclusion ties in with the fact that the disk causing the light is in just that location in which dust from the collision of asteroids and meteors and the disintegration of comets might be expected.

Interplanetary Dust

Beyond its probable appearance and effects in zodiacal light, the interplanetary dust has been intensively investigated in recent years because of the effects it might have on satellites at a great distance from the earth and on unmanned or manned spacecraft. The larger particles might pit and hole the shells of space vehicles, while the smaller ones might etch or erode the skins; both could be very dangerous for human occupants, particularly on longer trips, and might quickly disable the instruments or the operations of manned or unmanned craft themselves.

The interplanetary dust, including that found in the generalized-disk area of the zodiacal light, probably originates primarily from comets as they come close to the sun in their perihelion circuit. The cometary nuclei are heated by solar radiation, giving off the dust particles as well as larger particles and gases and producing the comae and tails, which are boiling, buffeted, and turbulent and lose many particles. These are dispersed as interplanetary dust by means of such forces as light pressure, solar radiation drag, the gravitational fields of planets, and collisions among themselves. Asteroids and meteoroids probably produce only a small proportion of the total interplanetary dust. The possibility must not be overlooked that some of the dust particles may be formed by accumulation from the atoms and molecules in space, as well as by the breaking down of larger bodies.

Planets and satellites gradually collect the dust, and much of it may be drawn in toward the sun itself. Current estimates as to how much dust the earth collects from space vary widely—from 1,000 to 10,000 or more tons a day, and 3 to 5 million tons a year. Once the dust has entered the ionospheric level of the earth's atmosphere, it may take a month or more to drift to the surface. Most of the interplanetary dust near the earth is very fine.

The tiny particles of interplanetary dust entering the earth's atmosphere may serve as the nuclei around which water vapor can condense, or condensation nuclei, as these are called. This dust may then explain how the noctilucent clouds form at such great heights (Chapter 6). The effects of lunar gravitation on this dust in the upper atmosphere have also been suggested as the explanation of lunar-phase rainfall. Some scientists believe that most of it must be concentrated clouds of dust blasted into cislunar space when meteorites crash into the moon. Laboratory studies have been made of the rock and sand particles sprayed out in a jet reaction when high-velocity projectiles strike. When meteorites strike the moon, some of the surface material may be sprayed up at greater than escape velocity, most of it going into orbit around the sun, but part of it, up to a ton or more of lunar dust a day, falling into the earth's gravitational field and being attracted toward it, or being collected again by the moon. Some of this dust sifting down to earth, then, may well be material from the surface of the moon, and it merits all the study being given it.

A "Venus fly-trap" rocket experiment, named for the flower that captures flies by closing its petals, has provided some indication of the nature of this dust. A rocket was sent up by the Air Force Cambridge Research Laboratories with sections that opened when it reached an altitude of 55 miles, stayed open as the rocket soared to a maximum height of 104 miles and started to fall back, and closed again when the rocket reached 72 miles altitude on the way down. While the sections were open—for a total of 4 minutes—cosmic-dust particles were collected on plates within coated with a thin metallic film that indicated the nature of the particles grazing them. Most of the particles were smaller than 1 micron (0.000039 inch) in diameter. The greatest proportion (72 per cent) were irregular in shape, but some of them were almost round (16 per cent), and others were "fluffy" or smashed particles (12 per cent). Some of the particles were found to be clumped together.

While the concentration of these tiny dust particles is still far from accurately known, it has been found they do not furnish too great an obstacle for space vehicles. The dust in interplanetary space should not be so dense, in terms of its average concentration, that it will greatly erode the surfaces of vehicles passing through it, if the surfaces are fairly hard and thick. Actually, the effect should be greater near the earth and in cislunar space than out in the farther reaches of space. This does not include the more sizable meteoroids, small and large, which may be found in space.

Space has provided designers of satellites and space vehicles with a variety of other knotty problems as well. Space contains no air to carry away by convection currents the heat generated by motors and other operating equipment in the vehicles. All of the heat must be radiated directly away from the hot surfaces, or from specifically designed antenna radiators which may require a great deal of extra weight and take up much needed space in the satellites. Polished aluminum exposed in space to solar radiation heats up to about 575° F., as hot as most baking ovens, since aluminum is a surprisingly poor radiator of

First ultraviolet photograph of earth made from lunar surface by Apollo 16 astronauts shows hydrogen ball or corona around the earth out to about 50,000 miles.
—GEORGE CARRUTHERS, U.S. NAVAL RESEARCH LABORATORY, AND APOLLO 16 ASTRONAUTS, NATIONAL AERONAUTICS AND SPACE ADMINISTRATION

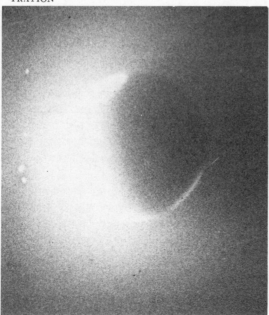

heat. A thin layer of silicon monoxide has been found to improve its heat-radiating capacity. The heat balance of Mariner 2 was not judged correctly, and its batteries began to heat up alarmingly as it came closer to the sun in its approach to Venus; fortunately, the batteries still functioned. The heat balance will be corrected in future designs.

The high vacuum of interplanetary space, containing only 10 to 1,000 particles per cubic centimeter, creates other design problems. The grease on bearings evaporates or sublimates into space very quickly. The coefficient of friction of these bearings increases because in space they lack two things present in a normal atmosphere: a thin coating of adsorbed gases and moisture from the atmosphere and a layer of oxide, both of which serve to form a shield keeping the metal surfaces from immediate contact. But the gases, moisture, and oxides evaporate into space, leaving the clean, raw metal surfaces, between which cohesion and cold welding may take place when they contact each other. They may weld tightly together even at temperatures as low as 50° to 75° F. Even graphite does not serve well as a lubricant in space. Consequently, units with rapidly moving parts and bearings often have to be completely enclosed in pressurized containers. Bearings have been developed of a solid lubricant, molybdenum disulphide, sintered or merged directly with the bearing metal. These and similar new products may eventually provide a solution for lubrication problems in space.

The x-ray, gamma-ray, and cosmic radiation in space, and the intense radiation of the trapped protons and electrons in the Van Allen belts, have crippled some radiation-sensitive devices, particularly electrical equipment, in space. Solar cells that generate electricity using sunlight have proved sensitive to the heavier Van Allen radiation. Many new designs are being developed to overcome this and give the cells a longer effective life. Transistors and other semiconducting devices have had to be placed in contained vacuums, since any gases around them are ionized by the radiation.

CHAPTER VIII

"Atoms or systems into ruin hurled,
And now a bubble burst, and now a world."

POPE

When the solar system was first formed, some of the ingredients must have been left over, for bits and pieces are still scattered helter-skelter through it, nearly 5 billion years after its birth. The result is by no means as neat and orderly as a perfectionist would desire. Over and beyond swirling masses of interplanetary gas and dust, there are thousands of so-called minor planets, or asteroids, and countless meteoroids and comets, large and small.

Minor Planets

At any time, day or night, a great asteroid 10 to 20 miles in diameter might suddenly darken the sky or black out the stars and crash into the earth unannounced, causing untold destruction over a wide area several hundred miles around its point of impact and leaving a great smoking crater, perhaps 50 miles across, blasted deep into the crust of the earth. Such a catastrophe is only an extremely slim possibility, however; it is much more likely that small asteroids passing near the earth will be put to use as space stations. Certainly they would make good ones—protecting men dug in under their surfaces from injurious solar radiations, furnishing plenty of room for observatories and research laboratories, and providing cheap and almost inexhaustible sources of nickel iron for the industries of the earth, the moon, and even of Mars, if and when it can be colonized.

The asteroids are rocky, subminiature bodies, in orbit around the sun like the nine major planets.

The first four asteroids discovered (Table 8) were the largest: Ceres (1) (the number assigned to the asteroid), with a diameter of

ASTEROIDS, METEORS, AND COMETS

about 480 miles; Pallas (2), 304 miles; Vesta (4), 240 miles; and Juno (3), 120 miles. Besides these, probably a handful of asteroids are more than 30 miles wide, and there are many no more than a mile or two in diameter. They are such small bodies that Vesta, with a variable visual magnitude of about 6.1, is the only one occasionally barely visible to the unaided eye; the rest can be sighted only through telescopes. Only a dozen or so are seen to have disks when viewed with the largest telescopes—the rest are simply points of light. As they move in direct orbits around the sun, they make trails on photographic plates in telescopes set to track the stars. This is the way most of them have been found, that is, by chance, and frequently no one has bothered to work out their exact positions and orbits.

The four largest asteroids were identified between 1801 and 1807 (Table 8). No more were added to the list until 1845. Since then, asteroid discoveries have been fairly continuous, becoming more numerous each decade. By 1890 about 300 had been identified. On a sampling basis, not by direct counting, it has been estimated that there may be an over-all total of 80,000 asteroids up to the 19th magnitude, but the majority are so diminutive that all put together their mass cannot be $\frac{1}{6}$ that of the earth and is probably less than the total mass of the moon. Roughly 3,000 of these tiny, glinting celestial objects have been identified as asteroids by their behavior. About 1,650 have had their orbits calculated, and have been given numbers and sometimes names and entered in asteroid registers. But they are such faint, fast-moving bodies that the five or more good observations needed to determine their orbits reliably have just not been made in most instances.

The courses of most asteroids lie in the great gap between the orbits of Mars and

Jupiter. They tend to have more elliptical, squashed orbits than the major planets; the average eccentricity is 0.15, with a few as high as 0.3 to 0.5. One asteroid, Hidalgo (944), discovered in 1924, with an estimated diameter of between 15 and 30 miles, has the longest known period of any of the asteroids—13.9 earth-years, an eccentricity in orbit of 0.656, and an orbital inclination of 42.5° to the ecliptic. In its tilted journey around the sun it sweeps beyond Jupiter and nearly out to the distance of Saturn. There are some others with orbits running out beyond Jupiter, and there may be minor planets swinging in dark, cold, lonely orbits out as far as Uranus.

The average inclination of asteroid orbits to the plane of the ecliptic (the imaginary plane formed by the earth's orbit) is 10°, and only a few of the orbits lie very close to it. Several asteroid inclinations run up as high as 20° or more from the ecliptic. Feodosia (1048), an asteroid identified in 1924 at the Heidelberg Observatory by Karl Reinmuth, who has more than 1,000 asteroid discoveries to his credit, has the largest known inclination of any of the asteroids—53.8°. Since, in the main, they are farther out from the sun than the earth and Mars, with some exceptions the periods of the asteroids range between 4 and 6 earth-years.

Only the four largest asteroids show definite enough disks for their diameters to be measured. The diameters of a number of others have been inferred: given a known distance from the earth in their calculated orbits and assuming an average asteroid reflectivity or albedo, their magnitude is measured and the size of the body which would produce that brightness under those conditions is calculated. Since the albedos of asteroids may vary considerably, this is only a rough approximation. Mercury and the moon have low albedos of 0.069 and 0.073, respectively; those of the asteroids Ceres and Pallas are very similar (0.06 and 0.07). Juno and Vesta have greater reflectivity; that of Juno is 0.12 and that of

Vesta 0.26, compared with Mars's albedo of 0.15.

Most of the asteroids are thought to be irregular masses of rock mixed with some iron and other metals; the smaller ones are probably odd-shaped fragments and splinters rather than neat spheres. These minor bodies definitely have the characteristics of the small inner, rocky planets, not those of the gas giants.

Variation in the track of the Sputnik 3 carrier case passing near Vega (bright star above trace) on July 31, 1958; the satellite's periodic fading from view and then brightening was probably caused by the tumbling of the cylindrical case. The light from some asteroids may vary for similar reasons. A timing device cuts the track of the satellite at regular intervals and the stars made arcs on the plate because the telescope followed the satellite.
—DOMINION ASTROPHYSICAL OBSERVATORY, VICTORIA, B.C.

Studies of the polarization of the light from some of the asteroids, of the color of their light with different filters and with photoelectric measurement, and spectrograms of the larger ones show that with some variation their reflected light is the color of sunlight or slightly yellower. While almost any type of rock would produce this effect, the asteroids are thought to be brown or gray-brown in color and craggy in character, with considerable dust on their surfaces. Whether rock on bodies with such small gravitational fields would have the texture and structure of rocks formed on the earth is open to question. When rocks from the moon can actually be studied, more can be inferred about this.

With their small masses, the minor planets have very little gravitational effect. On Ceres (diameter of 480 miles), weights would be only $\frac{1}{25}$ that on the surface of the earth. An average car would be easy to lift on Ceres,

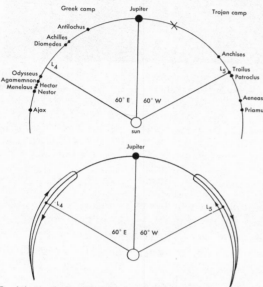

Position of the Trojan asteroids (above) on June 20, 1960, in the Greek camp east of and preceding Jupiter in the L_4 Lagrangian position, and west of and following Jupiter in the Trojan camp in the L_5 position. The "x" west of Jupiter indicates the closest that Anchises approaches the planet. The fairly large oscillation of Trojan asteroids (below) about the L_4 and L_5 positions is produced in part by perturbations caused by Saturn and other planets, and in part because the Trojans are not in exactly the same plane as Jupiter's orbit. Some of Jupiter's outer satellites may be Trojan asteroids that it has captured.
—ADAPTED FROM SETH B. NICHOLSON, No. 381, *Astronomical Society of the Pacific*

since it would weigh under 150 pounds, though one would not want to carry it very far. On an asteroid with a diameter of 10 miles things would weigh only $\frac{1}{1600}$ what they would on the earth. With such negligible gravitational fields, the minor planets could not hold atmospheres.

The light from many of the asteroids varies in brightness, often in a regular manner. Although they may be luminescent, reacting to the solar radiation with which they are pelted, they can have no light sources of their own. Their light pulsation must be caused by their own motion. Eros (433), which sometimes comes within 15 million miles of the earth, has a period of rotation of 5 hours and 16 minutes, during which the brightness of its reflected light shows two maxima and two minima. The generally accepted explanation is that it has an oblong shape, roughly that of a brick, and rotates on an axis perpendicular to its greatest length. The light variations would be due to the angle at which its rotation was visible from the earth; just the ends and sides, also part of the top and bottom, or only the top or bottom might be seen. Some of the artificial earth satellites have developed similar odd motions, such as tumbling from end to end about an axis.

Certain cloudlike satellites have been reported preceding and following the moon in its orbit of the earth, located in the Lagrangian positions called L_4 and L_5 (see diagram, Chapter 4), forming 60° angles between themselves, the moon, and the earth. These positions offer relatively stable locations for bodies to orbit. Much earlier, in 1906, the asteroid Achilles (588) was discovered in the Lagrangian position L_4 preceding Jupiter in its orbit. Achilles has a period of 11.98 earth-years, very similar to that of Jupiter's 11.862 years, but with a greater inclination of 10.30° to the ecliptic than Jupiter's 1.3°. Fifteen similar asteroids, collectively called the Trojans, have since been identified. Two of them, Achilles and Priamus, are shown in the photographs.

The Trojan orbits generally have small eccentricities and a variety of inclinations to the ecliptic. They tend to wander about the Lagrangian positions, principally because of the gravitational effects of Saturn out beyond them. They oscillate around the apex of the 60° triangle formed between them, the sun,

Two of the Trojan asteroids photographed with the 40-inch reflector of the U.S. Naval Observatory. Achilles (588) (left) was the first Trojan asteroid discovered, located in the Greek camp east of Jupiter, near the L_4 Lagrangian position in its orbit. Priamus (884) (right) is in the Trojan group west of Jupiter, following it by about 60° in the L_5 position. Achilles (estimated diameter, 35 miles) was photographed on November 9 and Priamus (diameter, 28 miles) on June 22, 1963, with 10- and 30-minute exposures, respectively.
—Elizabeth Roemer, U.S. Naval Observatory, Flagstaff, Ariz.

and Jupiter (see diagram), but still remain in fairly stable orbits. Probably some escape entirely from, and others are captured in, these Lagrangian positions mainly by the effects of Saturn, or possibly of other asteroids passing in the vicinity. Some of the Trojans are rather large, up to 50 miles in diameter, but their magnitudes run only up to about +12, and they are not visible to the naked eye.

A case has been made for the existence of several families or natural groups of asteroids, on the basis that calculations into the past from their present orbits, weighing in the various major perturbations which have affected them, reveal that they may have a common origin. One investigator came to the conclusion that nearly 200 asteroids appeared to fall into 5 such families. Another, working on the orbits of a greater number of the asteroids, found the same families and possibly 24 more closely related groups. Could such families, however vague the relationships of the members, indicate a common origin in space?

One obvious and popular way to explain the origin of the asteroids is to say they are the remnants of a small planet or primitive protoplanet which was in orbit between Mars and Jupiter hundreds of millions or billions of years ago. Then this planet broke up in some cosmic catastrophe and was gradually fragmented and pulverized by further collisions into the small bodies we call asteroids.

A planetary catastrophe of this sort is possible. It could be brought about by explosion, violent rotation, tidal disruptions when a large satellite or a planet like Jupiter is too close, or collision with some other body, such as an unusually dense comet. If two or more small planets were at one time in close orbits between Mars and Jupiter, as might be inferred from the five or more families distinguished, the collision theory would fit best. A crunching impact between two might lead to other collisions, the formation of more families, and the gradual breakup of the original protoplanets into the small, rocky bodies known today. The larger asteroids might even have been among the original small planets frequenting the gap between Mars and Jupiter, and two or more meeting with a calamity could have produced the remainder of the fragmentary smaller asteroids as they have been identified today. Or the asteroids may be clotted chunks that never formed a planet because of the disruptive effects of massive Jupiter.

A number of asteroids have orbits passing within that of Mars, some sheering fairly close to the earth on occasion (Table 9). Eros (433), the first of this type to be discovered, was found by G. Witt of the Urania Observatory in Berlin in 1898. Since it sometimes came within 15 million miles of the earth, its angle as observed from opposite sides of the earth varied by a whole minute of arc. So on its close approaches, in 1900–01, and 1930–31, it was used to redetermine the length of the astronomical unit. Amor (1221), discovered in 1932, comes within about 6 to 10 million miles of the earth, but because of its faintness (15th magnitude at its nearest) it cannot substitute for Eros.

Several other asteroids have rather eccentric orbits (Table 9), moving well out beyond Mars at their aphelia, crossing the earth's orbit, and passing within the orbit of Venus at their perihelia, as they zip around the sun. Apollo (1932 HA) passed within 1.9 million miles of the earth, Adonis (1936 CA) was only 900,000 miles away, and Hermes (1937 UB) was a mere 500,000 miles from the earth at its closest approach in 1937, about twice the distance to the moon. Hermes was expected to be in a good spot for observation in June, 1943, but no one could find it. Relative to the vast distances in interplanetary space, these asteroids were all but brushing the earth, the gravitation of which strongly perturbed their orbits.

Is the earth or the moon likely to collide with one of these wandering asteroids? The possibility has inspired some excellent fiction, but the truth is that it is very remote. The chance that bodies coming within 600,000 miles of the earth, as Hermes did, will strike the earth, which has a target diameter of only 8,000 miles, is calculated at only about 1 in 30,000. It is not likely to happen more often than once in an estimated 100,000 years. Some

of the alleged terrestrial meteor craters, as well as certain lunar seas, may have been produced by the impact of large asteroids in the past.

Meteors and Meteorites

On September 1, 1962, above Covington, West Virginia, and then again on September 5, over Clarksburg, West Virginia, great fireballs struck down through the atmosphere. The sky was ablaze with their bursting fragments, and sonic booms rattled dishes, furniture, and windows. Newspaper advertisements asked for eyewitness accounts to help plot the paths of these fireballs so they might be traced to the spots where they fell.

An amazing number of meteors hit the earth—at which time they become meteorites—in a 24-hour period, yet no one has ever been killed by one as far as is known. Only one human injury has been definitely recorded when, on November 30, 1954, "stars fell on Alabama." A small, stony meteoritic fragment weighing about 9 pounds crashed through the roof of a house in Sylacauga, Alabama, where Mrs. E. H. Hodges was resting on a sofa after lunch. The fragment ricocheted off a radio and struck her on the upper thigh, causing a slight bruise.

Called meteoroids in space, meteors as they penetrate the earth's atmosphere, and meteorites if they strike, very few of these stony, stony-iron, or iron bodies are ever recovered. For the most part, they land in the oceans, at the poles, or in such sparsely inhabited regions as forests and deserts. The bright but transitory streaks they make in the sky, which give them the name shooting stars, are due to friction with the atmosphere. The streaks are usually described as white, and sometimes as greenish, reddish, or yellowish. Very large and brilliant meteors are called fireballs, and when, rarely, these fireballs explode, they may be known as bolides. The term meteoroid, or meteoric body, is reserved for any fairly small object in space, smaller than an asteroid and considerably larger than an atom or molecule, before it enters the earth's atmosphere, which only a very small proportion of them ever do. Meteoroids in space are nonluminous and thus not detectable by optical telescopes, but eventually some may be picked up by radar telescopes. Because of the threat to space probes and ships, meteoroid-detection satellites with 50-foot wings are being launched to measure the probable extent of meteoroid penetration to which spacecraft on long flights may be exposed.

Here is the history of one meteorite that fell in Alberta, Canada, just a few years ago and was observed in some detail. It entered the earth's atmosphere at 1:06 A.M., Mountain Standard Time, on March 4, 1960. It was an evening meteor, entered the atmosphere at a velocity close to 8 miles a second (29,000 miles an hour), and detonated when about 20 miles above the surface. Fragments from the explosion were discovered across an ellipse-shaped area about 3.3 miles long by 2 miles wide, near Bruderheim, 30 miles northwest of Edmonton, Alberta. The flash of the detonation was visible for some 200 miles, and the noise audible over an area of 2,000 square miles (the equivalent of a square 450 miles on a side). Many fragments were subsequently recovered. Some were picked up on the snow over which they had bounded and rebounded, and farmers plowed up others that spring. The fragments had not had time to disintegrate, as many do. Some 188 sizable chunks were collected, weighing a total of 670 pounds. The Bruderheim meteorite was a chondrite, one type of stony meteorite, gray in color, with a low iron content.

Meteoritics, as the science of meteors is called, has made great headway during the

Elliptical distribution of the individual stones recovered from the Bruderheim (Alberta) meteorite of March 4, 1960. The fragments of the gray chondrite showered down from the west in a nearly west-east line, the larger chunks heavier than 9 pounds (4 kilograms) being carried farther than the smaller ones by their momentum.
—R. E. FOLINSBEE AND L. A. BAYROCK, *Journal of the Royal Astronomical Society of Canada*

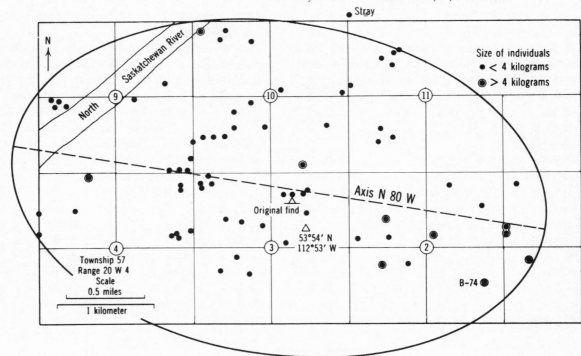

twentieth century, first by means of visual, photographic, and spectroscopic observational techniques, and then with radar. Amateur meteor observers around the world cooperate in the work. On a clear night an average of about 10 meteors an hour can be seen streaking across the sky. When there is a meteor shower, the number greatly increases. A star map is reproduced, designed by Peter M. Millman of the National Research Council of Canada. This is used to accurately plot and record the paths and magnitudes of visually observed

Meteors plotted (numbered arrows) by one observer at Springhill Meteor Observatory near Ottawa in 5 hours on the night of August 14–15, 1963, with their estimated magnitudes (20 indicates +2 magnitude, 25 indicates +2.5 magnitude, etc.). The radiant of most of the meteors was in the northern part of the constellation Perseus, the date falling near the end of the Perseid shower. A few sporadic meteors not related to the shower appear. The whole group of eight observers saw 500 meteors on this night, explaining the range of the numbers.

—PETER M. MILLMAN, NATIONAL RESEARCH COUNCIL, OTTAWA, CANADA

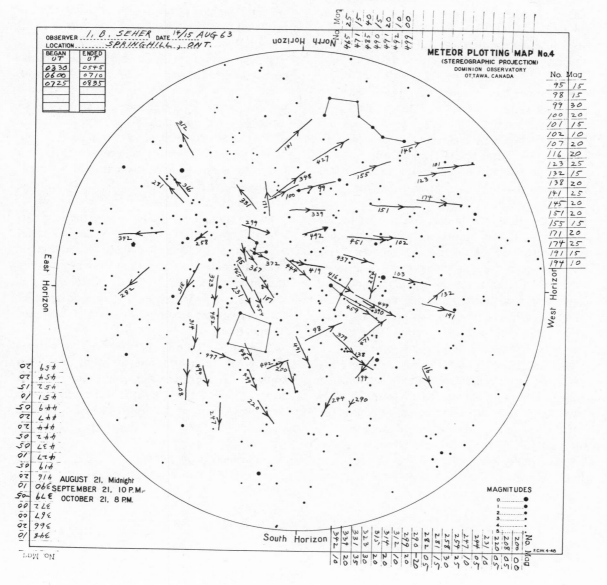

meteors. The paths of many meteors spotted over a period of about 5 hours on the night of August 14–15, 1963, are plotted on the map.

It has been estimated that about 24 million visible meteors pass through the atmosphere of the entire earth in 24 hours. Observations with telescopes up to the 10th visual magnitude indicate that 8 billion meteors must plunge into the earth's atmosphere a day. Added to this are the much more numerous micrometeorites, objects with a diameter of less than a millimeter (0.03937 inch), and the cosmic debris or dust (Chapter 7).

Only very rough estimates have been made of the mass which the earth must accumulate each day as a result of this infall from interplanetary space. The greatest meteorites which gouge out craters in the earth's crust may add from 0.5 up to 1.0 ton a day; fireballs, visual meteorites, and the dimmer telescopic and radio meteorites add from 1 to 10 tons daily. The largest share by far must come from the micrometeorites and interplanetary dust, depositing some 1,000 to 10,000 tons of cosmic debris on the earth each day. It has been calculated that all this must add from 5 to 10 pounds a year to each square mile of the earth's surface. We can stop hoping that smoke control will eliminate all of the dust and grit with which housewives do battle, for some of it has sifted down from the far reaches of interplanetary space, leaving the very stuff of the universe on our doorsteps.

Meteors penetrating the earth's atmosphere often belong to meteor streams or showers. A stream is a group of meteoric bodies in space with nearly identical orbits around the sun. A shower is a number of meteors with approximately the same trajectories actually entering the earth's atmosphere. Sporadic meteors, not linked with any recognized showers, greatly outnumber the shower meteors, though the latter sometimes come so thick and fast that they put on a much more spectacular scene, very like the grand finale of a fireworks display. The meteor radiant is that point where the

path of the meteor intersects the celestial sphere. A spray of meteors appears to come in a shower from the radiant, arching out from it as the ribs of an opened umbrella curve out from the center.

The more prominent nighttime meteor showers of both Northern and Southern hemispheres are listed in Table 10. Although there are variations, most of these appear consistently each year as the earth in its orbit intersects the meteor stream producing the showers. Sizable daytime showers occur quite often, but usually they are observable only by radio techniques. Great showers like the Leonids have sometimes been brilliant enough to be seen clearly in the daytime.

Many meteor streams follow closely in the orbits of comets. Thus the Southern Taurid stream is in the orbit of Encke's comet, and the Draconids are associated with the periodic comet Giacobini-Zinner. On the other hand, there are meteor showers unrelated to any present or previously known comets. A recent estimate is that perhaps 90 per cent of visually observable meteorites are of cometary origin. Most meteoritic orbits, however, have low inclinations to the ecliptic, with no apparent relation to the asteroids.

Opinions have clashed over the question of whether some meteors have hyperbolic orbits and velocities of their own over 26.159 miles a second. If so, they are likely to have come from interstellar space. The tendency now is to consider nearly all meteors members of our solar system. One estimate is that interstellar meteors constitute no more than 1 per cent of the total.

The calculated orbits of the bulk of meteors show that their motions are direct, though some are retrograde. Many of them come in from the direction toward which the earth is moving, and it is quite natural that the earth should sweep up a greater concentration of meteors from this direction. They usually become visible at heights of from 75 miles down to 55 miles in the atmosphere; the

Microphoto of a chondrule in a thin section of a meteorite. Chondrules are spheroidal bodies, usually about 1/25 of an inch in diameter, the origin of which is unknown. They may have been produced by the heating of carbonaceous chondrites, by direct condensation from a cool dust cloud among the earliest bits of matter in the solar system, or by metamorphic processes on larger bodies, although none have been found in terrestrial rocks.

—AMERICAN MUSEUM OF NATURAL HISTORY

brightest, flaunting their fiery and sometimes smoky trains, may plunge down to altitudes as low as 10 to 25 miles.

The actual velocities of meteorites range all the way from about 7 miles a second or less to 45 miles a second. Meteors in space move in elliptical orbits of the sun, with velocities of less than the 26 miles a second which would be escape velocity from the solar system. The earth's own velocity in its orbit is 18.5 miles a second. Thus meteors overtaking the earth have the earth's velocity subtracted and enter the atmosphere at 7.5 miles a second or less; those meeting the earth have the earth's velocity added, so they may enter the atmosphere with velocities as high as 45 miles a second. Meteorites of less than a ton in weight lose almost all their own velocity in atmospheric passage and collide with the earth at velocities of only 0.06 to 0.124 mile a second. Those with masses of 10 or more tons may

retain a good share of their initial velocity on impact.

Meteorites have been classified and analyzed intensively for what they can reveal about the interplanetary space through which they have traveled. The major meteoritic groups that have been identified are the stony meteorites (aerolites), the iron meteorites (siderites), and the intermediate stony-iron meteorites (siderolites). The stony meteorites consist of siliceous minerals in the main (magnesium-iron silicates), and resemble rather soft, light rocks, somewhat like volcanic rock. They may be either chondrites or achondrites. Chondrites are stony meteorites with fibrous, banded, or glassy inclusions, called chondrules, in their mineral matrix. Achondrites consist basically of magnesium-iron silicates without much nickel-iron and usually without chondrules. The stony irons are part mineral and part iron in composition (averaging half nickel-iron and half silicates). Iron meteorites, composed of masses of nickel-iron (averaging 90 per cent iron and 10 per cent nickel) with infrequent silicate inclusions, are very heavy. Some stony chondritic meteorites are called carbonaceous chondrites, since they contain amorphous carbon, graphite, and other carbon inclusions, notably complex hydrocarbons.

Among the over 1,500 meteorites actually recovered, the stony type predominates, with over 900; then come the irons with about 550; and, finally, the very scarce stony-irons, of which only 67 are known. Of the total number of meteorites recovered, 680 were picked up or located after actually being observed falling. A greater number, 860, were discovered and identified merely by their characteristics.

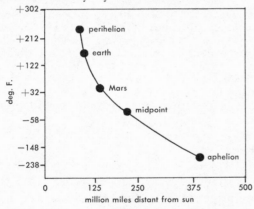

Temperature of the Pribam iron meteorite calculated for various points in its orbit on the basis of laboratory measurements of heat absorption and emission of meteoritic iron. Its orbit ranged out at aphelion to the asteroid region between Mars and Jupiter and almost in to the orbit of Venus at perihelion.

—C. P. Butler and R. J. Jenkins, *Science*

Recently, the temperature of an iron meteoroid in interplanetary space has been calculated. Visualized as bright and shiny as stainless steel, about 10 per cent of its surface might be pitted and cracked, absorbing more heat than the shiny, reflective surface. On this assumption and on the basis that the meteoroid followed the orbit of the Pribam meteorite, one of the well-observed falls, the temperature of the

meteoroid would be low (about —110° F.) at aphelion in its orbit, some 375 million miles from the sun. But when closest to the sun in perihelion, its temperature would have risen to about 255° F. At the earth's distance from the sun, the meteoroid's temperature would be about 195° F.—hot, but well below the temperature of boiling water. If it were an evening meteor, passing through the earth's shadow in driving toward the earth, its temperature might range down within half an hour to below the freezing point of water, depending on its size and the length of time it remained eclipsed by the earth. The temperature of the artificial satellite Explorer 4 dropped to about this extent in a half-hour on the nightside of the earth. So meteors may be either warm or cold as they enter the earth's atmosphere.

Meteoroids are traveling at many miles a second when they dash into the earth's atmosphere, but meeting the resistance of the gases high in the atmosphere begins to slow them down. Their great energy of motion must be dissipated, and by the time they have reached a height of 75 to 50 miles they are glowing in the sky because of the heat that has been produced. They become white-hot and their surfaces molten, streaming back from the direction of their travel, with drops flaming off and sometimes exploding or fragmenting and forming a number of wakes or trains in the sky. They lose fluid and vapors to the atmosphere in the process known as ablation, by which heat is rapidly carried away. The same ablation effects have been used to advantage in the design of spacecraft or missile nose cones, which must reenter the atmosphere with as little destruction as possible. The intense heat produced by the friction of their passage through the air must run off or be shucked off, as it were, with the molten nose-cone material.

As they pass through the atmosphere the larger meteors produce loud booms, the result of shock waves formed by their supersonic speed of entry. Few meteors are large enough to survive the atmosphere. Those that reach the earth's surface have been so slowed down that they have lost most of their surface heat, have formed what is called a fusion crust on their surfaces, and are barely warm, or may even be cold, to the touch. They cannot possibly start fires, as one might assume. The Hoba iron meteorite of southwest Africa probably weighed 100 tons when it fell, and a number of others weighing from 10 to 30 tons are known. Meteorites of over 100 tons will probably never be found, since their impact would be so explosive that they would be entirely vaporized or fragmented. Meteor craters constitute the sole evidence that such massive bodies have fallen from the sky.

A dozen well-authenticated meteoritic craters have been identified on earth, and many others are candidates. Their locations and principal features are given in Table 11. Iron or stony meteorites and many meteoritic fragments have been found in or near most of these craters. The Siberian formations were made as recently as 1908 and 1947, perhaps by clusters of meteorites or small comets. Their effects on seismographs and other instruments were noted at these times, and the air blasts of the earlier event felled forests and killed herds of reindeer. All but the Barringer Crater were identified in the twentieth century, yet most are thousands of years old.

The so-called fossil craters probably range from millions to billions of years old. In 1961 Peter M. Millman of the Canadian National Research Council and others thoroughly investigated the site of an alleged meteor crater in Algonquin Provincial Park, Ontario, and certified it as a fossil crater. Called the Brent Meteor Crater, it is about 2 miles in diameter, roughly circular, and dips down over half a mile deep beneath the material which has filled it in. Paleozoic rocks found within it are evidence of its antiquity. Judged to be of the Precambrian era, it must have been formed some 600 to 900 million years ago.

Evidence other than meteoritic fragments has been adduced as proof of meteor craters

and impact zones, particularly of the fossil variety, in which the meteoritic material as such may be entirely dissipated. Shatter cones have been found at a number of older, suspected meteor crater sites, such as the Sierra Madera circular structure near the W. J. McDonald Observatory near Fort Davis, Texas, or the huge Vredefort Ring in the Transvaal of South Africa, southwest of Johannesburg, which has a diameter of some 140 miles. Another type of evidence, the mineral coesite, has been found in the rock at the Ashanti Crater, Lake Bosumtwi, in Ghana, and in the Ries Kessel basin, 17 miles in diameter, located in southern Germany. Are these definite proofs of meteorites?

Shatter cones are conical chunks of rock with grooves or fissures (striations) radiating from their apexes. They vary in length from less than an inch up to many feet, and have been found in many kinds of rock, particularly in limestone and sandstone. They appear to be shock-wave products, and have been produced experimentally by firing small pellets at a velocity of 18,000 feet a second into limestone. Some think they may be the product of great volcanic stresses; others, that massive shearing forces such as those involved in the production of geological faults may have formed them. That they are solely the result of the heavy impact of meteorites and thus good evidence of meteor craters has not yet been proved.

The mineral coesite was first discovered in 1953 in the laboratory of the Norton Company in Worcester, Massachusetts, by Loring Coes, Jr., and was named after him. Coesite, a superdense form (or polymorph) of silica (SiO_2), was produced by Coes at pressures exceeding 300,000 pounds per square inch, pressures that meteorites striking the earth could initiate. Coesite has, in fact, been found in small quantities at a number of sites, including the well-known Barringer Meteor Crater, near Flagstaff, Arizona, the Holleford Crater in Ontario, Canada, at the less authenticated

Ries Kessel basin in Germany, and at the Wabar Crater in Saudi Arabia. Most remarkably, coesite was identified at the Teapot Ess Crater at the Yucca Flats, Nevada, nuclear test site, produced by the tremendous pressures from an underground nuclear explosion. The high pressures at which coesite was created in the laboratory occur at depths greater than 40 miles under the earth's surface, but it is possible that greater upheavals nearer the surface might produce such pressures, as well as meteorite impacts. Together with shatter cones, the presence of coesite may document in part the proposition that "a meteorite was here," but neither proves it.

Two stony meteorites of the type called chondrites were analyzed to determine their ages since their separation from a larger body and they were set quite reliably at 100,000 to 200,000 years, remarkably young for even the stony meteorites. Where did these stony meteorites come from? The iron meteorites with great cosmic-ray-exposure ages may come primarily from the asteroidal belt out between Mars and Jupiter, and it would take millions of years for their gradual perturbations in orbit to finally bring them to the earth. However, one iron meteorite only about 800,000 years old has been found, reaching the earth in a remarkably short time. Similarly, it has been believed that the stony meteorites with young exposure ages may have come from the moon, splashed or blown up from its surface by impacts of sizable bodies such as asteroids that may have dug out some of the lunar seas. Carbonaceous chondrites tend to have shorter average exposure ages than the other chondrites, and a lunar origin for them is not implausible. Still, the reliably calculated orbit of one chondrite appeared to indicate that it came from the asteroidal belt, and the velocities of many of the chondrites as they approach the earth bolsters this theory.

Ejection from the earth itself, or from Mars, in a collision with an asteroid in the remote past is another possibility. The place of origin of meteorites, particularly the stones, and the paths and times taken to reach the earth, are up in the air, both literally and figuratively. The theories now current hold that the meteorites came from the breakup of an original planet-sized body, occupying the space between Mars and Jupiter, from a number of bodies the size of the moon, from comets, or perhaps from a variety of smaller asteroidal or planetesimal bodies.

Glassy Enigmas

An absorbing story of detection is unfolding from studies of the small, glassy stones called tektites, found strewn in fields around the earth. Tektites range in size from about 1 to 4 inches, and are usually translucent yellowish-brown, but sometimes green. Silica-rich, containing 70 to 80 per cent silica, they also have high proportions of alumina, potash, and lime and small amounts of magnesia, iron oxides, and soda. They may be spheres, dumbells, teardrops, disks, or buttons. Early theories attributed their origin to terrestrial mineral or earth-soil fusion caused by flashes of lightning or volcanic eruptions. Currently, debate centers on the question of whether they sprang from terrestrial forces, were produced on earth by meteoritic, asteroidal, or cometary impacts, or may have been splashed up from the moon or other bodies in similar cataclysms, flying through space to be deposited finally on the earth. Tektites have not been seen falling, unfortunately, but they are to be found in Australia (australites), Indochina (indochinites), southern China, the Philippines (philippinites), Malaya (malaysianites), Thailand, Borneo, Java (javaites), Tasmania, the islands of Banka and Billiton (billitonites) in the Java Sea, the western African Republic of Ivory Coast, Czechoslovakia (moldavites), and Georgia and Texas (bediasites) in the United States.

One single fragment of a pale brownish-green tektite was reportedly found in a wash in the cliffs at Gay Head, Martha's Vineyard, Massachusetts, in 1960, below the only Miocene layer remaining on the New England coast. At least three groups of investigators have since combed the vicinity without finding any more. Perhaps some sailor off a whaling or coastwise ship dropped this memento of his voyages to the far corners of the earth, or to Georgia or Texas. Or perhaps this was a fragment separated by a detonation from other tektites as the parent body sped into our atmosphere.

Tektites have been dated by comparing the radioactive decay of the potassium 40 found in them to that of argon 40, the proportion of the radioactive elements in them roughly indicating their ages. The age of tektites in the United States has been set at about 34 million years, that of the tektites in Czechoslovakia at 15 million years, and those of the Australia-East Asia groups at about 700,000 years. But not enough clear-cut evidence has yet been marshalled to prove whether tektites originated in great meteoritic impacts on the moon or on the earth.

Only the Australian tektites appear to have been melted and then remelted. They might have been melted when splashed up from the moon and then remelted on passing through the earth's atmosphere; or they might have

(A) An australite "button" tektite that shows evidence of two periods of melting. The flanged edge indicates high-velocity flight through the atmosphere in which, in the second melting, the ablating material from the front melted, ran back, and formed the flaring flange (× 2.9). (B) Two bediasites from Texas (× 1.8). (C) The hollowed side of a Georgia tektite. (D) A tektite from Brunei, Borneo, with a surface flow pattern indicative of its internal structure.
—VIRGIL E. BARNES, DIRECTOR, RESEARCH ON TEKTITES, UNIVERSITY OF TEXAS

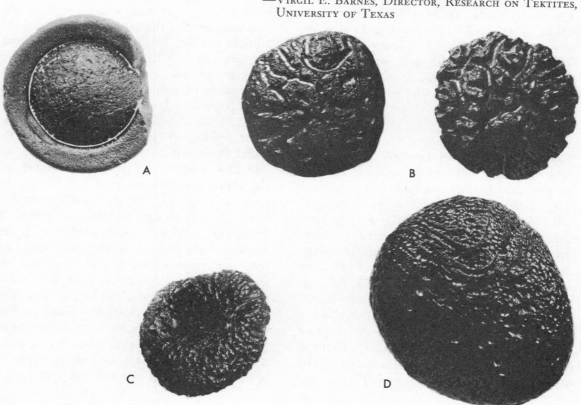

been melted when a huge meteorite impact on earth flung them up through the atmosphere at high velocities and then remelted on their reentry into the atmosphere after an orbit or part of an orbit of the earth.

On the one hand, tektite materials may have come originally from the earth, since they are related in chemical composition to igneous rocks, unlikely to occur on the surface of the moon, and not akin to the composition of meteorites. The analysis of certain tektites from Thailand, for example, has revealed the presence of mineral grains in some layers, most likely formed from earth materials. On the other hand, nickel-iron spherules have been found recently in a Philippines tektite; these spherules are closely related to, if not a kind of, meteoritic material. And, while they have been challenged, various calculations have led investigators to believe that the speed (7 to 8 miles a second) and angle of entry of the Australian tektites into the atmosphere indicate the moon as their most probable place of origin. Tektites may have come, in one instance, from the Cyrillic meteor shower of 1913, which orbited the earth in a nearly circular fashion and may have consisted of debris from meteor impacts on the moon.

Comets

Comets may be great luminous bodies with long streaming tails, clearly visible in the sky, or they may be tiny specks or smudged blurs when reflected in even the largest telescopes. Men may be filled with wonder at their loveliness or struck with terror of the doom comets are believed to portend. They suddenly emerge from the darkness at the fringes of the solar system and streak in toward the sun from almost any direction. Their orbits may be entirely changed by the mighty pull of Jupiter or the incandescence of the sun may rupture them. Their tails change direction and shape in the sky, and they vanish into the depths of space. Some are periodic, coming back year after year; others may be seen only once, perhaps never to appear again, or to return thousands of years hence. Few generalizations can be made about comets; they seem to behave as randomly or irrationally as frantic moths attracted to a light.

Sizable comets near the sun have a distinguishable head and tail. The head consists of a bright, hazy mass, called the coma (from the Latin for "hair"), and usually a bright, more or less central condensation or nucleus. The nucleus is believed to be a kind of loosely gathered body of metallic grains and frozen gases, perhaps averaging half a mile in diameter.

Comets shine by reflected sunlight, not by their own radiance, becoming brighter as they approach the sun. The average diameter of a comet has been estimated at about 70,000 miles, largest at about 1.4 a.u. from the sun, decreasing in diameter as perihelion is approached, and decreasing again when out beyond 1.4 a.u. The tail, which grows as the comet approaches the sun and may extend for millions of miles, becomes oriented away from the sun, just as a flag flutters away from the wind, except that cometary tails are influenced by streams of solar particles and radiation.

On September 1, 1961, Milton L. Humason, of the Mount Palomar Observatory, California, discovered Comet Humason (1961e), the fifth comet discovered or rediscovered that year. At that time it was a very dim, diffuse comet, of the 14th magnitude, with no tail, moving in toward the sun very slowly from a great distance, in an orbit with an estimated period of 2,000 years. When it had reached a point between the orbits of Jupiter and Mars, in July and August, 1962, it had developed an irregular tail about 9 million miles long, with an area judged to cover about twenty times the size of the full moon in the sky. Suddenly the comet collided with something, either solar particles or a radiation storm caused by a solar flare or a stream of gas or dust particles in space. Or possibly it had been affected by in-

Marked changes in the coma and tail of Comet Humason (1961e), photographed on July 9 and August 8, 1962, with the 40-inch Naval Observatory reflector. Discovered on September 1, 1961, by Milton L. Humason of the Mount Palomar Observatory, the comet was still at a distance of some 250 million miles from the sun, between the orbits of Mars and Jupiter, when these "explosive" effects occurred.
—Elizabeth Roemer, U.S. Naval Observatory, Flagstaff, Ariz.

ternal explosions. Whatever disaster befell it, its parts disintegrated and its tail rapidly changed form and volume, demonstrating the extremes of metamorphosis that comets can undergo.

On the average 5 or 6 comets are discovered or rediscovered each year, although in 1947 there were 14; two-thirds, or about 4 out of the 6 are usually new. They are named for their discoverer or discoverers and are identified by the year of discovery, followed by letters giving the order of discovery in that year—1956a, 1956b, and so forth. If they prove to be periodic when their orbits are calculated, they are identified by the year of their passage through perihelion about the sun (which may differ from the year of their discovery), and their order among the other comets in perihelion during that year is given in Roman numerals. Thus, the first periodic comet passing perihelion in 1964 was known as 1964I, the second as 1964II, and so on.

Comet Arend-Roland (1957III), discovered by Sylvain Arend and Georges Roland at Uccle, Belgium, on November 8, 1956, developed an odd spike directed toward the sun in April, 1957. At first this appeared fairly broad, then it became sharp and pencil-like (April 26), and then faint and stubby (April 29)—all within a week's time. One suggestion was that it was actually very thin but fan-shaped, and when the earth crossed the plane of the comet's orbit it was seen edgewise as a pencil-thin (less than 10,000-mile) jet directed toward the sun, even though it was millions of miles wide.

The existence of many faint comets, so small and dim that they are not discovered as they move into and out of their perihelion passage in the vicinity of the sun, was implied by the small, diffuse image found on photographs taken by scientists from the Goddard Space Flight Center of the National Aeronautics and Space Administration in a 20° area around the sun during its total eclipse of July 20, 1963. A blue-green filter was used to reveal any carbon molecules, which are abundant in comets. The image appeared on 7 of the 40 photographs taken, and while its identification as a miniature comet is not final, it might be profitable to follow up this lead in eclipses to come. There may be many more comets than has been supposed.

Whether any comets have definitely para-

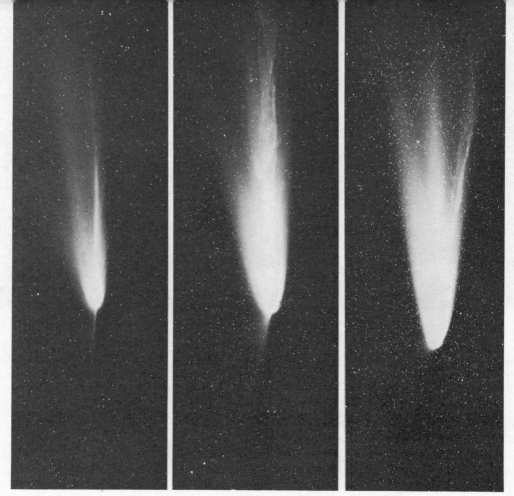

Photographed with the 48-inch Schmidt telescope on Mount Palomar in 1957, Comet Arend-Roland's (1957III) odd "spike" points toward the sun, changing radically in form and fading from April 26 to April 27 and 30.

—Mount Wilson and Palomar Observatories

bolic orbits with a velocity which will take them out of the solar system never to return is an open question. Many comets have nearly parabolic orbits, with an eccentricity very close to 1, in that portion of their orbits which can be observed near the sun. But comets have very little mass and can be strongly affected in their orbits by the sun, by huge Jupiter, and even by Saturn and other planets. In addition, their orbits are hard to determine since they are pulled tightly through perihelion near the sun; the exact positions of such large, diffuse bodies are difficult to pin down. However, most, if not all, comets are thought to be in elliptical orbits over the long run, but the wavering paths of these orbits near Jupiter and the sun makes calculation difficult.

Some comets have such great ellipticity in their courses that it will be thousands or even hundreds of thousands of years before they will return to the sun's vicinity, for they lose much of their velocity, and must move very slowly, near and through their aphelia. Other comets have periods not exceeding a few to a few hundred years; the Schwassmann-Wachmann 1 Comet (1925II) is an example. Revolving in a 16-year orbit entirely within the space between Jupiter and Saturn, it is visible in telescopes even at aphelion, when farthest from the sun. Occasionally it flares up to become 100 times as bright as usual—comets are known to do odd and inexplicable things of this sort. The sequence of photographs shows the changes in its appearance between October

Discovered in 1900 and definitely associated with the October Draconid meteor shower, periodic comet Giacobini-Zinner (1959VIII) is viewed on October 21, 1959, in photographs on the same scale in blue light (left) and yellow light (right) with exposures of 15 and 45 minutes, respectively. The blue-light photograph shows the distribution of material strongly emitting in blue light (by a process of fluorescence, activated by sunlight)—principally, CN, C_3, and NH of the head, and CO^+ and other ionized molecules on the tail. The yellow-light photograph shows the distribution of a few neutral molecules of C_2, but the principal contribution is of small solid particles which simply reflect and scatter sunlight.
—Elizabeth Roemer, U.S. Naval Observatory, Flagstaff, Ariz.

12, 1961, October 18, when it had become much brighter and extensive, and November 3, when diffuse material is visible in a kind of haze around its small, central bright area. One astronomer has suggested that it may be composed of antimatter, annihilating the matter of large meteors colliding with it in great explosions which cause its increase in brightness and the diffusion of material around it. Periodic Comet Encke (196II), a faint comet normally, now and then increases suddenly in brightness for causes unknown. This occurred in 1957 and again in 1960.

The first periodic comet identified (1682) was Halley's Comet, named for its discoverer, Edmund Halley, an English astronomer con-

temporary with Isaac Newton. It has reappeared about every 75 years. It returned in 1759, again in 1835, and last passed perihelion in 1910. In 1948 at aphelion it was invisible, far out beyond Neptune, and must now be moving back slowly in its retrograde orbit toward the sun. One of the most brilliant and impressive of comets, with an enormously long tail, it is expected to reappear in May, 1986.

When Halley's Comet approached the sun in 1910, the earth passed into or very close to its tail, and a good many people prepared to meet their Maker. However, nothing very much happened. Its tenuous tail may have caused very small variations in illumination in the earth's atmosphere, but they were not seen

One of the strange "outbursts" of periodic comet Schwassmann-Wachmann 1 (1925II), photographed with 30-minute exposures on October 12, October 18, and November 3, 1961. The envelope expanded with a linear velocity of about 200 miles an hour. Although the comet generally appears stellar or nearly so, as on October 12, it experiences such eruptions, which are not related to its distance from the sun, on the average of once or twice a year. The brightness may increase by as much as five magnitudes in a single day, the light becoming 100 times more intense.
—ELIZABETH ROEMER, U.S. NAVAL OBSERVATORY, FLAGSTAFF, ARIZ.

in the bright moonlight conditions at the time. When Halley's Comet then passed directly between the earth and the sun, the most meticulous observations with the best telescopes of the time revealed no trace of it against the sun's disk. Comets have been known to dim the light from bright stars slightly when they pass in front of them, as seen from the earth, so they are not completely transparent.

Theories about the nature and origin of these cosmic blowtorches are not easy to construct. Their behavior is often strange and seemingly contradictory. Conditions in comets are so different from familiar terrestrial conditions that few analogies can be drawn. Perhaps the most widely accepted theory of the

Periodic comet Encke (1961I) undergoing a great increase in brightness between (left) September 26, 1960, and (right) January 6, 1961, during approach to perihelion. The photographs, each 30-minute exposures with the 40-inch reflector of the U.S. Naval Observatory, are on the same scale.
—ELIZABETH ROEMER, U.S. NAVAL OBSERVATORY, FLAGSTAFF, ARIZ.

nature of comets is the one proposed in 1950 by the American astronomer Fred. L. Whipple. Called an "icy-conglomerate model" of comets, it visualizes the cometary nucleus as a porous mass of very finely divided mineral particles, perhaps in loosely related matrices called dustballs, immersed in an ice of solidified gases. The outer portion of these masses is heated and vaporized as the comet nears the sun, giving rise to the jets, the expansion of the coma, and to the gases streaming behind it in the tail.

The principal molecules in the nuclei of comets, inferred from spectroscopic evidence, are ammonia, methane, and possibly acetylene, with water and small amounts of such metals as iron, nickel, and sodium. These may be held as an agglomeration of ices frozen solid at great distances from the sun—though hydrates have been suggested in another model, and free radicals in still another, formed and trapped by radiation, perhaps when the comet is distant from the sun in its orbit. As comets near the sun, various ions of these substances appear in the coma and tail, while the metal traces are identified when the comet comes very close to the sun.

A theory was developed from bits of evidence that comets probably have a halo or "atmosphere" of gas atoms around their heads. To check this theory, observations of two bright comets were made in 1970 from the U.S. Orbiting Astronomical Observatory (OAO 2) and the Orbiting Geophysical Observatory (OGO 5) satellites. Sure enough, spectrographic and photometric instruments on these observatories reported the head halos around the comets that had been predicted, although they are invisible in naked-eye or optical telescope viewing.

Analysis of the satellite data showed that the comets had great halos of hydrogen encircling their heads, extending out about 6 million miles. Hydroxyl (OH) radicals were also found in these halos, so perhaps water molecules from the icy comet nuclei had been broken down in space into the hydrogen and hydroxyl particles. Perhaps 100 tons of comet nucleus

per second vaporizes to produce these halos. These observations tend to confirm the icy-conglomerate model of comets.

Some of the dustballs in the nucleus may be ejected or parts of the head may disintegrate and give rise to meteor streams or showers, meteoric dust, and some of the smaller, porous, stony meteors, as well as causing change in shape and brightness. Many meteor streams have been found to follow the orbits of comets. As the outer layers of the comet evaporate, the mineral particles collecting at the surface would be heated and might give rise to the explosive jets seen in the coma and the gas streams of considerable velocity seen in the tails. Such jets might also explain why the nuclei of many comets rotate; in similar fashion spacecraft are rotated around their axes by small jets of stored nitrogen or hydrogen peroxide gases on their surfaces, in this case to maintain a stable position or to effect a change in attitude.

Many theories of the origin of comets have been proposed, but none is considered adequate or accepted by most scientists today. Earlier discredited theories held that comets had been expelled from the earth, the other planets, or from the sun. It was thought that the sun's prominences might sometimes be ejected and become entirely detached from it. More recently, comets have been ascribed to masses of gas and dust about the outer rim of the gravitational field of the solar system, 100,000 to 200,000 a.u. from the sun; or to interstellar dust and gas clouds through which the sun may carry the solar system, scooping up some accretions of these materials to form the comets.

Plans are being drawn up to launch a space probe to pass through a comet's head. Technically this should be quite easy, since many comets cross the plane of the earth's ecliptic somewhere between the orbits of Venus and Mars, closer to the earth's orbit than either of these planets. Fifteen periodic comets are being studied to select the most promising target.

CHAPTER IX

"There was the veil through which I might not see."

A casual observer, a low-powered telescope, and a momentary viewing are all that is required to establish that the planets Venus, Jupiter, and Saturn are surrounded by those gaseous envelopes known as atmospheres. Mars's atmosphere, too, makes its presence fairly obvious, but Mercury does not so easily yield up its secrets in this respect, and the presence or absence of an atmosphere on the moon is, surprisingly, an even more vexing problem and the source of endless, heated debate.

One rough indicator of an atmosphere around a body is its reflectivity, or albedo (Table 5). The moon and Mercury have very low average albedos—Mercury 0.056 and the moon 0.067. Venus' albedo, in contrast, is 0.76, Mars's is 0.16; the earth's has been estimated at about 0.36, Jupiter's is 0.73, and Saturn's is 0.76. On the basis of albedo alone, the moon and Mercury are immediately set apart. It is a great temptation to attribute their low reflectivity to the absence of an atmosphere and thus of clouds to reflect the light.

Needle in the Lunar Haystack

The controversy over the presence or absence of an atmosphere on the moon has exhibited the all-too-human proclivity to adopt extreme points of view and to dabble in theories that go far beyond what is actually observed. It has been habitual for most astronomers and other scientists with lunar interests to think of the moon as utterly devoid of activity or atmosphere—a dead place. The American astronomer Fred L. Whipple has vividly expressed this point of view: "The moon's surface . . . is a

ATMOSPHERES OF THE MOON AND PLANETS

sublime desolation. The lunar plains are more barren than rocky deserts. The lunar mountains are more austere than terrestrial peaks above the timber line. . . . There is no weather on the moon. Where there is no air there can be no clouds, no rain, no sound. Within a dark lunar cave there would be eternal silence and inaction excepting possibly moonquakes. A spider web across a dim recess in such a cave would remain perfect and unchanged for a million years." The only question remaining is: Whence came the spider?

On the opposite side of the fence are those scientists, most prominently those who have devoted lifetimes to patient observation of the moon, who report that they have seen changes taking place on the moon and even definite indications that it has an atmosphere, however slight.

During an eclipse of the sun it would seem that any lunar atmosphere should appear as a delicate, luminous fringe along its edge. Peer as they will, however, astronomers have not seen this fringe—which proves only that the moon does not have a heavy atmosphere. A very thin one might well not appear against or near the glaring and mobile surface of the sun or its corona. The moon's edge as seen from earth, or limb as it is called, simply appears dark, clear-cut, and scalloped with mountains under these circumstances. But as a test for an atmosphere this is much too crude and hardly conclusive. Bright comets will disappear completely near the sun, swallowed in its light as they approach it in their orbits.

A second method of checking out the moon's atmosphere is observing it as the lunar limb occults, or passes in front of, one of the brighter stars. When an occultation of a star by Venus or Mars is observed, the star dims perceptibly and quite rapidly as its light passes through thickening layers of the planets' atmospheres. Lunar observers report to the contrary that the point of light from the bright star remains steady and then is suddenly cut off as it passes behind the limb of the moon, "snapping out" as an electric light does when the switch is flicked.

The presence of even a very thin atmosphere, perhaps down to one ten-thousandths of the density of the earth's at sea level, should produce a twilight effect by scattering the sunlight falling on it. This effect should be noticeable in the extension of the sharp horns, or cusps, of the crescent moon along the limb of the moon beyond the terminator, the line dividing the lighted side of the moon from the dark side. It might also be observable in the scattering of light across the terminator itself into the dark side on the disk of the moon, when gases were present in the area. What has been discovered on this score?

Not at every crescent moon, but on rare occasions, several skilled observers have noted the twilight effect in the extension of the cusps of the moon. This effect, clearly and frequently seen when Venus is a crescent, may, however, be hard to distinguish on the moon from the well-known earthshine, the dim reflection by the moon of light from the earth. Earthshine should be visible across the face of the darkened moon as a widespread dim glow, not concentrated at the cusps or along the terminator of the moon, and astronomers who have seen the cusp extension maintain it is not earthshine. High mountains range around the south pole of the moon, with sizable peaks also near the north pole. It has been suggested that the glinting of the light from these peaks after the sun has set for the lower levels produces this extension of the cusps, rather than a lunar atmosphere or a haze produced by gas emission. A long lunar ridge or mountain will often glint in the sun's light while the lower levels are still in darkness. So the twilight technique is also not a very sensitive one for distinguishing effects of a lunar atmosphere.

Observers often tell of mists or clouds in or near craters on the moon, and the walled-plain of Plato, just above the Mare Imbrium, has been notorious for this. Now and then a colored spot or a glow is seen in a crater area and interpreted as emanating from the crater itself. If this indicates volcanic activity, this could be at least one source for gases on the moon. The crater Alphonsus in 1958, and possibly again in 1959, may have emitted gases. A number of clouds or mists were said to have appeared within this crater, which is in the south-central portion of the moon, and in 1963 there were similar appearances in and near the crater Aristarchus, near the northwestern limb. The moon may well emit quite a number of such atmospheric blobs of residual gases, quickly diffused. Whether or not they form any very tenuous general atmosphere before they are swept away or escape from the moon's low gravitational field is another matter.

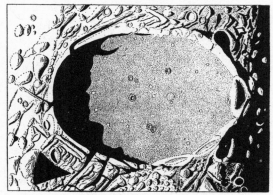

Plato, a walled plain some 60 miles in diameter, situated in the mountainous area between the Mare Imbrium and the Mare Frigoris toward the north of the moon. This careful drawing, showing a number of the tiny crater pits within the very smooth and level plain, was made by the British selenographer H. P. Wilkins, on April 3, 1952, using the 33-inch Meudon, France, refractor.

—H. P. WILKINS AND PATRICK MOORE, *The Moon*

During the last 15 years, radar echoes picked up by radio telescopes from signals bounced from the moon have shown unexpected drops in intensity, which implies that some of their strength was absorbed by passage through ionized gas. The ionized layers in the earth's atmosphere may have done some of the absorbing, and a slight lunar atmosphere may have absorbed the rest, but we know too little about the ionosphere to rely very heavily on speculations of this sort.

However, within the past few years radio astronomy has at last turned up some definitive evidence that there is a lunar atmosphere. The discovery was related to the fact that an atmosphere refracts radio waves in much the same way as it refracts light. In 1956 the moon occulted radio sources in the constellation Opiuchus. Again in 1959 it occulted radio sources in the Crab Nebula. Observations of these two events showed refractions of radio waves that could only be caused by a lunar atmosphere. The atmosphere responsible is estimated to be only about one ten-trillionth (10^{-13}) as dense as the earth's. It might prove to be a bit denser if argon could be measured by this method—which it probably was not. In any case, however tenuous it may be, the atmosphere is unquestionably there, and a long-standing debate is settled, leaving the field wide open for speculations about how this atmosphere was generated.

With only $\frac{1}{80}$ of the earth's mass and $\frac{1}{6}$ the earth's gravitational potential on its surface, the moon cannot have retained very much of its original atmosphere, if it had one, or of any atmosphere that may have evolved in the course of its development. Furthermore, a goodly proportion of its gases would congeal during the moon's night, when the temperature plummets to about −250° F.

There are several possible sources for a slight lunar atmosphere. One is the production of some relatively heavy argon gas by the radioactive decay of potassium in the rocks at or close to the lunar surface. Most of the minute amount of argon in the terrestrial atmosphere was probably produced in this way. Another possible source is the outgassing from volcanic or subsurface bubbles or domes of gas—too many gassy mists have been seen to discount

this possibility. The gases would probably include sulfur dioxide, carbon dioxide, hydrogen, and possibly water vapor, constituents of volcanic gas. Water vapor might also come from the layers of ice that some astronomers believe lie fairly close to the moon's surface, particularly near the poles.

Depending on the porosity of the dust and rock on the lunar surface, some of the gases would soon be absorbed. More would be jostled and swept into escape from the moon by the piercing solar wind, a formidable and fairly constant feature of interplanetary space, with a mean velocity of from 250 to 300 miles a second. In the face of such a wind, the unshielded moon would lose the battle to retain any atmosphere worth the name. Because of its probably very weak magnetic field—estimated at 100 gammas or less—it would lack the protection from the solar wind that the Van Allen belts give to the earth. There is also the distinct possibility that the lunar electrostatic field is sufficiently high to repel the ions even of heavy gases like argon, xenon, and krypton so forcefully that they would escape from the moon.

After Surveyor unmanned craft and Apollo astronauts actually landed on and studied the moon, most of these questions of a lunar atmosphere were answered. There is no water on the lunar surface, very little water in lunar rocks, and no permafrost, or frozen water, has been indicated below the surface. No signs of recent volcanic activity have been found in the few spots sampled. The solar wind does sweep the lunar surface and many of its particles are embedded at or near the surface. On one occasion, instruments left on the moon by astronauts did signal the presence of water vapor. But it was soon decided that the water must have leaked from containers in the base of the Lunar Expeditionary Module (LEM), left on the moon as the astronauts took off from it to return to the earth.

The possibility remains of local spots of gases briefly emitted from domes, rills, or craters, probably following moonquakes. The lunar atmosphere has been largely, but not quite completely, eliminated.

Veils over Mercury

Mercury is small. Its estimated diameter is 3,030 miles, a bit larger than the moon's and a little more than one-third of the earth's. Yet its density is nearly as great as that of our own planet, which makes it considerably more compact than either the moon or Mars and probably slightly more dense than Venus. This compactness is puzzling. Is Mercury composed largely of iron, or ferro-silicates, as has been suggested? No one knows.

One would expect such a dense body to have a high gravitational field, but because of Mercury's small size the escape velocity from its gravitational field is only 2.2 miles per second, in contrast to the earth's 7.0 miles per second. The acceleration of gravity is only 11.8 feet per second per second at Mercury's surface, compared with the earth's 32 feet per second per second. Thus atmospheric gases would not be so tightly bound in Mercury's gravitational field as they are in the earth's.

Early astronomers concentrated on Mercury's transits, which occur every 5 to 10 years. At such times it is seen as a tiny black dot crossing the disk of the blazing sun and looking smaller than some sunspots. Some of the early observers saw a light ring appearing around the dark disk of the planet when it passed in front of the sun. This gave rise to the belief that Mercury had a huge atmosphere, and this belief persisted for a long time. Other observers saw a gray or a dark-colored ring instead. It was finally concluded that the ring was an illusion caused by the extreme contrast of the black planet and the sun's flaming surface. Contemporary observation of Mercury in its transits shows no such misty ring. Photographs taken on May 11, 1937, as Mercury appeared to be passing through the corona only a few minutes of arc from the sun's limb, proved this conclusively.

At other transits during the nineteenth century, Mercury's black disk appeared to show a very bright spot, which caused a great furor in the astronomical world. This bright spot was interpreted as a high mountain, a seething volcano, and even as a Mercurian attempt to communicate with the earth—on the theory that intelligent Mercurians could deduce that we would be watching the transit! The 1937 photographs put an end to these flights of fancy. The spot was simply an instrumental error. The need to cut down telescope apertures so that Mercury could be seen in front of the sun was causing a diffraction effect.

Physical conditions on Mercury bespeak very little, if any, atmosphere. With its small size and low gravitational attraction, gases on its surface must escape very quickly. Temperatures on Mercury's daytime side must run high, perhaps between 400° and 750° F., while on its nighttime side they drop to −300° F. or below. These figures are based on few data, because Mercury is a difficult object to observe, staying close to the sun. Mercury rotates so slowly that it may have little, if any, magnetic field to yield a magnetosphere that would protect an atmosphere from the fierce solar wind battering it with particles. These conditions imply very little atmosphere.

The question is sometimes raised whether or not Mercury, and the moon as well, may have what is called an equilibrium atmosphere of a heavy gas such as argon—in equilibrium between what is lost into space, into condensation, or by photochemical decomposition under the sun's violent radiation, and what is added to the atmosphere from space or from sources within the planet itself. An atmosphere of this sort would not, in all probability, be much denser than interplanetary gas.

Argon gas has no lines in the visible part of the spectrum and could not be identified in spectroscopic analyses of the light from Mercury. However, it does make up some 0.9 per cent of the volume of the earth's atmosphere, probably originating in the main from radio-active decay of potassium 40 in the earth's crust, as has already been described. There is reason to believe that argon could have been produced on Mercury in a similar way. With an atomic weight of 40, argon is heavy enough not to escape readily, perhaps not even from Mercury. Actually, its high temperature on the side of Mercury facing the sun would place it on the borderline of escape, while its proba-bly very low temperature on the dark side might just about freeze it out of the atmos-phere. However, there is speculation that a slight atmosphere of argon or other heavy gases might result in enough circulation to keep the dark side less cold and the bright side less hot, so that an equilibrium atmosphere could be maintained.

The great Greek-born French astronomer, E. M. Antoniadi, a student of Mercury, was ambivalent on the question of its atmosphere. He wrote in the 1920's that Mercury does not have a gassy envelope in the usual sense. Its atmosphere, he wrote, is "absolutely invisible." Yet he was an astute observer of Mercury, found a number of faint markings on its sur-face which he was convinced were permanent, and wrote very persuasively of the "cloudy veils" or "clouds" that appeared to move over the permanent markings, sometimes dimming them or causing them to vanish. He believed that these obscurations implied that Mercury had winds in which clouds moved about, and that the veils might be composed of fine dust particles raised from the surface and blown about by the winds of the "invisible" atmos-phere. A contemporary French astronomer, Audouin Dollfus, has made most scrupulous observations of Mercury with the 24-inch re-fractor at the Pic-du-Midi Observatory and has not been able to confirm the existence of the cloudy veils so definitely seen by Antoniadi.

The polarization of light from the cusps of the moon failed to turn up indisputable evi-dence of even a very thin lunar atmosphere. Studies of the polarization of light from Mer-cury show an effect intermediate between that of the waxing and waning phases of the moon and very close to that of the full moon, so their surfaces must be very similar to each other. But Mercury has a stronger gravita-tional field than the moon, and other studies of the polarization of its light, compared with that of the moon, have revealed that it in-creases more rapidly at the cusps of the cres-cent phase than at the center, with an increase in the angle of the incident light. Dollfus, who made these polarimetric studies, concluded in 1953 that the stronger polarization of the light at Mercury's cusps, not noticeable for the moon, must be due to a weak atmosphere on the planet, perhaps three-thousandths of the earth's atmosphere (1 to 2 millimeters of mer-cury pressure as opposed to the 760-millimeter standard pressure of the earth's atmosphere at sea level). He also commented that "some-times local regions of the planet seem to show departures in polarization. . . . These results, if real, may indicate temporary veils in the at-mosphere. For instance, a dust cloud could weaken the local polarization." The "cloudy veils" have enwrapped us again!

In February, 1961, at the Crimean Astro-physical Observatory, the Russian astronomer, N. A. Kozyrev, obtained 20 spectrograms of the light from Mercury when it was high in the sky during a total solar eclipse. He re-ported emission lines in these spectrograms which he tentatively interpreted as caused by luminescence in an atmosphere on Mercury, attributing the lines to hydrogen gas, and sug-gesting they indicated an atmosphere $\frac{1}{100}$ as dense as that of the earth, of perhaps 7–8 millimeter mercury pressure at the surface. He believed Mercury might have an equilibrium atmosphere of hydrogen, steadily losing it by escape into space, with fresh hydrogen from the solar wind steadily replenishing it. A more definite spectroscopic identification of carbon dioxide above Mercury was announced in 1963 by V. I. Moroz of the same observatory. A much heavier gas, carbon dioxide would be much more likely to be retained by Mercury than hydrogen.

Perhaps, then, there are traces of an atmosphere, even a thin equilibrium one, on Mercury, composed of lighter gases like hydrogen or helium, or more probably of heavier, relatively inert gases such as carbon dioxide, argon, krypton, and xenon. Perhaps the veils come from fluorescing free radicals, bits of broken molecules and ions dancing in the radiation from the sun. The hazes may be simply diffusions seeping from large gas-filled domes beneath the planet's surface, or blown out by ranges of volcanoes, or bubbling out in the heat of the surface itself. Or they may be dust raised electrostatically above the surface by repulsion, not carried in a gaseous atmosphere at all.

Models of Venus

At one time the American astronomer Percival Lowell thought that Venus' atmosphere was nearly transparent and that its blurred markings were on the surface; others have believed they could see through the clouds to the surface on occasion, but it is accepted now that Venus has a heavy atmosphere. Venus' albedo of visible light is about 0.76, making it more brilliantly reflective than any of the other inner planets. Only the great outer planets have albedos as large or larger than that of Venus (Table 5). Venus' high albedo must be due to reflection from its clouds.

Venus is swathed, then, in a dense, almost opaque, yellowish-white atmosphere of clouds. The bright rim of this atmosphere has been viewed on those rare occasions when Venus makes a solar transit, and the atmosphere definitely affects the light from bright stars when Venus occults them.

Distinct elongations of the horns of Venus' crescent occur frequently. This type of phenomenon has often been looked for in the crescent moon, but rarely, if ever, actually seen. On Venus, the horns or cusps are often greatly extended in thin arcs, sometimes forming a complete narrow ring around the whole dark portion of the disk. This is caused by the scattering of the sun's light in the Venusian atmosphere and proves it to be considerable. The fuzziness or duskiness of the terminator line across the disk of Venus also reveals its atmosphere, demonstrating the scattering through the atmosphere of the light falling on the sunlit side of the disk toward the twilight and nightside of the disk. Often the terminator appears much more fuzzy and less clear-cut than the limb, or lighted edge, of the disk of Venus.

Either as the Morning or the Evening Star, gleaming brightly in the sky just before sunrise or after sunset, Venus is rather difficult to observe. The earth's atmosphere is ordinarily seething near the horizon, and while Venus' phase can be clearly identified by its shape, usually the planet appears to be nothing more than a blank, cottony surface, with such intense brightness that it causes trouble for the observer. Many of the best observations of Venus have therefore been made during the daytime when the planet is high in the sky, protecting the telescope's objective with a shade. Great effort has gone into trying to penetrate Venus' mysteries by prolonged visual and photographic observation.

Venus' atmosphere has certain features remarked by most persistent observers. Dusky, dim shadings can be viewed across the disk; these have been photographed quite clearly. Bright areas may also appear and persist for some time. Venus has cusp caps which seem to wax and wane in intensity, and on rare occasions a strange "ashen light," somewhat like the earthshine reflected from the earth on the moon, has been viewed on its darkened side.

Some observers have believed that the dusky shadings or markings on Venus observed in violet light are more or less parallel bands, belts, or stripes, somewhat like the very strongly marked belts on Jupiter. Others observing in

attrack

Venus photographed in ultraviolet light by Mariner 10 in 1974 just after the spacecraft had passed the planet. The central dark area, marking the subsolar point, and the long banners of clouds swirling around the planet are typical of its striking atmospheric effects, but the surface is not visible. —Jet Propulsion Laboratory, NASA

yellow light have asserted that the markings usually radiate from what is called the sub-solar point on Venus, directly beneath the sun's rays at the center. Since the bands or stripes seem to be seen most often through ultraviolet filters, and the radial markings or spots in visual light, it has been suggested that different atmospheric levels may be represented, so that both the radial and the parallel markings may actually exist distinct from each other, or at different times.

Other observers have maintained that in these dusky markings they are seeing the surface of Venus itself. Probably this is not the case, for the shading or mottling of the disk of Venus appears to be transient; it is not as if information were gradually being pieced

together about some permanent objects which are revealed in numerous dim glimpses. Even the net result of very persistent observations of Venus has not permitted astronomers to say with assurance, "Here is what I think I have been viewing from different angles and under varying conditions." The same impermanence is a feature of the bright patches. Like cloud or storm effects on Venus, they seem to be transient, not something beneath the atmosphere on the planet's surface. The bands, radial markings, and patches may well manifest the movement of large cloud and wind systems in Venus' atmosphere, differing from those that Tiros satellites have photographed in the earth's atmosphere because of Venus' lack of, or slight, rotation.

Observations of Venus with more effective techniques (such as studies with filtered light, of the polarization of the light, and by spectroscopic and radar instruments) reveal that a considerable amount of carbon dioxide may compose part of the atmosphere of Venus, greatly exceeding the proportion of this gas in the earth's atmosphere. There have been spectroscopic indications of carbon monoxide in the Venus daytime atmosphere, and nitrogen in the nighttime atmosphere, but these constituents are only tentative. It has also been asserted that there is water vapor there. Measurements of sunlight reflected from Venus, made in November, 1959, from a balloon at a height of about 81,000 feet above the earth, definitely implied water vapor, and lots of it, above the clouds in the Cytherean atmosphere but these measurements have been questioned.

Three distinct theories of Venus' atmosphere were called the greenhouse, the aeolosphere, and the ionosphere models. They implied very different surface conditions. According to the greenhouse model, the surface would be a dusky, almost dark, and very hot desert—the sun might be seen dimly through a heavy layer of clouds, and the air probably would be calm. Under the aeolosphere model, the surface would be entirely dark, with gales buffeting a dust-laden and hot atmosphere; the ionosphere model implied that the surface would be more like that on earth—reasonably warm and possibly entirely covered by an ocean. These three models were based in large measure on microwave radio emissions from Venus, which were discovered at 3- and 10-centimeter wavelengths. These indicate extremely high temperatures of from 500° to 800° F., higher than the heat of the average oven. The models of Venus' atmosphere were proposed to explain where these temperatures occur and why they are so extreme.

On December 14, 1962, the American space probe Mariner 2 flew past Venus at a distance of $21,648 \pm 10$ miles from its surface. Its few, carefully selected instruments reported unexpected data, which went far toward proving that the planet is unique in many ways. While some of the data have not been finally interpreted, they definitely imply that Venus is not a sister planet of the earth.

The dust detector on Mariner 2 strangely did not record a single dust-particle impact in the vicinity of Venus; near the earth it would have been bombarded with the particles in the rather dense dustball which enwraps it. The significance of this lack of dust near Venus is not yet clear.

In another negative result, Mariner's magnetometer, sensitive to a magnetic field of 0.5 gamma, about $\frac{1}{100,000}$ of the magnetic field of the earth, gave no indication of a magnetic field at its distance from Venus. This implies a dipole magnetic field, if any, no stronger than $\frac{1}{10}$ to $\frac{1}{20}$ of that of the earth and probably much weaker. It is possible that Venus has a very weak multipole magnetic field, or that its magnetic field was compressed by the solar wind in its vicinity more than that of the earth, so that Mariner 2 did not penetrate it.

This also implies that very weak, if any, radiation belts similar to those of the earth's Van Allen belts exist on Venus, and probably very few, if any, of the auroras associated with such belts. In fact, the high-energy radiation experiment on Mariner II confirmed the lack of this sort of radiation in Venus' vicinity.

But the reports of the two radiometers aboard Mariner 2 showed that they had picked up emissions from Venus' atmosphere and surface and further opened the lid of its Pandora's box of mysteries. They implied very little water vapor in Venus' atmosphere, less than $\frac{1}{1000}$ of that in the earth's atmosphere. The temperatures previously inferred from radio-echo studies of Venus were confirmed. Finally, there is little, if any, carbon dioxide in Venus' upper clouds, apparently, though its concentration at the lower levels may be considerable. In addition, over the regions scanned by the radiometers no breaks in the cloud cover were found.

A second Mariner space probe flyby of Venus and three Russian probes dropped from passing spacecraft into Venus' atmosphere, the last two of which broadcast surface conditions for 23 and 50 minutes, have confirmed the results from Mariner 2. Venus has a very heavy atmosphere, 90 to 100 times the density of that of the earth at the surface, it has been shown, largely composed of carbon dioxide.

The evidence from these space flights to Venus has discredited the ionosphere model of the planet on many points. No direct indications of the dense ionosphere of this model have been found. Limb darkening occurs, rather than the limb brightening required by an ionosphere. The surface temperature may run, with variations, up to 900° F., again counter to the ionospheric model. Since the illuminated hemisphere seems somewhat warmer than the dark side, the evidence favors the greenhouse model—the aeolosphere model implies little if any difference in temperature between the two hemispheres. Beneath a heavy layer of clouds with considerable carbon dioxide and possibly water vapor, Venus' lower atmosphere would be at the high temperature indicated.

The greenhouse model should not be assumed to have been fully verified, however. Other evidence may come in to force a full revision of the thinking about Venus' atmosphere, requiring a radical adjustment of the greenhouse model, some combination of these models, or an entirely new model. But it is very unlikely now that enormous oceans of water, hydrocarbons, or other liquids cover the surface of Venus or that it has a temperature much lower than 800° F. Great progress has been made in understanding the atmosphere, although its exact composition and structure are still in doubt.

Gases detected spectroscopically in Venus' atmosphere are carbon dioxide and monoxide, water vapor, ammonia, and hydrogen chloride. Astronomers have had trouble identifying the

material in the visible clouds of Venus' upper atmosphere. They have discarded proposals of very small silicate or salt particles, ferric chloride, pure water or ice, and mercury compounds. Carbon suboxide (C_3O_2) and hydrochloric acid (HCl), which would be yellowish in combinations or dissolve iron from dust particles to yield brownish colors, have recently been proposed to explain the straw colors in Venus' clouds. But carbon suboxide and hydrochloric acid do not fit the evidence perfectly, and while not absolutely excluded, the likelihood for either is not high. A new look is being taken at the Venus cloud composition

Mixed Martian Clouds

Anyone who has observed the planet Mars for even a short time can assure you that it has an atmosphere; the frequent variations reveal it. Sometimes the planet's entire disk is shrouded in clouds or mists; then on occasion most of the vapors have cleared away and portions of the surface can be viewed distinctly. The atmosphere of Mars has been susceptible to the usual proofs. Photographs of Mars in red or infrared light penetrate deeply into its atmosphere and the details of its surface are clear, while plates taken in the violet or ultraviolet record the outer portions of the atmosphere and therefore a much larger sphere.

On the basis of spectroscopic examination, there can be very little water vapor in Mars's atmosphere. According to recent polarization studies there is less than $1/1000$ as much water vapor in the Martian atmosphere as in the earth's. Mariner-Mars reports have shown that its polar caps consist largely of frozen carbon dioxide, but with some water. And in polarization studies, some of the whitish clouds above Mars have been shown to contain ice crystals. Theoretical calculation of the abundance of water vapor has shown it to be present in very small amounts, and this has recently been confirmed spectroscopically. If there is so little water vapor, Mars' surface temperatures are too low to allow much liquid

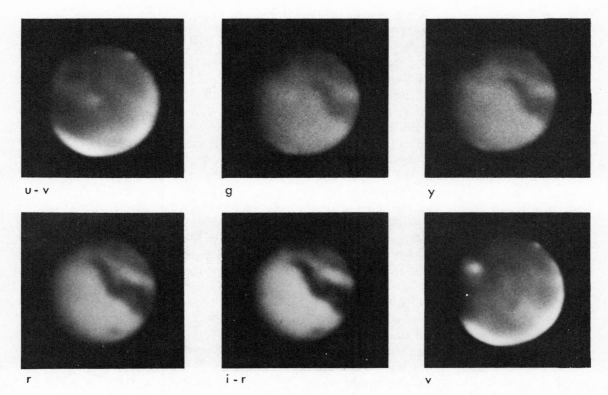

u - v   g   y

r   i - r   v

Mars photographed in six colors: ultraviolet (u-v), green (g), yellow (y), red (r), infrared (i-r), and violet (v). Made at the Lick Observatory with the Crossley reflector, these photographs show how much more plainly the dark markings of its surface stand revealed in the red and infrared.

—Lick Observatory, Mount Hamilton, Calif.

water, except perhaps in absorption in the ground, and the water present on Mars would change from water vapor to ice or back again directly by sublimation.

Some observers believe the circulation of the water vapor to be found on Mars is very simple. It may, they think, sublimate from one polar cap directly into vapor in the atmosphere, diffuse in the white clouds toward the equator, and then pass down to the other polar cap, where it freezes as a kind of hoar frost a fraction of an inch thick on the surface.

In 1963 at the Jet Propulsion Laboratory, Lewis Kaplan, Guido Münch, and Hyron Spinrad made critical spectroscopic observations which gave an estimated atmospheric pressure on Mars of about 30 millimeters of mercury (compared with 760 millimeters atmospheric

pressure at the earth's surface). On the basis of similar observations, Gerard P. Kuiper, of the Lunar and Planetary Laboratory of the University of Arizona, concluded tentatively that the atmospheric pressure on Mars was only about 7.6 millimeters of mercury, $\frac{1}{100}$ of the earth's sea-level pressure and about the pressure to be found at an altitude of 100,000 feet.

Three Mariner spacecraft have flown by Mars, collecting data as they passed, and one American and one Russian craft have orbited the planet, with a Russian craft soft-landing on Mars but returning no photos or data from the surface. Among many other features of Mars, they have confirmed the earth observations that the planet's atmosphere is very thin, with a pressure of between 6 to 7 millimeters of mercury. Also they have indicated that the atmosphere is largely carbon dioxide, with a little water vapor on occasion, and with only a little oxygen in the upper levels of the atmosphere.

Such an atmosphere is not livable for man, of course, both in terms of pressure and in lack of oxygen. Astronauts landing on Mars would need to be space-suited as on the moon.

Clouds, mists, and hazes have been the clearly observable features of the Martian atmosphere and corresponding attention has been focused on them. The composition of the yellow and white clouds on Mars may have been fairly well identified, but the so-called blue or violet hazes or veils have not.

The yellow clouds seen near the surface of Mars are probably the effect of dust storms. Large portions of the Martian surface are colored yellow-ochre or buff; they are thought to be dry regions somewhat like the vast desert areas here on earth, providing the dust for the yellow clouds. During Mars's opposition to the earth in 1956, its surface was largely obscured by such clouds, veiling the planet for a great proportion of the time. Surface markings cannot normally be viewed through these clouds, so the opposition was a great disappointment to astronomers.

The dust of the yellow clouds, if it be dust, does not usually reach great heights, for well-defined white clouds are sometimes observed above the yellow ones. In 1956, an opaque, bright yellow cloud appeared above the surface and soon had extended to form an almost continuous belt around Mars in the temperate zones. Later the south polar cap was also covered, and when the yellow cloud had cleared the cap appeared to have dimmed markedly from its previous brightness. Even after the dust seemed to have largely settled, the sunset limb of the planet showed a distinct light yellow border, or halo. The polarization of the light from the yellow clouds has been studied intensively and shows their composition to be that of the Martian deserts.

The varying white clouds of Mars predominate. They are usually large clouds, last-ing sometimes for several days or even weeks, and can run up to 1,250 miles long. Ordinarily, they move slowly over the surface features, at rates up to about 20 miles an hour. On occasion, they appear on the limb of Mars as bright prominences. They may be surrounded by lighter hazes and can sometimes be resolved with good seeing into smaller, composite formations. They tend to appear above certain regions frequently, particularly the darker areas of Mars's surface.

The polarization curves for the white cloud formations on Mars are very similar to those for fogs of ice particles, and it is most often assumed that they consist of such particles floating fairly high in Mars's atmosphere, as the ice crystals composing the cirrus clouds do above the earth. However, these clouds might be made up of crystals of carbon dioxide (dry ice), if the temperature of the Martian upper atmosphere is low enough (about $-200°$ F.) to make it condense and crystallize. The polarization curves definitely show some ice crystals to be present, though their proportion of the total is undetermined.

Among other types of white clouds in Mars's atmosphere are small, bright, and often isolated clouds, which may appear each day in about the same place, though with varying shapes. These clouds may form over surface elevations or other formations that might cause updrafts, for example, similar to the bright formations often seen around the polar caps where elevations are probable. Other white-cloud formations are definitely hazes or fogs which appear locally on the limb of the planet each morning and disappear a few hours after sunrise. On occasion, they form in the evening near the sunset limb. Bright and fairly permanent clouds seem to form over the polar regions during their fall and winter seasons also. Hazes do not appear as frequently over what are called the desert regions. All in all, the observation of its white clouds makes the Martian atmosphere seem very similar to that of the earth, although they do not shroud it to the same extent as do our clouds, which often cover large portions of the earth.

Although the yellow and white clouds were not observable in the photographs from the earlier Mariners, they were brilliantly confirmed by Mariner 9, which began its orbit of Mars as the planet was shrouded in yellow clouds. These do consist of fine dust, mostly silicates like sand, raised to great heights in the Martian air by fast-blowing gales. And waves of high white clouds were visible, mainly along the limb and terminator of Mars, but in other places as well. Earth telescopes were not so bad!

Almost a complete mystery, the blue or violet haze high in the Martian atmosphere is definitely there, though no one knows of what it consists. One interpretation of the Martian haze is that it is nothing more than morning or evening fogs of water vapor or droplets of water, sometimes connected with the white clouds or seen above them. Most observers, however, are pointing to a much more widespread and permanent feature when they refer to the Martian blue haze. Usually it covers most of the planet and obscures the surface markings to a greater or lesser extent, though "blue clearings" of a very striking nature have been observed. Two phases, or types, of this blue haze have been distinguished: a diffuse, low-contrast haze, spreading over great regions of the planet; small, higher contrast blue "clouds" which move with the surface features and drift about from day to day.

Actually the blue haze is not blue in color —it is called blue because it absorbs the blue component of solar light, so that the blue part is extinguished, leaving the actual reflected light brighter in the red than in the blue end of the spectrum. Thus we call Mars the "red planet," though this is for its ochre or somewhat reddish colored desert regions as well.

One of the currently favored explanations of the blue haze is that it consists of a layer of ice crystals at a considerable height above the surface—perhaps somewhat similar to, though higher than, the earth's noctilucent clouds. The crystals would have a very small diameter (about 0.3 micron), which might tend to absorb the blue wavelength of light but would be transparent at longer wavelengths. Again, it has been suggested that the haze may consist of very fine suspended particles; if so, how could the haze clear as quickly as it sometimes does? The haze has also been ascribed to carbon dioxide crystals and to carbon smoke. A recent explanation for the haze and its clearing is simple: the condensation of water vapor on dust nuclei during Martian summer periods of highest humidity, resulting in precipitation, thus clearing the air. Perhaps this mystery has been solved.

Over the last 50 years, continued observations by telescopes and then by spacecraft have constantly whittled down the elements in Mars' atmosphere similar to those of earth. Thus, the possibility has become smaller and smaller that the Martian air is of the type likely to nurture and protect life as it is known on earth, although this question is considered in greater depth in Chapter 12. The first Mariner spacecraft seemed to show that Mars was a dead, almost unchanging, body, like the moon. But the data from the orbiting Mariner 9, showing dust storms and cloud systems on Mars, have proved many changes occur there.

Atmospheres of the Outer Planets

Jupiter's light is on the whole more yellowish than the sun's. It has yellow zones with a series of gray belts parallel to its equator. Some of the darker gray has a brownish cast, and traces of pink or blue can be seen occasionally. There are some very white zones that stand out in contrast to the yellowness. And there is the great Red Spot, which has recently grown much fainter than its striking hue in the late nineteenth century. The names adopted for Jupiter's zones and belts are shown in the chart. Many of these can be picked out in the photograph taken with the 200-inch Hale telescope, in which the great Red Spot is also clear.

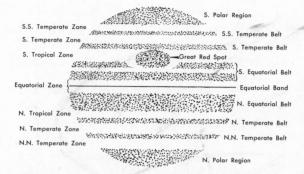

S. Polar Region

S.S. Temperate Zone
S. Temperate Zone
S. Tropical Zone

S.S. Temperate Belt
S. Temperate Belt
Great Red Spot
S. Equatorial Belt

Equatorial Zone {
Equatorial Band

N. Equatorial Belt
N. Tropical Zone
N. Temperate Zone
N.N. Temperate Zone

N. Temperate Belt
N.N. Temperate Belt

N. Polar Region

Chart of Jupiter's dark belts and lighter zones with the names that have come to be used for them and the position of the great Red Spot.
—Bertrand M. Peek, *The Planet Jupiter*

The light zones and dark belts around Jupiter must trace currents in its atmosphere that are maintained by its rapid, 10-hour rotation. While the belts are semipermanent, their position, size, and form vary, and sometimes the variations occur very rapidly. The Red Spot in Jupiter's south tropical zone has maintained its elliptical shape over the years, but it has moved up and down in latitude and wandered in longitude. While it measures about 25,000 miles long and 8,000 miles wide, its size has varied. Unique in the solar system, no one knows just what it is.

Jupiter has a strong magnetic field, perhaps 6 to 8 gauss compared with the earth's average 0.5 gauss. On the basis of Jupiter's microwave radio emissions, Van Allen belts must encircle the planet as they do the earth, in conjunction with the magnetic field. Beyond this, decameter radio waves (several meters long) are emitted by Jupiter in strong, sharp bursts. Their explanation has been a hard nut to crack. They have finally been related to the position of Io, a satellite close to Jupiter, in relation to the planet and the earth. Io apparently triggers the radio bursts from Jupiter's Van Allen belts. As it moves through the belts, which act like a huge particle accelerator, at a given angle Io causes the radio bursts directed toward the earth. If Pioneer 10 space

probe survives to fly to Jupiter in 1973, much more should be learned about this.

Within its strong magnetic field lies Jupiter's swirling atmosphere. Infrared scans were made of its disk for the first time in 1963 with a very sensitive detector in conjunction with the 200-inch Palomar telescope. Jupiter's temperature averaged about —228° F. across its disk, with extremely small variations and a drop of only 2° to 3° toward the edge of the disk. Jupiter's dark belts, its bright zones, and the great Red Spot did not vary in temperature by more than 1° F., and no change of more than 2° was discovered with latitude. This temperature was a little lower than that previously determined by measuring the massed radiation from all parts of the disk.

Jupiter's atmosphere would hardly be exhilarating to breathe. Spectroscopic studies have shown that hydrogen is its main gas, with quantities of helium, methane gas, and ammonia as well. The clouds and belts visible above Jupiter probably are frozen ammonia crystals in the main.

Jupiter's great quantities of hydrogen and helium suggest a mixture similar to that of the sun. Further study may show that Jupiter has been keeping its gases in "cold storage" since the early days when the solar system was formed, some 4.5 billion years ago, and reveal much more about how the sun and other stars form solar systems and planets develop.

Since Jupiter's clouds show varying colors in brown, orange, and yellow, it is inferred that ammonium hydroxide and sulfides, as well as water and hydrogen sulfide, are likely to be present in Jupiter's atmosphere in small quantities. And given these compounds, other simple organic compounds similar to those present in the primitive atmosphere of the earth may be found eventually in Jupiter's atmosphere. The same basic pre-life chemical reactions that occurred on the earth may at present be taking place above the planet Jupiter.

Saturn's light is yellower than that of the sun. Yellow and green colors predominate in

Jupiter's Red Spot, and a shadow of the moon, Io, plus Jupiter's cloud structure are shown in this photograph taken in late 1973 by the Pioneer 10 spacecraft from a distance of about 1,580,000 miles from the giant planet. (The picture has been reversed to conform with the view of the planet as seen through a telescope.)

—National Aeronautics and Space Administration

its atmosphere. Belts band it, too. They are parallel to its equator and much less distinct than Jupiter's. The belts also change in form and appearance. Rare white spots may indicate eruptions or storms at lower levels on the planet.

Methane, ammonia, and hydrogen have been identified in Saturn's atmosphere, and it must also contain large quantities of helium, with perhaps nitrogen, neon, and other inert gases in small amounts. Again, the visible clouds must be frozen crystals of ammonia in turbulent rotation with the planet, with perhaps a cooler zone of lighter gases above them. Like the sun, both Saturn and Jupiter have differential speeds of rotation of their equatorial atmosphere and the atmosphere at high latitudes.

When viewed with binoculars or a small telescope, Saturn looks lopsided; small won-

der that the early observers of the planet drew its shape in many different ways. Finally, it became clear that Saturn's odd protuberances were delicate rings encircling it in the plane of its equator. Observers have marveled ever since at their geometric beauty. While Saturn's diameter is about 75,000 miles, the outer of the three distinguishable rings ranges far out around the planet, with a maximum diameter of 171,000 miles. The middle ring is the brightest and densest, while the inner, "crepe" ring is rather hard to distinguish and, like the outer ring, is so tenuous that bright stars can be seen right through it.

The rings have been calculated recently to be not more than 8 and possibly only 4 inches thick, rather than the 10 miles previously estimated. In 1965, and at about 15-year intervals, Saturn's rings are oriented exactly edge-on toward the earth; relatively paper thin, they

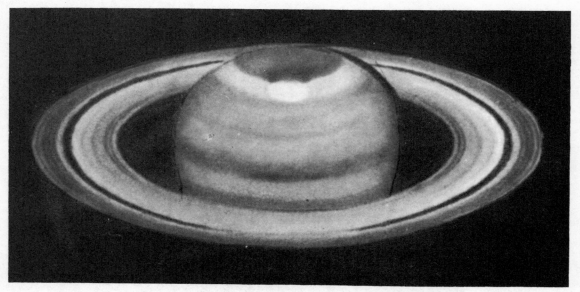

Saturn clearly reveals its dark belts and lighter zones (north at top), and its outer, middle, and inner or "crepe" rings in a photograph made with the 24-inch Pic-du-Midi telescope on April 27, 1960. One of the rare white spots appears in Saturn's atmosphere at 60° north. White spots were recorded before in 1876, 1903, and 1933. By May 5 the spot had faded, forming a bright regular zone around Saturn at this latitude.

—A. Dollfus, Observatoire de Paris, Meudon, France

then disappear completely for a day or two, unresolvable from the earth. The rings are composed of finely divided, probably ice-coated, grains or particles of dust with perhaps some ice crystals, somewhat like the earth's noctilucent clouds but much more dense.

Methane gas has been observed spectroscopically on Saturn's large satellite, Titan, which has a diameter of nearly 3,000 miles, but no other constituents have been identified above its surface, on which some dim markings are apparent.

At 1,775 million miles from the sun, Uranus is not an easy object to study. In a sizable telescope, however, it shows a small blue-green disk. Very dim bands can be picked out parallel to its equator when Uranus' equatorial plane is directed toward the earth, but when its polar axis is so oriented the bands are not visible.

Spectroscopic observations of Uranus have demonstrated the presence of methane and hy-drogen in its atmosphere, but at its prevailing temperature of about —280° F. any ammonia must have been frozen out entirely. Its major gases are probably hydrogen and helium, as with the other gas giants.

Neptune shows a slightly greenish disk in the telescope and only vague, irregular markings, unlike the bands of Saturn and Uranus. The constituents of its atmosphere are probably similar to those of Uranus. Molecular hydrogen has been directly observed in it with the spectroscope.

The theoretical average temperature at Pluto's distance from the sun is a crackling cold of about —300° F. Spectroscopic examination of the light from Pluto has shown no absorption lines indicative of an atmosphere, although its size might well enable it to retain a thin atmosphere of the heavier gases. Variations in Pluto's light suggest that its surface bears dark markings; a rough rotation of 6 days has been inferred from these markings.

CHAPTER X

". . . As when by night the Glass
of Galileo . . . observes
Imagin'd Lands and Regions in the Moon."

MILTON

Despite centuries of dedicated attention by countless observers, professional and amateur, and despite the observations of astronauts and the analysis of lunar rocks by countless scientists on earth, many features of the lunar surface are still open to dispute and speculation. Before space vehicles or astronauts landed on the moon, a leading lunar authority, Harold C. Urey, urged that its investigation should be made with ". . . the realization that the moon is very different from the earth and that naive analogies are useless. The surface of the moon will not be like that of the earth. The surface origin is quite different. . . ."

The best existing photographs of the moon from earth give resolutions of about 0.4 to 0.5 second of arc. This means that objects much smaller than 0.4 to 0.5 of a mile (2,000 to 2,500 feet) in at least one dimension could not be distinguished. Direct observation of the moon with a telescope is somewhat better. A visual resolution nearly four times greater than the photographic has been claimed, so that objects only 500 to 1,000 feet in size can be picked out. Reports from the 200-inch telescope are that craterlets only 240 feet across and 15-foot-wide clefts or cracks can be seen.

Then the Ranger series of spacecraft took photographs up to the last second as they neared impact with the lunar surface. Suddenly craters one to three feet in diameter were visible! Surveyor soft-landing space vehicles settled on the ancient surface and took pictures all around them, even scooping trenches in the surface. Lunar craters of all sizes, rocks, and soil, all appeared in the photographs Surveyors sent back to earth, including the first

SURFACE OF THE MOON

pictures of earth photographed from the moon!

Finally, astronauts took that last "giant step" and loped about on the lunar surface, picking up rocks, then hopped on the lunar "buggy" or Rover to explore the moon. We will trace this whole development here.

Lunar Topography

When the moon sails high in a clear sky, the naked eye easily distinguishes its bright areas from its principal dark regions. The dark areas, which reflect on the average only 5 to 10 per cent of the sunlight falling on them, seem to predominate, particularly across the middle of the moon and to the north and west. The bright areas, seen primarily in the southern and eastern parts and far to the north around the lunar pole, reflect up to 20 to 30 per cent of the light they receive from the sun.

The moon's main features near full are displayed in the small photographs made with the 15-inch refractor of the Dominion Observatory in Ottawa. The sun's light probes directly into some of the large craters, which become brilliant spots on the moon at this phase, and the ray systems around them are most striking.

The major lunar features apparent at the first and last quarter appear in the two photographs forming a general map of the moon, made with the 36-inch refractor of the Lick Observatory. Many of the most significant features are named in the accompanying table and can be located easily on the photographs by means of the coordinates. With the exception of a few tiny points, all of the lunar features referred to in this chapter can be found on this map. In these photographs, north is at the top, south at the bottom, east to the right, and west to the left, following normal cartographic (or, in this case, what is called "astronautical") convention, just as prominent features of the moon are seen in the sky. By tradition, and for convenience, astronomers follow another convention, showing the moon (and other celestial objects) with directions reversed—north at the bottom and south at the top.

The dark areas, called "maria" (the Latin for seas), are relatively flat, with very gradual slopes. They seem to have quite a different composition and origin than the bright areas, as do the related "sinuses" or gulfs off the seas, and "paludes" or smaller dark areas that look like marshes at first sight. If the moon is seen with binoculars or in a small telescope, it appears to have a leaden, pewter, or silver color —duller when the sun is low, brighter and even scintillating when it is high above the surface features. The bright areas resolve into rough, mountainous, or broken regions and into pockmarks or holes called craters.

A number of great mountain ranges cut across the moon and some regions are jumbled mountainous and hilly areas. The Apennine Mountains slash down at an angle from northeast to southwest just above the center of the moon's disk, which is near a rather dark spot called the Sinus Medii, or Central Bay. The Apennines tower up to 18,000 feet above the Mare Imbrium (Sea of Rains), the vast plain to their northwest. The Caucasian Mountains (reaching up 19,000 feet) and the Alps (12,000 feet) lie still farther north.

The moon's southern hemisphere is very mountainous and cratered. Near its south pole are the Leibnitz Mountains, with peaks up to 19,600 feet. Although this is 10,000 feet short of Mount Everest's height and about comparable to Kilimanjaro in Africa or Mount Logan in the Yukon Territory, Canada, the second highest peak in North America, the moon is so much smaller than the earth that its greatest mountains are at least relatively higher than ours.

The variety of types and sizes of moon craters has given birth to a multitude of terms to distinguish them. The larger ones are called

Change in the appearance of the moon in a four-day interval from just before full moon (above) to just after full moon (below), revealing the prominent features apparent in this phase with a 15-inch refractor. The bright and dark areas are sharply contrasted, and some of the larger craters and their ray systems stand out. (North at bottom in the astronomical convention.)
—DOMINION OBSERVATORY, OTTAWA, CANADA

walled plains, ringed plains, and mountain rings; then come crater plains, craters, and craterlets; the crater pits and crater cones are smaller still. "Crater" is often used in a general sense for all these ringed shapes and does not imply a volcanic origin. In a more technical sense it is reserved for the smaller rings, perhaps 4 to 12 miles in diameter, often with a cone or central peak rising within them and usually sunken well into the surrounding surface. It has been estimated that from 200,000 to 1 million craters of all types can be seen on the visible surface of the moon with large telescopes. The photographs relayed by the Ranger 7 spacecraft in 1964 showed many more tiny crater pits as small as three feet in diameter in the Mare Nubium.

Forty miles wide, with walls rising to over 12,000 feet, the great crater Copernicus, perhaps the most striking of all, stands all by itself northwest of the moon's midpoint. It is conspicuous for its multitude of rays, those unexplained long, whitish streamers which radiate out from and around craters like the points of a star. Tycho in the south central region has the most pronounced system of rays on the moon. Hundreds of them radiate out to great distances and are separated from Tycho itself by a dark ring around it. Tycho is believed to have been one of the last great features produced on the moon. Whatever its origin, the explosion that created it must have been enormous. One of its rays can be traced across the Mare Serenitatis in the moon's northeastern region, and out to the limb, where it disappears onto the far side. It is thought that it may traverse the entire far side of the moon and that one of the rays appearing on the limb in the southwest may be its continuation.

A prominent crater system just south of the midpoint of the moon consists of the craters Ptolemaeus, Alphonsus, and Arzachel, lying close together on a north-south arc. Ptolemaeus is a great walled plain over 90 miles across, with a very dark, gray interior, full of detail that has been a focus of interest. The 70-mile-wide Alphonsus has walls up to 7,000 feet, a strange series of dark spots or streaks near the rim on its floor, and frequent signs of activity in its interior. Arzachel, the smallest of the three, has a central peak and much interior detail.

Far to the north of the Mare Imbrium, just below the moon's north pole, is another object of much speculation, the great walled plain of Plato, 60 miles in diameter with walls rising to 7,000 feet. Mists, vapors, or obscuring gases sometimes seem to veil Plato's floor and have been reported at the same time in the United States and in England. Bright flashes have been seen there, and from time to time light and dark spots or streaks appear in it. One astronomer has described Plato as "one of the most continuously active volcanic regions on the moon"; others deny that anything unusual ever takes place there.

Numerous "domes" have been discovered on the moon's surface. Domes may be somewhat like the bulges or uplifts caused by lava beneath the earth's surface, representing arrested stages of - volcanism. Sometimes circular, though usually irregular in shape, and sloping or bulging up by perhaps 700 to 1,000 feet, these domes may run 1.5 to 2.0 miles in diameter. There are occasional larger ones, as wide as 50 miles, which may have been volcanic shields, uplifts due to bubbles in the lava of the plains which did not burst through.

A great many clefts, sometimes called rills or rilles—many of them may be lunar faults—meander along the surface of the moon. They may be 1 to 2 miles or more in width, sometimes narrower. Some are fairly straight, others crooked or curved; they may extend for tens or even hundreds of miles, occasionally cutting through hills or the walls of craters. The great Cleft of Hyginus, for example, lying just to the northeast of the moon's midpoint, cuts through the small crater pit called Hyginus. This cleft runs a rather jagged course and patient scrutiny has revealed several small branches that are lesser clefts, or cracks. Parts of the cleft are composed of crater chains.

The moon's maria probably never contained any water, nor were in any other respect like

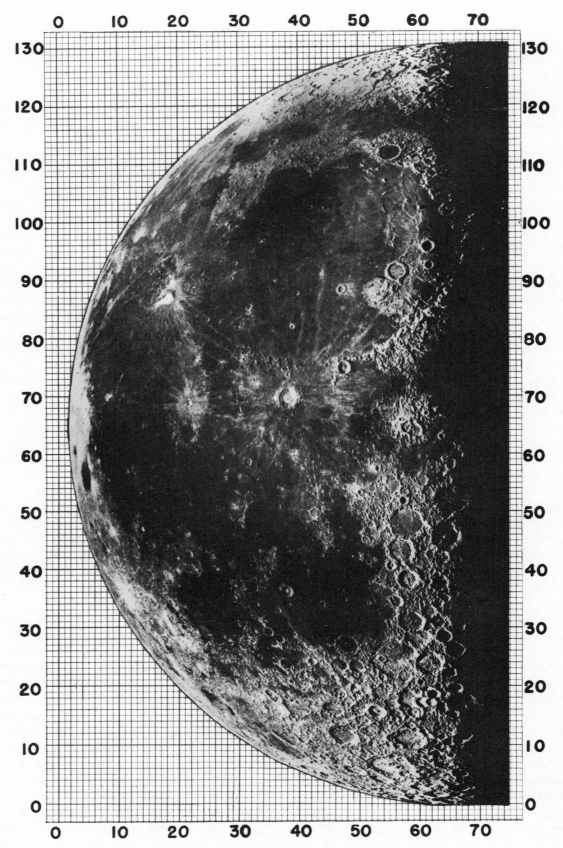

A general map of the moon in the normal cartographic or astronautical convention. As on any map of the earth, and as the moon is seen in the sky, north is at the top, south at bottom, east to the right, and west to the left. The grid is arbitrary, with x coordinate numbers running horizontally toward the right and y coordinate numbers vertically toward the top from the left-hand corner. (Note that the two photographs, and consequently the x coordinates, overlap somewhat in the middle.) The x and y coordinates for 204 named features are given in the table (pages 110–111). *To find a given feature whose name is known*, obtain its x and y coordinate numbers from the table, locate its y number on the grid vertically, and move

the eye or place a ruler across the map horizontally until the x number is reached, the feature being located where lines from the x and y numbers cross. *To identify a feature seen on the map,* obtain its x and y numbers by placing a ruler vertically and horizontally, respectively, from it to the grid at the edge, and use these numbers to find the name of the feature in the table. The two half-moon photographs were taken in a series by J. H. Moore and J. F. Chappell (1937–38) with the 36-inch refractor of the Lick Observatory.

—DINSMORE ALTER, *Pictorial Guide to the Moon;*
LICK OBSERVATORY, MOUNT HAMILTON, CALIFORNIA

Name of feature	x	y	Name of feature	x	y	Name of feature	x	y
Abenezra	73	34	Dionysius	79	60	Manilius	70	73
Abulfeda	74	42	Doppelmayer	24	29	Mare Australe	105	15
Agrippa	72	61	Endymion	95	115	Mare Cognitum	35	48
Albategnius	64	46	Eratosthenes	47	75	Mare Crisium	115	80
Aliacensis	65	25	Euclides	29	50	Mare Foecunditatis	112	55
Almanon	75	39	Eudoxus	74	105	Mare Frigoris	30–85	116
Alphonsus	56	43	Euler	32	85	Mare		
Alpine Valley	63	109	Fabricius	95	17	Humboldtianum	99	118
Alps, mountains	63	110	Flammarion	55	55	Mare Humorum	25	34
Altai Scarp	83	32	Flamsteed	17	55	Mare Imbrium	40	95
Anaxagoras	59	124	Fracastorius	94	35	Mare Marginis	127	80
Anaximander	43	122	Fra Mauro	40	51	Mare Nectaris	95	40
Apennine Mountains	58	80	Furnerius	110	25	Mare Nubium	47	36
Arago	83	64	Gärtner	81	117	Mare Serenitatis	78	88
Archimedes	55	91	Gassendi	22	40	Mare Smythii	129	62
Ariadaeus	80	64	Geber	75	36	Mare Spumans	122	63
Aristarchus	18	87	Geminus	108	95	Mare Tranquillitatis	90	68
Aristillus	61	96	Godin	72	59	Mare Undarum	122	69
Aristoteles	74	110	Grimaldi	4	56	Mare Vaporum	65	72
Arzachel	57	38	Guericke	44	46	Marius	16	74
Atlas	73	108	Gutenberg	104	49	Maurolycus	73	15
Autolycus	61	92	Haemus Mountains	70	80	Menelaus	77	75
Azophi	74	33	Hell	53	25	Mercurius	104	109
Bailly	45	4	Hercules	90	108	Mersenius	17	37
Ball	53	21	Herodotus	17	87	Messala	107	98
Barrow	66	124	Herschel	57	52	Messier	109	57
Beaumont	91	38	Herschel, Caroline	33	97	Metius	97	18
Bessarion	23	76	Herschel, John	44	120	Meton	69	125
Bessel	78	82	Hesiodus	45	27	Milichius	28	70
Birt	51	34	Hind	68	48	Moretus	63	2
Bond, W. C.	64	120	Hippalus	32	33	Mösting	53	57
Bonpland	41	49	Hipparchus	65	52	Newton	60	0
Brayley	24	83	Hortensius	30	66	Oceanus Procellarum	10	95–
Bullialdus	38	36	Huygens, Mount	57	81			45
Bürg	83	106	Hyginus	66	65	Olbers	3	71
Byrgius	12	36	Jansen	90	74	"Oval Rays"	44	66
Carpathian			Janssen	92	16	Palus Nebularum	60	99
Mountains	35	75	Julius Caesar	76	67	Palus Putredinis	60	88
Cassini	65	105	Kepler	22	69	Palus Somnii	105	76
"Cassini's Bright			Kepler A	23	68	Parry	42	50
Spot"	57	24	Lacaille	61	33	Petavius	114	34
Catharina	85	38	Lambert	39	88	Phocylides	32	11
Caucasian			Landsberg	31	59	Picard	113	76
Mountains	68	98	Langrenus	119	51	Piccolomini	91	27
Censorinus	95	57	Laplace, Prom.	42	108	Pickering, W. H.	108	57
Clavius	55	6	Leibnitz Mountains	75	0	Pico, Mount	54	107
Cleomedes	110	90	Le Monnier	91	88	Pitatus	47	27
Conon	62	81	Licetus	67	11	Pitiscus	81	13
Copernicus	37	70	Lick	112	74	Piton, Mount	59	102
Cuvier	69	9	Linné	72	88	Plato	54	112
Cyrillus	86	43	Longomontanus	46	12	Playfair	69	31
D'Alembert			Lyot	58	49	Plinius	85	75
Mountains	3	54	Macrobius	105	83	Posidonius	88	94
Delambre	79	55	Mädler	92	45	Proclus	107	78
Deslandres	56	24	Maginus	58	11	Ptolemaeus	56	48

Name of feature	x	y	Name of feature	x	y	Name of feature	x	y
Purbach	58	31	Schröter's Valley	17	88	Thales	86	120
Pyrenees Mountains	102	43	Seleucus	8	86	Thebit	55	40
Pythagoras	40	122	Sinus Aestuum	50	72	Theophilus	88	45
Pytheas	39	82	Sinus Iridum	37	108	Timocharis	46	88
Rabbi Levi	82	21	"Sinus Iridum			Tobias Mayer	31	77
Ramsden	33	25	Highlands"	40	110	Triesnecker	64	61
Regiomontanus	59	28	Sinus Medii	58	60	Tycho	52	16
Reinhold	35	62	Sinus Roris	26	107	Vitello	28	28
Rheita	100	21	Stadius	44	70	Walter	60	24
Rheita Valley	99	18	Stevinus	107	26	Wargentin	27	15
Riccioli	3	60	Strabo	88	120	Werner	63	28
Rook Mountains	20	22	Straight Range	47	109	Wrottesley	112	34
Rosse	96	38	Straight Wall	52	34	Yerkes	109	77
Scheiner	50	6	Taruntius	108	66	Zach	66	5
Schickard	26	17	Taurus Mountains	100	95			
Schiller	37	11	Teneriffe Range	51	109			

Map of first-quarter lunar surface temperatures by isotherms of equal temperature on a simultaneous photograph of the moon. Temperatures are shown here in centigrade degree. A portion of the Mare Crisium north of the subsolar point seems to show as a hotter region, as it did also in other thermal maps. (North at top.)
—ANN R. GEOFFRION, MARJORIE KORNER, AND WILLIAM M. SINTON, *Lowell Observatory Bulletin*

the seas or oceans of the earth. A century ago, a few astronomers did think that the maria might actually be the dried-up beds of ancient oceans, their dark color due to sediments that had collected on their floors, from the time when the moon had liquids on its surface and an atmosphere above it. Though it has recurred recently, this has always been a minority view. The major lunar seas that are immediately distinguishable include the Mare Crisium (Sea of Crises), the dark, rather elliptical area on the east-central side of the moon above the larger and broader Mare Foecunditatis (Sea of Fertility). The Mare Crisium is one of the few isolated seas on the moon; it is surrounded by mountains and unconnected with other seas. Probably the most beautiful of the moon's seas in the telescope, its 66,000-square-mile area is visible to the naked eye as a small, dark, oval spot. Some observers report that it has a greenish tint, with many fine white streaks and bright spots on its surface. It is near the eastern limb of the moon, and appears unmistakably on some of the photographs of the far side of the moon made by the Soviet Lunik 3 in 1959.

The large dark sea just northeast of the moon's midpoint has been named Mare Serenitatis (Sea of Serenity), and between it and the Mare Foecunditatis lies the large Mare Tranquillitatis (Sea of Tranquility). To the west of the moon's central meridian are the Mare Imbrium (Sea of Rains) to the north, and to the south the Mare Nubium (Sea of Clouds). The Sea of Rains is a large, dark area measuring some 750 miles from east to west and 690 miles from north to south. The great smooth area extending far north and south and almost to the moon's western limb is known as Oceanus Procellarum (Ocean of Storms). Its area is estimated as some 2 million square miles.

An isolated portion of Mare Nubium within which Ranger impacted has been named Mare Cognitum (Known Sea), since Ranger photographed parts of it.

The moon probably never had enough of an atmosphere for the formation of any sizable body of water. Water vapor issuing from within the moon would very quickly vanish into space unless it were frozen near the poles, or in rare spots remaining in dense shade or under heavily insulating dust. Suggestions have been made recently that ice or water may exist at some distance beneath the surface, insulated by thick dust from the extreme temperature changes on the surface itself.

Surface Temperatures

Temperatures on the moon's surface have been measured since 1868, and today the very sensitive thermocouples used in scans of the moon's surface show detailed fluctuations in its temperature with changes in the angle of the sun as it passes from east to west in the moon's day. The temperature contour map shows isotherms (lines of equal temperature) on the moon in its first quarter as the whole surface was scanned over a period of three hours. The center of the moon, Sinus Medii, is indicated. Where the sun is full up, above the eastern part of the disk, the temperature is 230° F., 18° above the boiling point of water, while it ranges down to the boiling point at the next isotherm line marked. From there, the temperature drops off rather fast, reaching a thoroughly comfortable 68° at this phase just past the moon's center. Then it plummets in the moon's darkness to —140° F. The temperatures shown on the contour map are centigrade.

While the temperature at the moon's midnight point has not been established precisely, a number of measurements have placed it at from —250° to —300° F. When the earth's shadow covers the moon in lunar eclipses, the temperature goes down very rapidly. The moon's surface, however, seems to be a good insulator, and the temperature well below it should not vary a great deal between lunar day and night. With subsurface protection, not only from the sun's heat but also from its much more dangerous particles and gamma rays, human or other forms of life could be maintained very safely and comfortably.

Lunar Maps

Sometime before 1603, William Gilbert, the British discoverer of the earth's magnetism, drew the first known map of the moon, simply from naked-eye viewing. In 1609 (Galileo) and 1610 (Harriot), the first rough telescopic maps of the moon were made. For three and a half centuries thereafter, more and more accurate and detailed lunar maps were prepared.

Only in the last few years have complete maps of the moon been made, covering the farside, nearside, and the poles. These maps are based on photographs made from unmanned spacecraft and by astronauts in orbit.

Determination of the exact position of any point on the moon's surface must be made from the lunar coordinates of latitude (beta or β) and longitude (lambda or λ). On the moon, latitudes (north and south) are determined from the lunar equator and longitudes (east and west) from the principal meridian perpendicular to the equator. The inclination of the moon's equator to the ecliptic has been found to be about $1° 32' 40''$. The principal meridian is the meridian plane cutting the moon through the point at which the radius vector (or straight line) joining the centers of the earth and the moon is situated at the time of the passage of the moon through perigee or apogee.

The dead center of the moon's disk in the Sinus Medii, where the equator and central meridian intersect, is not marked by any prominent surface feature. The small but unmistakable crater of Mösting A nearby, on the western slopes of a rather dark plain called Flammarion, has been selected as the basic reference point. (See topographic lunar map of the United States Army Corps of Engineers.) Located to the southwest of the central

point, its walls rise 7,000 feet above the depressed bowl of its interior, and 1,600 feet above the surrounding area, making it an easily identifiable landmark. From this reference point, a group of five to nine other reference points are determined, and the whole system is then used to determine the positions of all of the other objects being mapped.

Artists have customarily portrayed the lunar craters, ridges, and mountains as dramatically sharp and rough. While it is true that shadows thrown by the sun on the moon's surface often give this impression, modern contour research, using microdensitometer tracings of lunar shadows, has proved that most of the lunar craters and mountains are undulating rather than jagged. The heights of crater walls are usually very small in comparison with their widths.

Most slopes or hills on the moon are not inclined more than a few degrees, and very few slopes run at angles of 10° or more from the horizontal. The typical vista on the moon, then, would be of a normally flat, only occasionally gently undulating or rolling surface, somewhat like the Great Plains of the West, unrelieved by sharp mountains and promontories.

Some observers have noted various hues and shades of color here and there on the moon, such as the greenish hue of portions of Mare Crisium. In the main, however, the surface appears to run through monotonous shades of gray. While other colors may often have been due to chromatic aberrations in optical systems, the existence of significant color contrasts on the lunar surface has been demonstrated by spectrophotometry of a number of areas. In general, the brighter areas studied appeared to be redder than others, and color contrasts were even found within the monotonous-looking Mare Nubium.

Contacts with the Moon

The first known direct human contact with

Section of topographic lunar map showing Flammarion, Mösting, and Mösting A southwest of the Sinus Medii, and below them the line of Ptolemaeus, Alphonsus, and Arzachel. Scale 1:2,500,000; contour intervals 1,000 meters (3,280 feet) and 500 meters (1,640 feet); north to top and east to right.

—Army Map Service, Corps of Engineers, U.S. Army

The western border of the Mare Tranquillitatis, photographed at the coudé focus of the 120-inch Lick reflector. (North at top.) Craterlets dot the surface of the mare, the crater Julius Caesar rises to the west with crater pits scattered on its floor, and to the south the Ariadaeus Rill, a distinctive valley cleft more than 150 miles long, runs almost east and west.

—Lick Observatory, Mount Hamilton, Calif.

the moon was the impact of the Soviet lunar probe Lunik 2, at 5:02:24 P.M., Eastern Daylight Saving Time, on September 13, 1959. According to reports in America, Lunik fell into the area of Palus Putredinis (Marsh of Decay) in the north-central region of the moon. The Marsh of Decay is an offshoot of the Mare Imbrium and the Sinus Aestuum (Bay of Billows), just north of the Apennine Mountains. The Russians have located the impact point a

little farther north and east, on the eastern wall of the crater Autolycus.

Ranger 4, of the American moon-probe series, launched from Cape Canaveral (now Cape Kennedy), Florida, with an Atlas-Agena B rocket, was scheduled to place a 727-pound capsule with instruments on the lunar surface. It covered the distance to the moon in about 64 hours and is believed to have crashed on April 26, 1962, on the far side—just around the limb of the visible moon. The second stage of the Agena B rocket passed on by the moon and into solar orbit.

On July 31, 1964, Ranger 7, which had functioned smoothly throughout its flight, plummeted to impact in the Mare Nubium near the crater Guericke in the south central quadrant of the moon. Starting at a height of about 1,100 miles above the moon, it transmitted some 4,300 television pictures back to earth, the last made at an altitude of about 1,000 feet above the surface, revealing details with a resolution of one to three feet. While some areas were heavily pitted with secondary craters, apparently caused by debris shot out when crater Tycho was produced, other areas appeared to be relatively smooth and uncluttered, covered with what was interpreted to be crunchy dust of an undetermined depth. It was estimated that the photographic resolution was some 2,000 times better than in the best photographs ever achieved through the earth's atmosphere.

Another significant lunar first was the taking of photographs of the far side of the moon in October, 1959, by the Soviet spacecraft Luna 3. After a series of hard-landing Soviet and American craft, many soft-landing Luna and Surveyor craft, balancing on their rocket jets, settled on the lunar surface. They panned their directed TV cameras around and furnished on-the-spot close-ups of the lunar soil, the strewn rocks and boulders, and the large and small craters pocking the landscape.

Then artificial lunar satellites called Lunar Orbiters spanned the moon in highly eccentric orbits, coming as close as 30 miles above the surface, to provide detailed photographs of areas of the moon from which the most favorable and instructive landing sites could be selected.

Men then came to the moon, first circling the moon in close orbits, and then on July 20, 1969, descending to the surface in the LEM. With the close of the Apollo series of flights, a half-dozen lunar areas had been sampled.

From all these photographic and direct contacts with the moon came a wealth of information. The jumbled peaks of vast mountain ranges could be observed from nearby. Astronauts walked, dug, and rode over lunar seas or maria. Small and large craters, were examined and wrinkled ridges and bright rays were studied, as well as groups of domes scattered on the moon's surface.

Astronauts of the Apollo 15 mission in 1971 explored the edge of the Hadley Rill, one of the long, meandering valleys of the moon. Faults and cracks where the lunar surface had broken and shifted look like those on earth, and dikes of lava had forced their way through the cracks. Selenology rapidly became a science like the geology of the earth, as solid knowledge began to replace speculation.

Early fears of spacecraft and astronauts sinking deep within fine lunar dust soon evaporated. The lunar soil, a layer from 10 to 30 or 40 feet deep, was similar in texture to soil on earth. It was made up of fragments of rocks and dust broken and churned by meteoritic impacts. Most of the soil was basaltic, a kind of igneous rock that has been heated and melted. Beneath the soil, a layered rock crust 40 miles thick was revealed by study of seismic waves from the impact of spacecraft and the rare lunar earthquakes. The upper 15 miles of crust is largely basaltic rock. Beneath this comes a 25-mile layer of anorthositic, older, but still igneous rock. Under the crust, the moon has a mantle, as does the earth, and probably even a differentiated core.

Lunar Seas

The vast dark seas on the moon and its hundreds of thousands of visible craters seem to be uniquely lunar. Scientists naturally have done their best to explain them, but to tie down an explanation of phenomena in so distant a place as the moon is well nigh impossible.

One type of theory hinges on a volcanic

Principal features of the farside of the moon as prepared from photographic records of Lunar Orbiter Missions I–V. Jumbled mountainous features cover about 90 per cent of the farside, as opposed to 60 per cent of the nearside, and no large maria are found on the farside. A number of great impact craters with several rings of upheaval around them form the most prominent regions on the farside, and where lava has welled up in their centers these are called maria.

—NASA Lunar Chart, Aeronautical Chart and Information Center, U.S. Air Force

origin, demonstrating how most of the moon's surface features can be explained completely in terms of internal, volcanic processes. Another type of theory explains everything on the moon in terms of the impact of meteorites and asteroids, on the assumption that the moon is an entirely inert and inactive body, without any apparent erosive forces because it almost entirely lacks an atmosphere. These impacting bodies, according to the theory, produced both

the seas as they now exist and the thousands of craters, which look as if some giant had poked the end of a blunt stick into the soft or plastic surface of the moon over and over again.

While many of the mountainous regions and ranges do appear to have been produced by the debris of impacts, there are bright areas in the southwest quadrant of the face which may be among the oldest, unaffected portions of its surface and may demonstrate what the original,

On the first manned lunar landing, via the Apollo 11 mission, in July 1969, the astronauts ventured outside their lunar module to explore the surface for only 90 minutes. Here one astronaut adjusts the passive seismic experiment package at the astronauts' greatest distance from the module of 300 feet.

—NATIONAL AERONAUTICS AND SPACE ADMINISTRATION

At the end of their third and final trip in the Rover (see tracks), the Apollo 15 astronauts stopped to make these panoramic photographs. Nearby, Mount Hadley (left); in the distance, higher peaks of the Apennines (center) look strikingly like foothills on earth.

—NATIONAL AERONAUTICS AND SPACE ADMINISTRATION

On the Apollo 15 mission, in July 1971, the astronauts explored the area near Mount Hadley (background) and the Hadley Rill (right) for 18.5 hours. They toured some 18 miles in the first car on the moon, the Lunar Rover, collected some 170 pounds of lunar samples, and discovered that the moon's crust was layered.
—National Aeronautics and Space Administration

In this portion of the panorama, Silver Spur is on the left, with St. George crater dark on its right flank. The lunar module, home for the astronauts on the moon, is some 300-feet distant (right) between the astronauts and the Hadley Rill.
—National Aeronautics and Space Administration

solidified crust was like. They appear to belong to a time when the shape of the moon was nearly spherical, before it was disorganized by impacts or internal explosions or eruptions, and are without the lava-type surface which covers a lot of the moon. The surface in such regions seems to be quite smooth, with gentle inclines, although the dark shadows look rugged.

The volcanic hypothesis assumes that the maria were produced by huge lava flows early in the moon's history when it was still very hot or had become hot, or when it had recently crusted over; that it then became hotter still, perhaps from radioactive elements under its surface, or was possibly disrupted by such a titanic event as capture by the earth, and great floods of lava or molten rock poured forth to create the seas. On the other hand, the impact hypothesis envisages a tremendous meteorite or small asteroid up to 100 to 200 miles in diameter arriving at the surface of the moon, gouging out huge gashes as it came in, striking with great shock waves and producing heat. With this, lava was released from below the surface. The lava may have already existed or been melted by the heat and streamed out at the impact time or later, when the dome produced by the impact cracked and subsided. Those who support the impact theory have pointed in particular to the Mare Imbrium in the northwest region of the moon, around which the ranges of the Apennines, the Caucasians, and the Alps curve in a full three-quarter arc of a circle, presumably raised as part of the quantities of debris shot forth by the impact of an asteroid. This highly dramatic explanation for Mare Imbrium has been developed in great and convincing detail. But does it apply to all of the large seas on the moon?

One lunar expert, Gerard P. Kuiper of the Lunar and Planetary Observatory of the University of Arizona, believes that the lunar seas are of two different types. The first is nearly circular in shape, with considerable symmetry. The Mare Crisium, the Mare Nectaris (Sea of Nectar) in the southeast quadrant of the moon, Mare Serenitatis, Mare Imbrium, and Mare Humorum (Sea of Moisture) are examples. He believes that these seas, which are usually surrounded by rings of mountains, could well be large impact craters. The second type of sea is asymmetrical, without a surrounding ring of walls or mountains. It has invaded many low areas around its circumference, producing numerous bays and estuaries. The eastern half of Mare Tranquillitatis, Mare Nubium, and most of the great Oceanus Procellarum (Ocean of Storms), fall into this second category. These are flooded regions in which the lava welled up from underneath the surface of the moon without a catastrophic impact.

Kuiper points out variations in the impact seas which have also been noted by the exponents of the impact hypothesis. In certain maria it appears that the impact from without occurred at one time and the lava flooding at a later date; among these are the Mare Crisium, Mare Nectaris, and Mare Humorum. The great splashes of lava so apparent about the Mare Imbrium do not surround them. In the Mare Imbrium masses of lava apparently were present under the surface and splashed about when the impact came. The lava in the Mare Crisium and others of its kind may have welled up at some time after the dry impact, and the lava may not have been produced mainly by the heating from the impact shock, but have come from the moon's stored internal heat.

Before the lunar surface and the minerals of its rocks had been studied at first hand, it was generally believed the moon had formed cold with the falling together of the condensing materials of the original nebula of gas and dust making up the solar system. But the lunar minerals soon showed chemical differentiation had taken place. This must have been during an early, molten period of the moon, shortly after it formed some 4.5 billion years ago. In a melted state the lighter elements

continued on page 153

COUNTDOWN TO SPACE

A Special Pictorial Supplement

Man's ventures into space during the last decade are displayed in the picture story on the following pages.

With the creative ideas, technical ingenuity, and cooperative efforts of hundreds of thousands of people around the earth, men finally walked on the moon and explored its surface.

After all these advances into space, however, the stars go on sparkling in all their brilliant colors in the night sky, still unapproached, only brought closer with telescopes and other instruments on earth and on probing spacecraft.

(Above) OFFICE OF MANNED SPACE FLIGHT, NATIONAL AERONAUTICS AND SPACE ADMINISTRATION.
(Below) G. C. ATAMIAN, TALCOTT MOUNTAIN SCIENCE CENTER, AVON, CONNECTICUT

Stars probably are born in the falling together of great masses of gas and dust like those in the Lagoon Nebula in constellation Sagittarius pictured above by the 200-inch telescope on Mount Palomar. Astronomer at the prime focus of the 200-inch (left) with the guiding eyepiece preparing to make an exposure. —Photographs © California Institute of Technology and the Carnegie Institution of Washington.

A spiral galaxy, NGC 4565, located in the constellation Coma Berenices, viewed edge-on with the 200-inch Hale telescope, just one of the billions of galaxies in the universe.
—MOUNT WILSON AND PALOMAR OBSERVATORIES

Astronomers have been extending our vision and knowledge far out into space during this time, to the galaxies like NGC 4565 (above) formed of 100 billion or more stars, to the very powerful energy sources like quasars, and to clusters and superclusters of galaxies, perhaps nearly to the limits of the universe.

Meanwhile, space scientists and engineers of the United States, the Soviet Union, and many other countries around the earth have learned how to launch unmanned and manned satellites and space ships into space near the earth, among the planets of our solar system, and finally, in 1972, with Pioneer 10 spacecraft, into deep space beyond our system.

Thus our knowledge of our immediate environment in near space and our remote environment in deep space has been gathered by the often integrated efforts of both astronomers and space scientists.

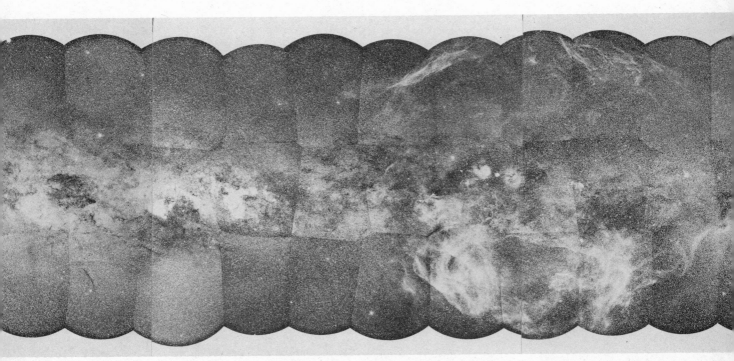

Part of our own Milky Way Galaxy, as we see it from inside, from
where the sun is located in one of the galaxy's vast spiral arms.
—The Mount Stromlo and Siding Spring Observatory
Australian National University

The Large Magellanic Cloud, a member of the Local Group of galaxies.
—The Mount Stromlo and Siding Spring Observatory
Australian National University

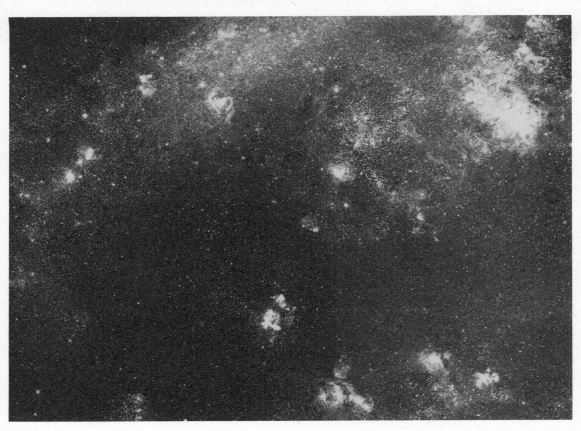

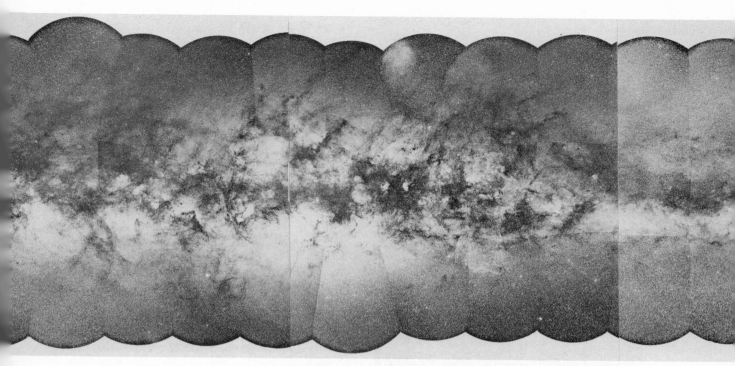

Black spots are dense dust in the Milky Way through which light
cannot penetrate; brightest patch is toward the center of our galaxy.
—THE MOUNT STROMLO AND SIDING SPRING OBSERVATORY
AUSTRALIAN NATIONAL UNIVERSITY

During the last decade, astronomers have
learned much more about our own galaxy, the
Milky Way, and the many occurrences going
on in it. Its shape and structure have been
mapped more precisely with radio telescopes.
Computers are also being used to simulate
models of how galaxies develop and to describe
their life histories. The tremendous events that
must be going on in the dense centers of galaxies are being explored.

The Large Magellanic Cloud is an irregular
galaxy, closest to our own at about 160,000
light-years. With its many variable stars, it has
been a very fruitful source for understanding
stars, their groupings, and their evolution.

At the edge of our Local Group of galaxies
lie Maffei 1 and 2. Buried beyond a dense
portion of the Milky Way, these galaxies were
only recently discovered and their distances estimated at about three million light-years. Spiral

arms can just be glimpsed in the plate of Maffei
2, as shown on the right.

With all this work, astronomers are beginning to sketch the backdrop of the universe
against which the nearby stars and our solar
system seem almost infinitesimally small.

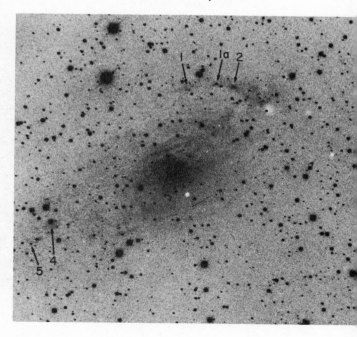

Recently discovered spiral galaxy, Maffei 2, beyond
Milky Way. Arrows point to hydrogen clouds.
—HYRON SPINRAD AND LICK OBSERVATORY

The Crab Nebula in Taurus resulted from a great supernova explosion observed in A.D. 1054.
—MOUNT WILSON AND PALOMAR OBSERVATORIES

In our era of change, pulsars were first discovered by radio telescopes and their pulsations, many times a second, confirmed by light telescopes. Rapidly rotating balls of neutrons, perhaps only 10 miles in diameter, probably resulted from supernovas like the Crab. They send out all kinds of radiation along the electromagnetic spectrum.

The more common nova explosions of stars cause the rapid brightening and then dimming shown for Nova Herculis. The causes of novas and supernovas are still being explored. They are only two among the models of many different kinds of stars astronomers are building.

Our sun, fortunately, limits its particles to the electrons and protons seen in its corona and extending far out beyond the earth in the solar wind, although it is now suggested that the sun has increased in luminosity by perhaps 40 percent over the 4-billion-odd years of geologic time. As the sun moves swiftly through interstellar space, it may make a streamlined hydrogen wake, as shown.

A pulsar, or pulsating neutron star (NP 0532), left by Crab supernova, pictured during flashes only (left) and nonflashes only (right).
—LICK OBSERVATORY, UNIVERSITY OF CALIFORNIA

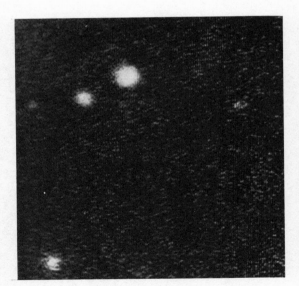

Stars almost tear themselves to pieces when they explode in novas. The explosion of this star, named Nova Herculis, late in 1934, resulted in its rapid brightening by March 1935 (left), to nearly the brightness of Saturn, then dimming rapidly by May (right).
—LICK OBSERATORY, UNIVERSITY OF CALIFORNIA

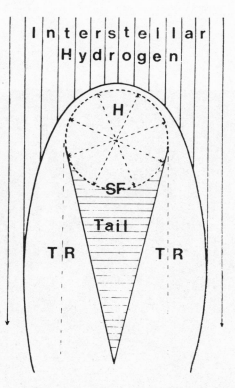

—MOUNT WILSON AND PALOMAR OBSERVATORIES
Bright prominences project from sun (above) in eclipse by moon and solar corona floods into space. Astronomers suggest (right) that corona forms heliosphere (H) around sun, with shock front (SF), a turbulent region (TR), and a tail as sun races through hydrogen of space.

—P. W. BLUM AND H. J. FAHR, UNIVERSITY OF BONN

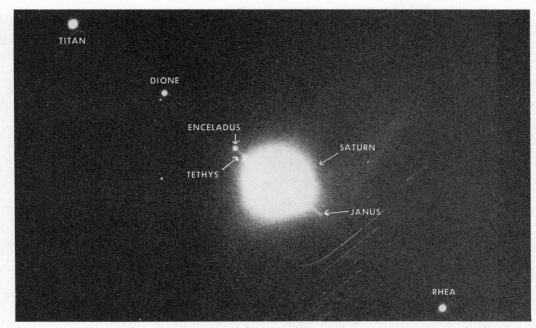

Janus, Saturn's tenth satellite, discovered in 1966. Rings were edge-on as viewed from earth.

—FLAGSTAFF STATION, U.S. NAVAL OBSERVATORY

Solar wind (streaks) forms streamlined shape around earth with a tail streaming behind.

—NATIONAL AERONAUTICS AND SPACE ADMINISTRATION

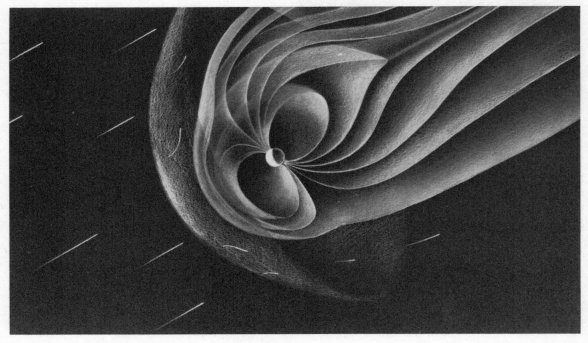

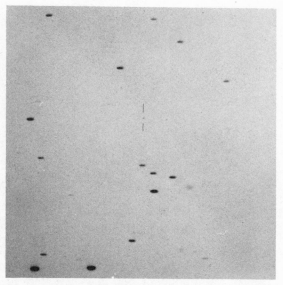

Minor planet Betulia (arrow), about one mile in diameter, at its closest approach of 14 million miles to earth in 1963. Streaks made by stars.
—FLAGSTAFF STATION, U.S. NAVAL OBSERVATORY

Jupiter's twelfth satellite, discovered in 1951, is about 4 miles in diameter and orbits the planet at a distance of 13 million miles.
—MOUNT WILSON AND PALOMAR OBSERVATORIES

Apollo 11 astronauts photographed earth (left) from a distance of 113,000 miles and moon (right) from a closer range, contrasting many of the differences between earth and lunar environments.
—NATIONAL AERONAUTICS AND SPACE ADMINISTRATION

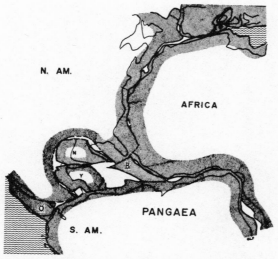

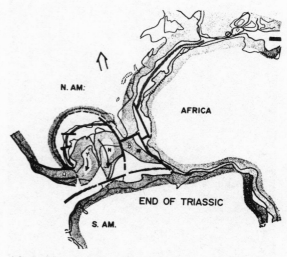

The drawings, left to right, show how the Gulf of Mexico and Caribbean Sea may have formed, under the continental drift theory. 200 million years ago (above), continents were one, called Pangaea. Y stands for Yucatan, N for Nicaragua-Honduras, O for Oaxaca, and B for Bahamas.

After 20 million years, North America moved (arrow), opening Atlantic; Y and N moved to leave Gulf of Mexico as a small ocean basin. Bahamas plate moved northeast, but S. America stayed where it was, joined to Africa. Oaxaca had slipped east.

The last decade has brought a revolution in ideas how the earth's crust changes as well as new knowledge of the solar system. The crust, it is believed, consists of a number of large plates that break up, collide, and drift under the pressures of convection cells in the mantle. An hypothesis how the Gulf of Mexico and Caribbean Sea formed with the drifting of the North and South American and African plates is shown in the maps above. If only volcanoes had not formed the Panama Isthmus, the Canal would not have had to be dug! The drifting of crustal plates offers a fruitful theory of how the continents, mountains, and oceans were shaped.

The Clearwater Lake Craters may be fossil remains of a meteoritic impact 2 billion years ago, and Wolf Creek Crater a much more recent event. Such explosive impacts from space must have played a role in the shaping of earth's crust, and perhaps even more on Mars and the moon. But erosion by wind and water have erased most of such craters on earth.

Wolf Creek Crater in Australia, about half a mile in diameter, in which one 400-pound meteorite was found, a fragment of a much larger impact.
—THE AMERICAN MUSEUM OF NATURAL HISTORY

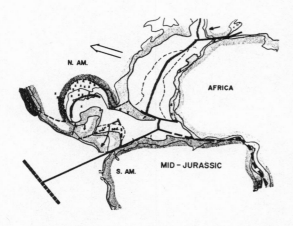

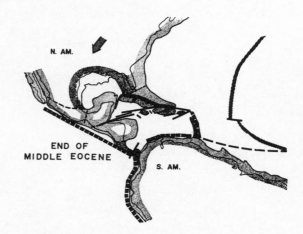

In another 30 million years, as North America moved west, the Atlantic Ocean had widened, and Central America was nearly formed with Yucatan and Nicaragua in place. Gulf of Mexico was formed and Caribbean was opening up. Salt deposits (x) formed in Gulf basin.

About 45 million years ago, South America had split from Africa and Atlantic Ocean widened to give essentially the present positions. Volcanoes had built up the Panama Isthmus.
—G. L. Freeland and R. S. Dietz, National Oceanic and Atmospheric Administration

Clearwater Lake Craters, near Hudson's Bay, Canada, 14 and 20 miles in diameter, may have been blasted out by a huge double meteorite.
—Carlton Beals and Royal Canadian Air Force

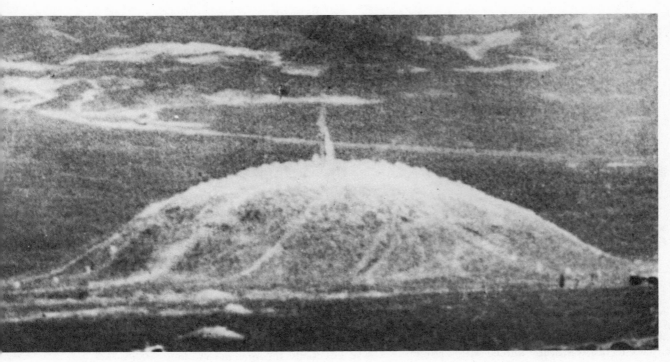

Man has learned how to make huge craters, too. A thermonuclear explosion 600 feet underground raised this earth bubble 300 feet high in 3 seconds at AEC Nevada Test site. Plume at top of bubble is sand used to plug hole for device.
—Lawrence Radiation Laboratory,
Livermore, California

Explosion above created Sedan Crater below, measuring 1200 feet in diameter and 300 feet deep. Such explosions creating craters like many on moon may someday be used for earth-moving projects.
—Lawrence Radiation Laboratory,
Livermore, California

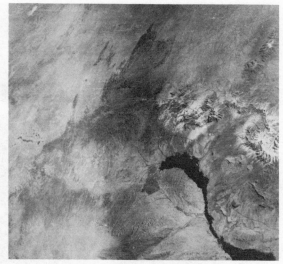

Summer (above left) and winter (above right) scenes as viewed by camera aboard a sounding rocket some 70 miles above earth in American Southwest. Black area (lower right) is a lake and dusky areas (lower halves) are forests. Note the earth's curve over the Pacific Ocean (top left) and the snow on the mountains (center right).
—U.S. NAVY AND JOHNS HOPKINS UNIVERSITY

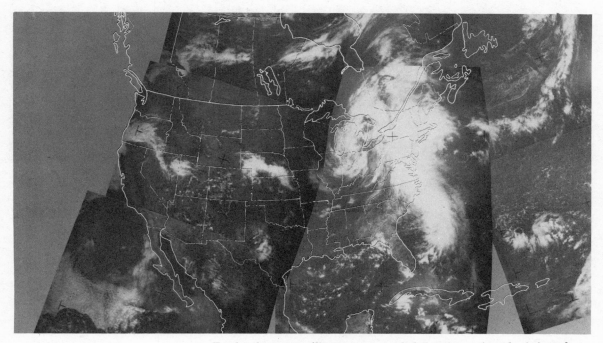

Earth-orbiting satellites soon extended the views of earth. A few photos made this view of United States' cloud cover on one day.
—NATIONAL OCEANIC AND ATMOSPHERIC ADMINISTRATION

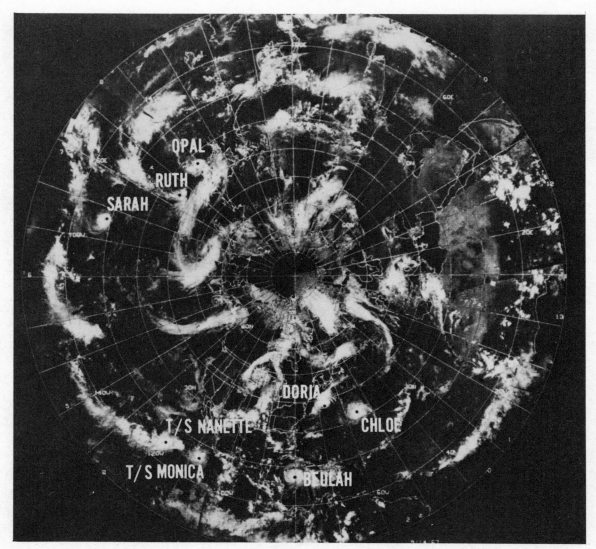

—National Oceanic and Atmospheric Administration

The ESSA 5 weather satellite reported these eight hurricanes or tropical storms (T/S) around the earth in the Northern Hemisphere on one day, September 14, 1967. Continents are traced lightly on the mosaic of photographs. Beulah, shown here south of Cuba, was the hurricane that eventually destroyed a billion dollars worth of property.

Weather satellites like ESSA 5 are now launched into polar orbits, crossing near both poles. Then they can send down observations of nearly the whole globe in one day, as the earth rotates under their orbits.

By observing the earth through the cameras and many other instruments carried by satellites, scientists can now stand off and view the earth from a distance in perspective. This information can then be interpreted and passed on in many applications for all mankind. Satellites also permit communication around the whole earth. An infrared TV camera on a satellite at about 900 miles above the earth takes the temperatures of the surface in October in the picture below. The cloud tops (white) are cold, and the land of the eastern United States (light gray) cool, while the Great Lakes are warmer (gray). Lighter shades in Canada show the land cooler there than in the United States.

The Atlantic Ocean is warm (almost black) where the Gulf Stream runs from the south, and the Atlantic is cooler along the East Coast.

Other infrared instruments on satellites can now measure the temperatures of the atmosphere right down to the surface. Giving results that compare precisely with those from weather balloons, satellites may begin to replace them.

—NATIONAL OCEANIC AND ATMOSPHERIC ADMINISTRATION

PTOLEMAEUS
ALPHONSUS
CENTRAL PEAK
DARK
MARKINGS
STRAIGHT WALL
MARE NUBIUM

—MOUNT WILSON AND PALOMAR OBSERVATORIES

A generation of hard-landing Rangers, spacecraft that returned photographs to earth as they plummeted into the moon, gave much new information about its surface. On March 24, 1965, Ranger IX hit its target in the crater Alphonsus, identified in the earth-made photograph above. Strange cracks, rills, and dark markings have been observed in Alphonsus, and scientists wondered if it might emit gases since red glows have been reported in the crater. The central peak rises a sheer 3300 feet from the crater floor.

The pictures on the right were made during Ranger's fall to the moon, the first at a distance of 775 miles, the last at about a half mile. The white circle in each picture corresponds to that in the next, each picture being a closer view of a small part of the view of the prior picture.

—NATIONAL AERONAUTICS AND SPACE ADMINISTRATION

A—Alphonsus (left) from 775 miles, with its central peak, rills, and dark spots on crater floor. Albategnius crater on the right.

D—From 12 miles up. Crater at left on Alphonsus' floor is 1.5 miles across and rills are chains of craters with sunken surface between.

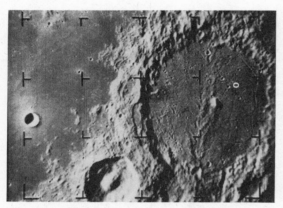

B—Alphonsus from 258 miles with many more rills and craters now visible. Alpetragius crater (left bottom) has been called an egg in nest.

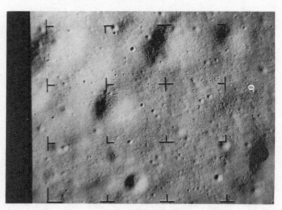

E—At 4.5 miles, Ranger nears its target. Many more craters can be seen, down to 40 feet in diameter, but no signs of volcanism.

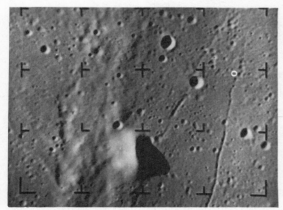

C—Central peak from 58 miles. Rills running up from peak's shadow and at right appear in picture D in much more detail.

F—At less than half a mile, 3-foot craters are visible and Ranger IX crashed down at white circle (top) next to a 25-foot crater.

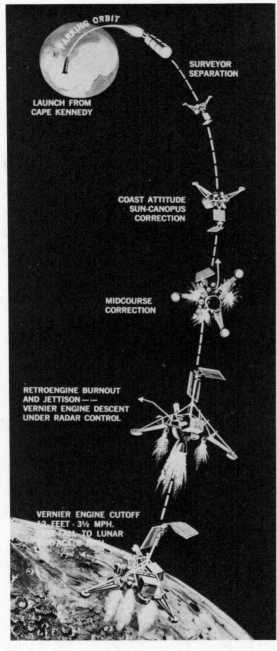

Rock-strewn lunar surface pictured by Surveyor VII on January 9, 1968, in highlands, 18 miles north of the large impact crater Tycho. Rocks cast long shadows in rising sun. Boom and omnidirectional antenna of craft at left.

— JET PROPULSION LABORATORY AND NASA

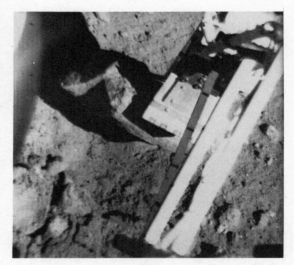

Surveyor spacecraft, soft-landing on the moon from 1966 to 1968, studied the surface to aid the selection of sites for Apollo manned landings and to reveal more than Rangers about lunar surface composition and load capacity.

— NATIONAL AERONAUTICS AND SPACE ADMINISTRATION

Surface sampler arm of Surveyor VII turns up a sizable rock fragment on lunar surface. Surveyors proved even dusty lunar surface strong enough to bear weight of incoming manned Lunar Expeditionary Module (LEM).

— JET PROPULSION LABORATORY AND NASA

Rocky landscape with craters (above) pictured by Surveyor VII was probably formed from debris of impact that blasted out Tycho. Rolling hills and ridges to horizon, 8 miles distant at center, look like parts of eastern United States. Rocky crater in foreground was gouged by block ejected from another crater. Rock in foreground is 12 feet across with 6-foot shadow. Plenty of material for moon colonists to play "duck on the rock," but dangerous in space suits!

White circle in center identifies rock in closeup from Surveyor (right photograph). Many of these rocks fragmented in crater are a foot or more across. Most lunar rocks are basaltic (maria) or anorthositic (highlands).

—JET PROPULSION LABORATORY AND NASA

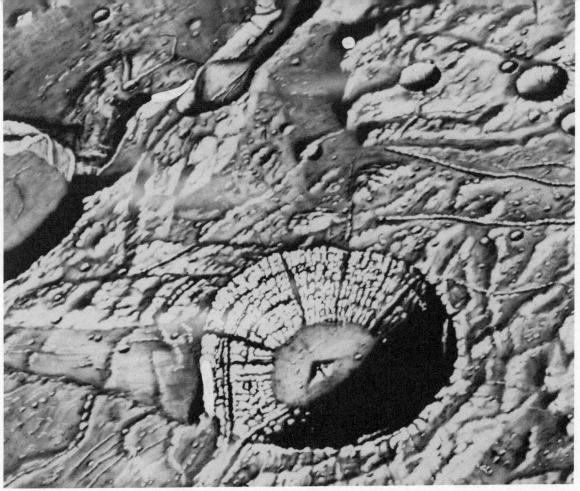

One of best drawings of Aristarchus made from telescopic study.
—U.S. Air Force Chart and Information Center.

Aristarchus is one of the brightest craters on the moon, located toward the northeast. Astronomers have watched it from earth and drawn it as shown on the upper facing page, because many red glows have been reported in and around the crater. Then Lunar Orbiter V took the photograph on the lower facing page of the southern wall of the crater, showing much erosion, and its floor (top right) seamed with tiny craters. Whether all the holes are impact craters or some are cinder cones of volcanic action, recent or in past, may not be decided until astronauts land at Aristarchus and explore it. Lunar Orbiters followed Surveyors with many such photographs from as close as 70 miles.

Lunar Orbiter V made the photograph above of the Alpine Valley in the Alps Mountains in the north central region of the moon. From earth, astronomers thought perhaps this slash through the mountains had been carved by a meteorite coming in at a low angle. But it turns out to be a wide irregular valley with a rill wiggling down its center, its cause still unknown!

—National Aeronautics and Space Administration
Astronauts began to explore the moon directly with the Apollo 11 mission. Here the Lunar Expeditionary Module (LEM) returns with two astronauts from the moon's surface to rendezvous with the astronaut in the Command Module, who took this photograph through its porthole. The return to earth (left) completed this first landing mission.

Two kinds of space exploration meet (above) with Apollo 12 astronaut taking off TV camera of Surveyor soft-lander III to return it to earth for examination. LEM (on horizon) landed within 600 feet of the Surveyor in the Oceanus Procellarum. Apollo 15 astronaut with Lunar Rover (below) looks down the winding Hadley Rill.
—National Aeronautics and Space Administration

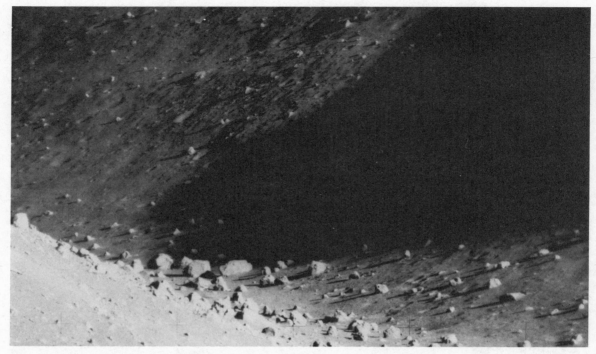

First view down the Hadley Rill by telephoto (above) by astronaut in last photo reveals rocky sides but no signs of flow along bottom, though faint horizontal layering. Mount Hadley telephoto (below) shows blocky horizontal layers that prove development of lunar crust here was complex, with many layers forming one above the other.
—National Aeronautics and Space Administration

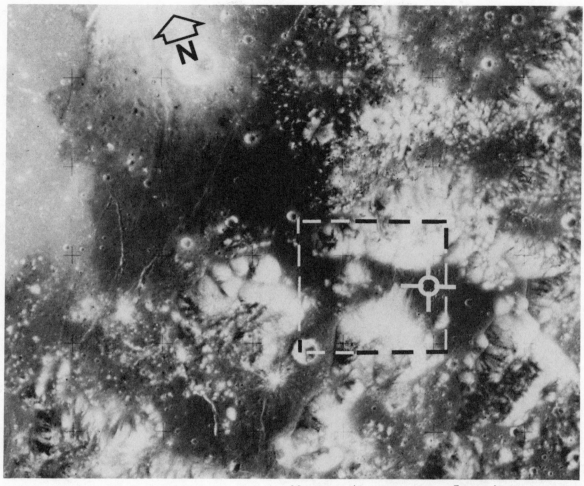

The mysterious Taurus-Littrow region in the northeast portion of the moon, in which astronauts of Apollo 17 landed, within the crossed circle. This photo from the Command Module of Apollo 15 shows the bright, old Taurus Mountains, whose steep sides rise 7000 feet above the dark plains, stretching out to the lighter Mare Serenitatis (left). Dark material spread across the mountains and in the basins may be volcanic ash and the small craters with dark haloes may be cinder cones. Here the astronauts may prove whether or not the moon was actively volcanic in the past.

Unless the Russians undertake manned lunar missions, a gap in such explorations has been reached with Apollo 17. But scientists of many disciplines will take years to digest the data already obtained from the moon. Data are still coming in, also, from instruments left on the moon by Apollo crews. Seismographs have revealed many moonquakes, for example. Without doubt, men will be back on the moon to explore it further and perhaps to set up a first colony there. Meanwhile, man's reactions to long-term weightlessness must be investigated in earth-orbiting space stations, while unmanned studies of the planets Venus and Mars continue.

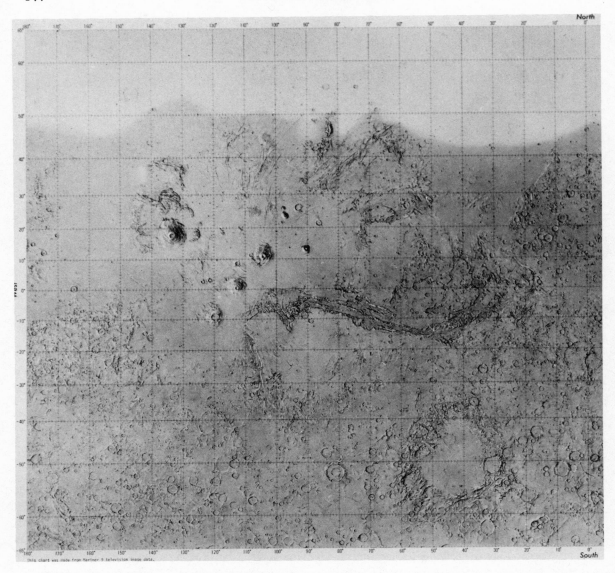

SCALE: 1 INCH = 1000 MILES, AT EQUATOR

This chart of the surface of Mars is the most complete and detailed one yet made. It is based on TV image data radioed to earth by Mariner 9 as it orbited Mars with its cameras scanning the planet. The positions of the surface features, such as volcanoes, canyons, impact craters, and smooth plains were taken from photo-mosaic charts like that shown at the top of the second following page. Some features may be inaccurately located by as much as 35 miles.

On this scale, features may be as much as 1/32 of an inch off. This chart shows only location of Martian surface features. A revised version will more accurately portray brightness variations on the surface. Comparison of this chart with that on page 169 (turning the book bottom side up, because the earlier chart follows the astronomical convention) reveals the great advances of Mariner 9.

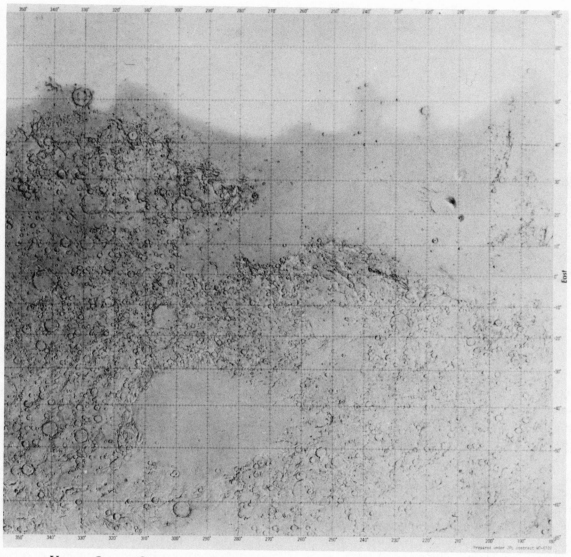

—United States Geological Survey and
National Aeronautics and Space
Administration

Note that this chart covers Mars from the equator to only 65 degrees north and south latitude, but includes whole Martian circumference to 180 degrees east and west of the Prime Meridian (0 degrees longitude).

Two unmanned Viking spacecraft are scheduled to orbit Mars in 1976 and release capsules for soft landings on Mars. The capsules will probe for signs of life on the Martian surface and study the winds, earthquakes, and surface chemistry of Mars. Five landing sites have been picked for the Viking capsules on low, smooth plains likely to be warm enough to favor any Martian life. The first two sites are north and south of the equator west of Nix Olympica, the third north of the central giant canyons, and the last two on the plains north and south of the equator to the east.

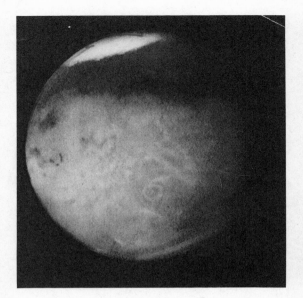 

—JET PROPULSION LABORATORY AND NASA
Far encounter photographs of Mars made by Mariner 7 spacecraft in 1969 from 280,000 miles. The south polar cap stands out, with ragged melting edge. The circle near center is volcano, Nix Olympica. Its motion in the 46 minutes between these photographs shows Mars' rotation.

TV photographs transmitted from Mariner spacecraft 4, 6, and 7 as they whipped past Mars into solar orbits in 1965 and 1969 raised more questions about Mars than they answered.

With Mariner 9 placed in orbit of Mars in late 1971, Mars could be thoroughly mapped and its features studied in detail. The two-page map of Mars, the mosaic along its equator on the facing page, and the photographs from as close as 800 miles are among Mariner 9 returns.

Data from Mariner 9 showed that although Mars' atmosphere and polar caps are largely carbon dioxide, more oxygen and water may be present in both than had been thought. Only running water could explain all the signs of carving and erosion of Mars' surface with deep canyons and branching valleys looking like river systems. Perhaps there were more atmosphere and water on Mars in the past than now. Slow changes in direction of Mars' poles of rotation might explain a cyclic gain and loss of atmosphere and water. Or perhaps Mars, at a great distance from sun, is only now warming up.

Mariner spacecraft have shown that much of the Martian surface is peppered with meteoritic impact craters, as is the moon. Mars is near the asteroid belt and its thin air would not burn up larger meteors coming in. See the mosaic (next page) for clusters of such impact craters.

Mars' many large volcanoes, however, like Nix Olympica and Nodus Gordii above, showed great activity in the past. When they erupted, on occasion, such volcanoes emitted vast amounts of carbon dioxide, water vapor, other gases, and dust. These gases probably made a denser atmosphere and more water on Mars, until its weak gravity let them escape. This may be why so many signs of erosion and running streams appear on Mars and why in some places the impact craters have been worn down almost flat, as they never have been on the moon.

With so many large volcanoes, Mars must once have had a hot interior, in which convection cells could make its huge canyon.

—Jet Propulsion Laboratory and NASA
Mosaic of TV photos of Mars from orbiting Mariner 9 in 1972 shows band of about one-third of Mars' equator. Group of giant volcanoes with summit craters appear in upper left and deep, crisscrossing canyons along equator in center.

The huge canyon undulating along Mars' equator is about 2500 miles long, 75 miles wide, and nearly 20,000 feet deep, 3–4 times the depth of the Grand Canyon. On earth, this canyon would extend from Los Angeles to New York. It shows signs of erosion along its serrated edges.

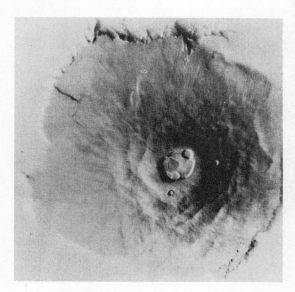

—Jet Propulsion Laboratory and NASA
A gigantic volcanic mountain, Nix Olympica (above) rises perhaps 6 miles high and appears on mosaic (far upper left). It is about 310 miles across its base and the summit crater is 40 miles across. It has given Mars plenty of volcanic action.

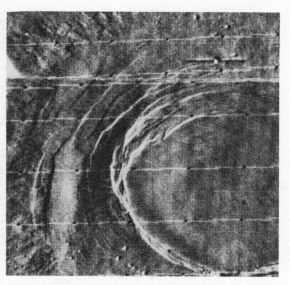

—Jet Propulsion Laboratory and NASA
Called Nodus Gordii (The Gordian Knot) as viewed from earth, this feature turned out to be a huge volcanic collapse crater, or caldera, 70 miles across. It appears dimly on the mosaic, below and to the right of Nix Olympica.

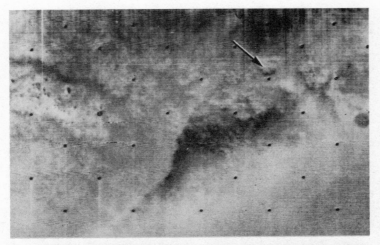

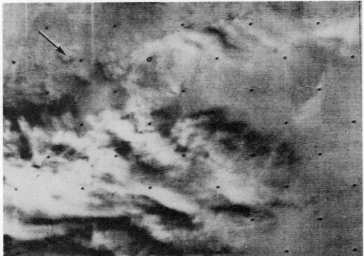

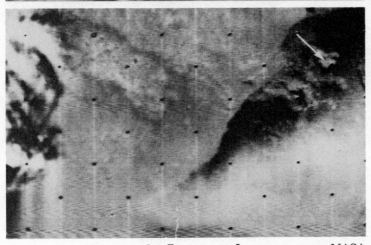

—JET PROPULSION LABORATORY AND NASA

Do these photographs from Mariner 9 as it orbited Mars in February 1972 offer an explanation for the puzzling changes in Mars' dark markings, which astronomers have sometimes attributed to the growth of vegetation on Mars?

The arrows in the photos point to the same crater in a region called Euxinus Lacus above the Martian equator. The top photograph shows normal conditions in the area, with a dark marking in the center. The middle photograph, taken 24 hours later, shows a large dust storm, 300 miles wide, sweeping from west (left) to east (right) across the region. The bottom photograph, taken two weeks later, shows that the dark marking in the top photograph has become blacker and much larger! A new dust storm is appearing (left) on the bottom photograph.

The winds in the first dust storm may have uncovered a dark area of the Martian surface under the lighter-colored dust, causing the enhancement of the dark marking. Perhaps the alternate covering and uncovering of a dark surface by wind-blown, light-colored dust causes the changes in the dark markings on Mars.

This explanation of the changes in Mars' dark markings is very similar to that offered years ago by Dean McLaughlin of the University of Michigan (see page 185 of text), and he should receive due credit for his early theory. He was far ahead of his time. Although vegetation no longer makes sense in the explanation of the dark markings, and no reports have been made of indications of organic molecules on Mars surface or in its air that have been received from Mariner 9, the chance of native Martian life at locations where water is available remains possible.

The network of great canyons (above), pictured by Mariner 9, may be seen below the equator on the left half of the Martian mosaic, ending in a "Badlands" at the equatorial canyon. Erosional forces must have been at work on these canyons, as well as on the long valley (below) running for some 250 miles and 3 miles wide. No branching "tributaries" like this are found in the rills on the moon.

—Jet Propulsion Laboratory and NASA

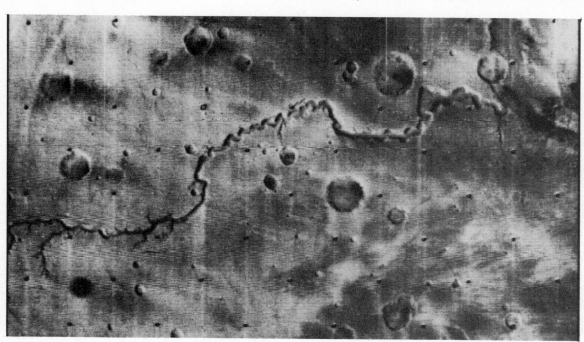

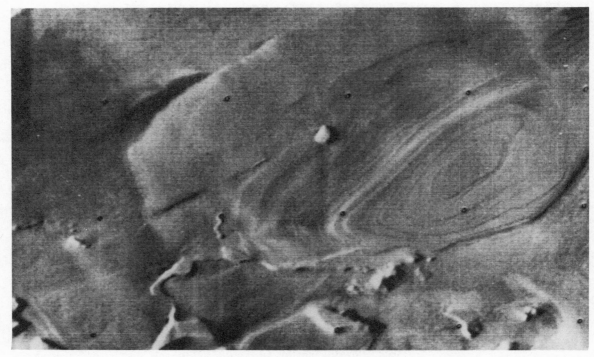

Mariner 9 photographed (above) what may be an oval glacier near Mars' south pole. The light layers may be frozen carbon dioxide and water ice between dark layers of dust or volcanic ash. Beneath these are jagged pits and grooves of an older deposit. Mysterious curves and deep depressions in photograph (below) on polar cap may result from interplay of Martian geological forces unknown on earth.

—JET PROPULSION LABORATORY AND NASA

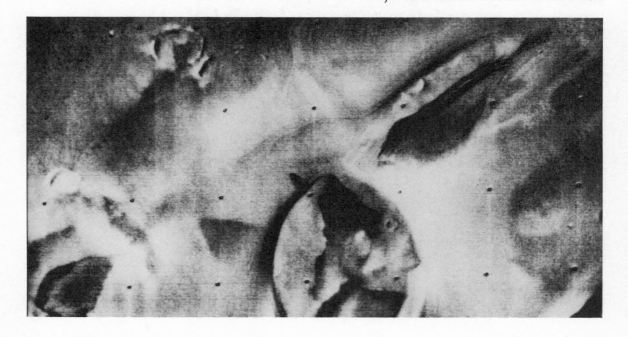

The manned, rocket-powered X-15 (above) penetrated near-space in 1960, after release from a
mother B-52 aircraft. In landing on its own at
Edwards Air Force Base, California, it was very
like the future space-shuttle craft (below).
—UNITED STATES AIR FORCE

One design for an earth-orbiting space station
(below) to be manned for long periods of space
and earth research. Space shuttle (left) will exchange astronaut crews, shot up by rocket and
gliding back to earth landing like X-15 above.
—NATIONAL AERONAUTICS AND SPACE
ADMINISTRATION

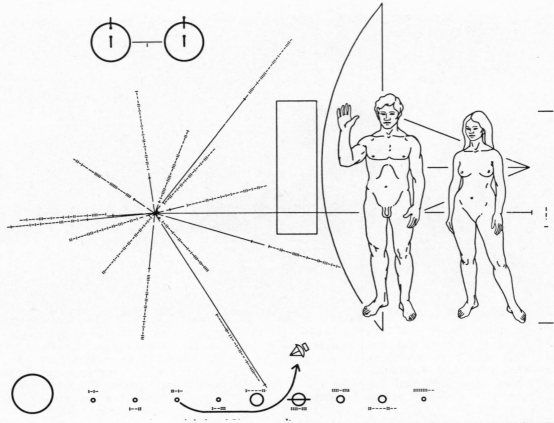

Message (above) on Pioneer 10 spacecraft, passing Jupiter in 1973 and then orbiting out of solar system, tells other cultures our location from 14 pulsars (left), our appearance (right) to scale of Pioneer, and our place in system (bottom). Such communication would be an historic event.

—CARL AND LINDA SAGAN AND FRANK DRAKE

Early astronomers thought moon's Straight Wall (below) was made by intelligent moon men. Lunar Orbiter photograph here shows wall to be a natural fault scarp, 75 miles long and 870 feet high. Much more solid evidence would be necessary to prove existence of extraterrestrial life.

—NATIONAL AERONAUTICS AND SPACE ADMINISTRATION

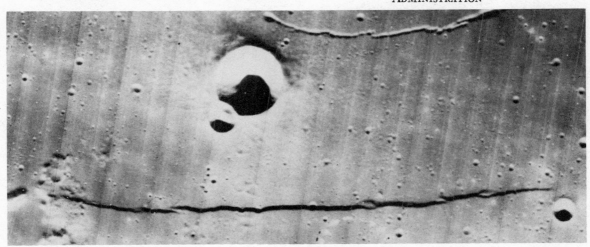

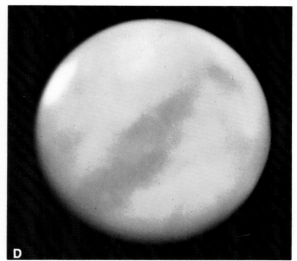

Above (left), NASA; above (right), MOUNT WILSON AND PALOMAR OBSERVATORIES
Earth photographed from space and Mars photographed from earth reveal both differences and similarities when compared side by side.

Jupiter, the largest planet, and Saturn, acclaimed the most beautiful, may tell us a great deal about the origin and character of the solar system when data from Pioneer 10 and the spacecraft that will follow it are analyzed and interpreted.
—MOUNT WILSON AND PALOMAR OBSERVATORIES

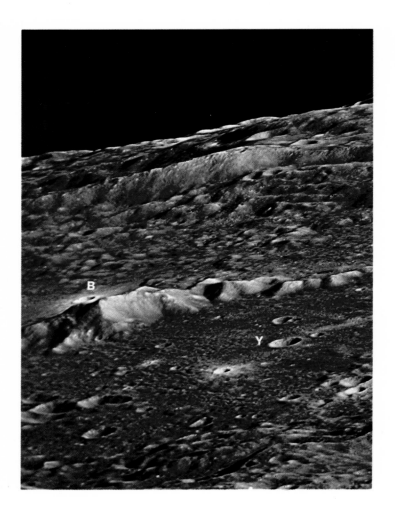

Features on the other side of the moon, always hidden from earth, photographed by astronaut. The only feature that would give an astronaut direction on the moon would be the sun and the shadow it casts. So an astronaut going from B to Y would head directly toward the sun.
—National Aeronautics and Space Administration

Four meteors made these trails as they ducked into (below) and out of (above) earth's atmosphere, photographed from a balloon at an altitude of 19 miles. Such trails leave dust in the atmosphere that affects our weather. Greater understanding of these and many other space events will offer opportunities for us to control and improve our environment.
—E. K. Bigg, Division of Radiophysics, CSIRO, Australia

rose to the surface, forming a quickly hardening crust. At this time the largest of the impact craters probably formed, such as the Mare Imbrium and the Mare Crisium.

It is believed that another period of melting occurred in the upper mantle of the moon after about 1 to 1.5 billion years, when the vast lava flows forming the maria took place. With all the layering found in the crust of the moon, probably there were hundreds or thousands of local floods of lava. These filled some of the older crater bowls and made the wrinkle ridges in maria. And during all this time, impact craters kept forming, as fragments of asteroids, comet remnants, and smaller meteorites came plunging into the moon.

When Lunar Orbiters first circled the moon, changes in their orbits indicated the presence of a number of large mass concentrations, called mascons, beneath some of the maria. Too large to be the remains of asteroids that hit the moon, it is believed they indicate pools of heavier materials melted in the upper mantle.

Lunar Craters

Robert S. Richardson, an American astronomer, comments that, "The chief difficulty in writing about the origin of the lunar craters is the appalling number of hypotheses that have been advanced to explain them. Apparently everybody who has looked at the moon has had a try at it. Some of these explanations sound like plain crank ideas. Others are undeniably ingenious. But today there is only one that seems acceptable to us." He refers to the meteoritic impact hypothesis. On the other hand, varied volcanic explanations of the origin of lunar craters have been developed, and many scientists, particularly some in the Soviet Union, have elaborated on the volcanic theory.

The meteor-impact theory received a great boost when Ralph B. Baldwin of the University of Chicago pointed out in 1949 that his graphs showed a continuous curvilinear relationship between the width and depth of craters made on earth by shells, bombs, and other explosions at the lower end and lunar craters at the upper end. He was able to formulate an equation for the relationship between the diameter and the depth of craters, holding for those on the earth and on the moon. But this does not appear to be final proof that meteorites alone blasted out the lunar craters—there may also have been internal explosions of some kind. Baldwin himself points out that, while "Almost every observed condition of the lunar crust may be completely explained by the meteoritic theory or associated subsequent processes," still there are some exceptions, "the very few formations obviously due to a mild igneous action."

The United States Atomic Energy Commission more recently has investigated the relationship of size of detonation, depth of detonation, and breadth and depth of the resulting craters in studies of chemical-explosive and nuclear-explosive craters produced here on earth. These preliminary studies should enable the formulation of a full-fledged technology of crater production, which might then be extrapolated to the conditions existing on the moon.

Two types of terrestrial craters, the maar and the meteoritic-impact craters, appear to be very much like certain craters observed on the moon. These two types are often so similar here that they are confused, just as those on the moon may have been. Maar craters are usually small, circular or elliptical, or quite irregular depressions, opened up by the slumping of the walls of volcanic vents. Debris is usually distributed widely around them by the gases rushing out of the vent, and a low rim forms around the crater, continually being engulfed and built up again as the crater widens with growth. Maars are found around the world, scattered in volcanic regions.

Eugene M. Shoemaker of the United States Geological Survey has attempted to establish criteria that clearly distinguish maar craters from impact craters. He considers size as a criterion, but concludes that maar and impact craters overlap in this respect, though the terrestrial impact craters run much larger than do the maar craters; shape does not distinguish them, as both types may be circular, elliptical, or sometimes polygonal (like so many of the

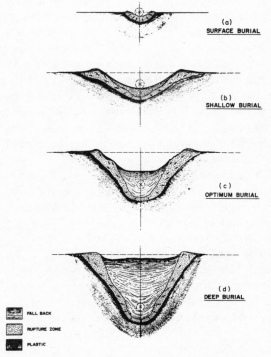

Typical crater profiles resulting from shallow to deep burials of nuclear explosives, based on experiments conducted at the atomic test site in Nevada.
—U.S. ATOMIC ENERGY COMMISSION

craters on the moon), depending in the main on the structure of the surrounding terrain. Both kinds of craters may have a central hill or peak within their depressions, a common feature of lunar craters. In terms of slumping of the sides and of faulting, the crater types may be similar. But because of their origin, the maar and meteoritic craters do seem to differ in terms of the debris thrown out about them. The outer slopes of the rims of maars are usually smooth; those of impact craters are rough and hummocky. Rays may be created around impact craters, which is not true of maar craters.

Fluidization is another process suggested for the formation of craters as the moon began to age. When gas moves through dust it picks it up and carries it along like a fluid. So it has been suggested that outgassings from hot depths below nearly inactive areas on the moon are the source of a number of craters.

Although astronauts have not yet explored them, photographs from Apollo and Orbiter craft have proved that volcanic openings do occur in some craters and even in the central peaks of craters, giving evidence of volcanic action by either lava or gas at some time in the past. The conclusion that all processes, impact, fluid volcanic, and outgassing with fluidization, have played a part in the formation of the moon's surface makes excellent sense.

Changes on the Moon?

Diametrically opposed views exist on whether any changes at all have occurred on the lunar surface during the period over which it has been observed with telescopes.

Proponents of the meteoritic-impact theory of lunar-crater formation are pronounced skeptics when it comes to any reports, past or present, of changes on the moon that might have been internally caused. It is true that changes were reported much more often in the past than nowadays. Before detailed maps of the moon had been drawn, it was quite natural to interpret an observed variation as an appearance of a new object or as a disappearance of a familiar one, when in reality it may have been only a changed reflection of light from an object due to the angle of the sun at the time of observation. Nonetheless, most selenographers who have devoted a great deal of time to the moon say they have seen activity or obscuration there. Proponents of the volcanic theory of lunar-crater origin are, of course, looking for evidence of continuing activity to buttress their theories. Any evidence for changes must be thoroughly evaluated.

There are numerous reports of veils, mists, and small clouds on the moon, which obscured known features in such craters as Aristarchus and Plato. So many of these reports have come in, in some instances confirmed by independent observers, that it is difficult to shrug them off as due to poor seeing, inferior instruments, inaccurate drawings, illusions, or memory lapses. Something akin to the escape of gas or the vaporization of solid gases or solids does seem to occur on the moon, though very rarely, despite the firm conviction of some that the moon was formed cold a billion years or more ago and has not changed one iota since. Just a few years ago one such event was recorded on a spectrogram—or at least was so interpreted by the astronomer who obtained it.

The event in question took place on November 3, 1958, in the crater Alphonsus, located just south of Ptolemaeus and to the west of the central meridian. Three famous dark spots along the floor of the crater near the edge, that can be distinguished even with small telescopes, have aroused a great deal of interest. But the story begins two years earlier when, on October 26, 1956, Dinsmore Alter, using the 60-inch Mount Wilson reflector, took photographs of the Alphonsus area and the adjoining craters of Ptolemaeus and Arzachel. Some plates were made without a filter, in violet-blue light and others were made with infrared filters. There was considerably less contrast and more blurring in the blue light than in the red. Alter believed that this was in the portion of the photographs showing the crater Alphonsus, not throughout the blue photographs. He said that the fissures in the eastern half of the crater appeared in these photographs to be much more blurred than those in the adjoining crater of Arzachel and took this to mean that a temporary emission of gas had occurred in the eastern part of Alphonsus.

Following up on the reports of the Alter photographs, N. A. Kozyrev, of the Pulkovo Observatory in Leningrad, U.S.S.R., began making spectrograms of the lunar surface near the terminator in the fall of 1958, using the 50-inch reflector of the Crimean Astrophysical Observatory. He made a special point of obtaining spectrograms of the interior of the crater Alphonsus. While taking the second of three 30-minute spectrograms of the area on November 3, 1958, he noted that the central peak in the crater suddenly appeared brighter and whiter than usual and then dropped to normal intensity. At this point a third spectrogram was taken, and this, as well as the first spectrogram taken an hour earlier, showed a normal spectrum. The second spectrogram, however, showed lines that were not normal for Alphonsus, which Kozyrev interpreted as a band indicating diatomic molecules of carbon (C_2) and representing an emission of volcanic gases containing carbon from the crater. From the timing of the spectrograms, Kozyrev estimated that this process could not have lasted for more than an hour.

While there has been disagreement about the interpretation of Kozyrev's spectrogram, particularly whether its bands manifest the presence of carbon or not, it is now generally agreed that a small residual gaseous emission must have occurred in Alphonsus on November 3. Kozyrev reported another similar occurrence on October 23, 1959, but the spectrogram does not show such striking bands as the earlier one and it is questioned whether any emission took place at that date.

Kozyrev definitely sides with those who believe that meteorites had little to do with the formation of the moon's surface. He concludes, for example, that his spectroscopic observations "go a long way toward establishing the fact that the moon, even at the present time, has sufficient internal energy for orogenic [mountain-building or upheaval] processes. This result clearly shows that the history of the formation of the lunar landscape may, in fact, be mainly the history of the internal processes of the moon's cosmic existence; while external influences may have been of secondary importance and, in particular, the role played

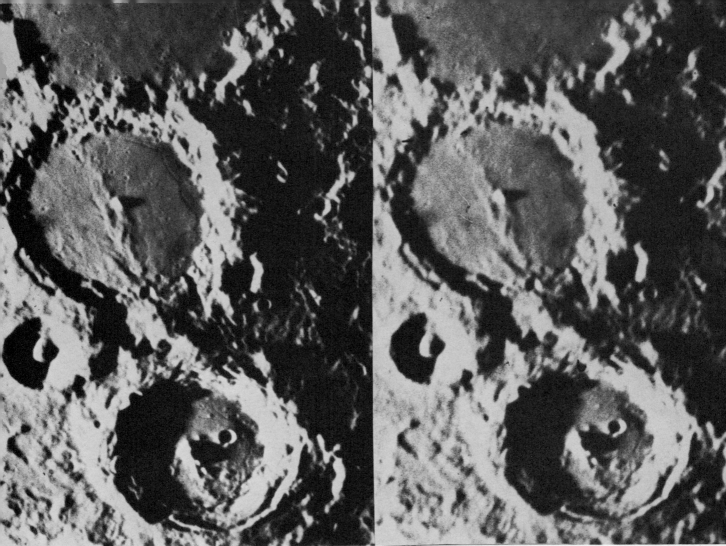

The crater Alphonsus photographed in infrared (left) and blue-violet (right) light on October 26, 1956, with the 60-inch Mount Wilson reflector, showing possible obscuration of the floor of Alphonsus at the upper right in the blue-violet. The rill clearly visible in the infrared near the edge of the floor of Alphonsus is barely discernible in the blue-violet, particularly toward the top. (North to top and east to right.)
—DINSMORE ALTER, MOUNT WILSON AND PALOMAR OBSERVATORIES

by meteorite impacts scarcely more significant than in the formation of the known terrestrial features."

The number of discoveries of activities on the moon may be as much a function of the amount of time large instruments are trained on it as a measure of actual events occurring there, for in recent years reports of lunar events have multiplied. On the evening of October 29, 1963, James Greenacre and Edward Barr were using the 24-inch refractor of the Lowell Observatory at Flagstaff, Arizona, in the lunar mapping program of the United States Air Force, and saw three colored spots in the vicinity of the crater Aristarchus, toward the limb of the moon northwest of Copernicus. Light ruby-red or reddish-orange in color, two of these spots (one covering an area of about 1.5 × 5 miles over a domelike structure and the other about 1.5 miles in diameter on a hilltop) appeared near the Cobra Head widening of Schröter's Valley, and a pinkish streak about 1.5 × 11 miles extended along the inner rim of Aristarchus itself. Within 25 minutes the spots had all disappeared.

On November 27, 1963, the same observers

and others saw a larger light-ruby-red streak. This one was about 1.5×12 miles in size, appeared in almost exactly the same place on the rim of Aristarchus, and lasted for 45 minutes. They notified two colleagues at the Perkins Observatory at Flagstaff, who checked the area in their 69-inch reflector and verified the observation, so the streak must have occurred.

These color flashes and red patches glimpsed on the moon have been explained in many ways. Perhaps the rising sun heats minerals that then luminesce; perhaps moonquakes uncover granite that flashes as its minerals, irradiated for eons by solar radiations, are exposed and warmed. The best explanation may derive from the moonquakes reported by instruments left on the moon by Apollo astronauts. Most quakes occur at times when the moon is most pulled by tidal gravitational forces from the earth. Gases trapped under the surface may be released by quakes and raise clouds of electrically charged dust that sparkle and glow.

In 1955 the Russian astronomer Kozyrev had noted a peculiarity in a spectrum he made of Aristarchus, which he interpreted as a luminescent glow. Again in 1961, working with the 50-inch reflector of the Crimean Astrophysical Observatory, spectra he made of the central part of the crater on November 26 and 28, and two on December 3, showed emission lines that he identified as those of molecular hydrogen gas, which he believed must have escaped from the moon's interior.

Lunar Surface Composition

Full moonlight is much too brilliant for visual comfort when it is gathered in the optical system of a telescope, yet here on earth we see an average of only some 7 per cent of the light that falls on the moon from the sun. The moon's over-all reflectivity is very low. What kinds of material would absorb so much of the light falling on them? For many decades, astronomers tried to answer this qestion. Speculation about the material of the moon's

surface has run riot. A few of the studies made before Apollo astronauts landed on the moon and gathered soil and rocks to return to the earth will give an idea of the wide range of explanations that were offered.

Since the advent of radio astronomy, radio waves were bounced off the moon and its effects on them were analyzed, furnishing some clues about the nature of the moon's surface. These studies indicated that it was generally very smooth and only slightly hilly, with rises up to about the order of 1 foot in 10, and with only about 10 per cent of the surface covered with small objects below the optical limits of resolution. In fact, the reflected radio signals have been compared to those received by an aircraft flying over dry, desert country on earth. The radio astronomy gradient results compared well with the evidence eventually obtained directly from the lunar surface.

Other lines of investigations led to more specific, but still inconclusive, results. During the 1920's, Bernard Lyot, a French astronomer, compared polarization curves of the light from the moon with those obtained by the reflection of light from a great many mixtures and compounds of materials here on earth, including vitreous and granular mixtures, sands and clays, powders, igneous rocks, volcanic ash, water droplets, and artificial fogs. Lyot concluded from these studies that the moon was covered almost entirely with a powdery material, probably in a thin layer, closely resembling volcanic ashes found on the earth. Others have found that the light polarization from the maria resembles that produced by volcanic lava, although that from the brighter regions does not.

In the 1950's, Lyot's work was followed up by Audouin Dollfus and his colleagues at the Meudon astrophysical section of the Paris Observatory. Their results tended to confirm Lyot's conclusions. Dollfus thought that the dusty surface of the moon could not have come from the pulverizing effects of solar ultraviolet and x-ray radiations, nor from the cracking of rocks under rapid temperature changes, since these would not produce enough dust. Rather, he believed, the dust was ejected when the meteorites which formed the craters on the moon plunged into its surface. The dust formed with the impacts would be shot out at the same time and spread quite uniformly over the moon's disk; with this, also, would be mixed the dust formed by smaller meteorites, and the fine micrometeoritic dust collected from space by the moon.

Rather different conclusions have been drawn from photometric studies of the moon's surface. These tend to show not a layer of dust on the surface, but an extremely porous, granular substance containing separate grains capable of reflecting light backward and of standing out separately enough to cast shadows in their immediate vicinity. The results of laboratory experiments with agglomerated stones and volcanic ashes did not fit in with the moon's photometric properties, which indicate similarity to surfaces pitted with deep-rounded, porous wells, as if the moon might be covered with small meteoric or other pits.

Other laboratory studies have indicated that the lunar surface might have a porous structure made up of tiny dust particles formed into what is called a "skeletal fuzz," with varying estimates of its possible depth—anywhere from less than an inch to several inches. Such a fuzz has been described as perhaps made up of "fairy castle," "tinker-toy," or "Christmas tree" configurations in order to explain the way in which the light reflected from the moon appears to come from a fairly open structure. These shapes might result from electrostatic surface forces which may be considerable on the moon, from sputtering with the impact of protons in the solar wind, or from the striking of micrometeorites raising spurts or clouds of dust on the surface; finally, some combination of these factors might actually be at work on the lunar surface.

In another approach, bulk densities of about 0.1 that of solid rock have been calculated for the upper layers of the lunar terrain. Its porous character has been described as perhaps having the consistency of fine, matted, hairy filaments, of a layer of loosely compacted dust, of spun-sugar candy, Cracker Jack, rusk, sponge, or moss.

With an eye on the design of landing craft capable of coping with lunar surface conditions, Krafft Ehricke conceived the surface as consisting of a soft layer, perhaps up to one or two feet thick, resembling sun-baked, cracked mud. The bombardment of cosmic rays, solar flares, micrometeorites, and cislunar dust would have produced fine particles on the surface over millions of years, he believed; these particles would tend to adhere to one another in the absence of an atmosphere, be baked together and cracked by further bombardment, and riddled with small meteoritic impacts. Such material would smooth out the surface over small pits and cracks but would not cover larger fissures, pits, and clefts. With the exception of the fissures and cracks, Ehricke's idea of the surface came closest to the reality that was discovered, perhaps, but most of the other theories were "in the ball park" too. Fortunately, the loose dust to great depths did not engulf astronauts and their vehicles.

As experience on the lunar surface soon proved, astronauts only sank a matter of inches into the dust, beneath which was a fairly well compacted soil. This could bear the weight of the astronauts and their buggy as they trudged and drove across it. Unmanned Soviet vehicles also trundled over the surface with ease.

The soil was composed of fine particles of basaltic rocks and contained many glassy spheres, created in the heat of impacting meteorites that churned the lunar surface over millions of years into the consistency of a well-plowed field. The footprints of no other visitors to the moon from space have been found. If they are there, they will wait for eons.

CHAPTER XI

"... or if they list to try
Conjecture, he his Fabric of the Heav'ns
Hast left to their disputes, perhaps to move
His laughter at their quaint Opinions
wide ..."

MILTON

PLANETARY SURFACES

The quality of information available about the surfaces of planets other than the earth is almost inversely proportional to their distances. Even so, we know little enough about Venus, which at its nearest is only 100 times farther away than the moon, and about Mars, which on rare occasions approaches within 150 times the moon's distance. As relatively close as they are, these bodies are still millions of miles away, and in view of the observational difficulties, even the scraps of knowledge that we have represent a remarkable achievement.

No matter how large the telescope, the unhappy fact is that no fine details at all can be seen on any of the planets, even the closest ones. With a rare resolution of one second of arc, the estimated smallest spots visible photographically on Mars at opposition (with the earth nearly on a straight line between Mars and the sun) are from 180 to 250 miles across. On Venus at inferior conjunction, the spots distinguishable photographically are about 150 miles across. Visually, under exceptionally favorable seeing conditions, the smallest spots discernible on Mars are 18 to 25 miles across and about 16 miles across on Venus (at a resolution of 0.1 second of arc). (Offhand, it seems doubtful that Martians ever dug "canals" or built anything as large as this!) Thus only large-scale conformations can be seen on the planetary surfaces—merged patterns reminiscent of the way objects on the far horizon look from the top of a mountain.

The degree of detail in planet photographs is illustrated by the prints. With a long and patient look at the images of Venus, studying them for perhaps five minutes and trying over and over to distinguish all the gradations in light and shade, bright and dark markings can gradually be picked out that were not immediately apparent. Astronomers have had to work on about this level of discrimination, although they can see somewhat more in their telescopes than a photograph can show. For comparison, the drawings of what they have seen of Venus appear later in this chapter. Often a number of drawings are put together and the composite drawing reveals more. Only a few features of the drawings can be picked out in the photographs.

Much more can be distinguished in the photographs of Mars. Drawings made on the basis of visual observations depict more fine detail than do those of Venus and Mercury and go far beyond photographs as well. Nonetheless, the vagueness of all the photographs and drawings demonstrate something of what astronomers have had to cope with in their efforts to map the surface features of the nearby planets with earthbound telescopes.

Mercury's Dim Markings

Mercury so closely precedes or follows the sun in the sky that it is a very elusive object to study and is often lost entirely in the sun's glare. Under the best seeing conditions, with powerful telescopes, astronomers have been able to observe on the Mercurian disk some faint shadings or markings that appear to be fairly permanent, though shifting somewhat with its large librations.

In a powerful 24-inch telescope under ideal conditions, Mercury resembles the moon as seen with the naked eye. However, the markings on its surface are usually less distinct and do not offer nearly as much contrast as those

Three views of Venus in ultraviolet light, taken in June and July, 1927,
with the 100-inch Mount Wilson reflector.
—MOUNT WILSON AND PALOMAR OBSERVATORIES

View of Mercury's northern limb made by Mariner 10 in March, 1974,
from a distance of 49,000 miles. Mercury's surface is very similar to that
of the moon, except for its many long, high scarps, or ridges (top center),
perhaps formed by compressive forces caused by crustal shortening.
—JET PROPULSION LABORATORY, NASA

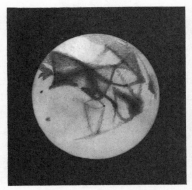 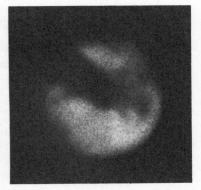

Comparison of a drawing and a photograph of Mars made at the same time during the planet's opposition in 1926.

—LICK OBSERVATORY, MOUNT HAMILTON, CALIF.

Photomosaic of part of Mercury's surface photographed by Mariner 10 as it passed the planet in 1974. The huge basin with concentric rings (left center) named Caloris is some 800 miles in diameter and similar to Mare Orientale on the farside of the moon.

—JET PROPULSION LABORATORY, NASA

on the moon. Polarization curves of the integrated light from Mercury are intermediate between those of the waxing and the waning moon when Mercury is at quadrature (when the line from Mercury to the sun forms a 90° angle with that from the sun to the earth). But when it swings closer to inferior conjunction with the earth (on a line between the sun and the earth), its polarization curves are almost indistinguishable from those of the moon. Sometimes the polarization of light from small regions of Mercury shows wide variations. This may indicate the presence of a local and temporary atmosphere, dust veils, or some other events. Polarization studies seem to indicate that in an over-all way the surface of Mercury may be very like the moon's.

Very recently, the first radar observations of Mercury were made by scientists in Russia and at the Goldstone tracking station of the Jet Propulsion Laboratory in the Mohave Desert, where an 85-foot parabolic antenna was used both for transmitting and receiving the signals. Both teams used strong transmitting power (100 kilowatts or more) in order to receive identifiable echoes back from the tiny, distant planet. The Russians reported that Mercury and the moon appeared to have similar reflection characteristics. The Goldstone scientists obtained striking confirmation of the length of the astronomical unit as determined from the earlier radar-echo studies of Venus. Radar bouncing from Mercury is similar to the echo from a dime at a distance of 10,000 miles. They also concluded that Mercury's surface is much rougher than that of Mars and perhaps twice as rough as Venus, more like that of the moon. Even more recent studies have revealed the presence of several big, rough surface areas and one smooth area on Mercury.

The heat radiated by Mercury, presumed to be entirely or largely reflected solar radiation, has been studied with thermocouples, and the surface temperature has been calculated for that portion of the disk directly beneath the sun, called the subsolar point. These measurements were made on a total of 26 days spread over a 2-year period. They had to be based on the entire image of the planet since it is so small. For the average distance of Mercury from the sun, this temperature proved to be about 640° F.; when Mercury is at perihelion, closest to the sun, it rises to about 780° F., and at aphelion drops to about 545° F., ranging over 200°. In addition, the phase-radiation curves for Mercury were very similar to those for the heat radiation from the moon.

The Italian astronomer, Giovanni V. Schiaparelli, who had directed attention to the Martian "canali" or "channels" in the 1870's, observed Mercury carefully with a small telescope during the 1880's. He reported what seemed to be permanent spots and linear markings on Mercury's surface. He reproduced these on the simple planisphere, or representation of a sphere on a plane, given here.

Schiaparelli also shook up the astronomical world with the announcement that Mercury rotated only once in a revolution of the sun, keeping the same face toward it except for slight veerings, or librations, as the planet moved in its orbit. Earlier observers had thought the rotation of Mercury was close to 24 hours, though they indicated slight variations in the turning. Apparently their preconceptions were father to the thought that the planet had spun full around in 24 hours, returning to nearly the same position, and so presenting the same face to the observer.

The drawings of Percival Lowell given here carry the notes "Rotation 88 days" and "Effect of Libration," in accord with Schiaparelli's report. Imagine the surprise when radar observations in the mid-1960's indicated that Mercury actually rotates rather faster on its axis, once in 59 days, about two-thirds of its 88-day-long period of revolution of the sun. Apparently the markings on Mercury are so dim, and the rotation so slow, that its 59-day rotation could be mistaken for periods of 88 days or of 24 hours.

With a surface like the moon's but much hotter, Mercury is not among the likely candidates for the establishment of a human colony in the near future. One of the unmanned spacecraft making a "grand tour" of a number of the planets might fly by Mercury as well, however. Knowledge of any other planets will help to explain the origin of our solar system.

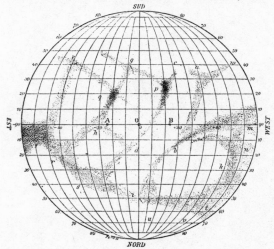

Map of the faint markings of Mercury, drawn by the Italian astronomer Schiaparelli on a planisphere of the planet. (North at bottom and west at right, according to astronomical convention.)
—E. M. ANTONIADI, *La Planète Mercure et la Rotation des Satellites*

After Schiaparelli, in 1897, the American planetary astronomer, Percival Lowell, drew an odd map of Mercury. Its diagonal slashing strokes and outline do not give the impression

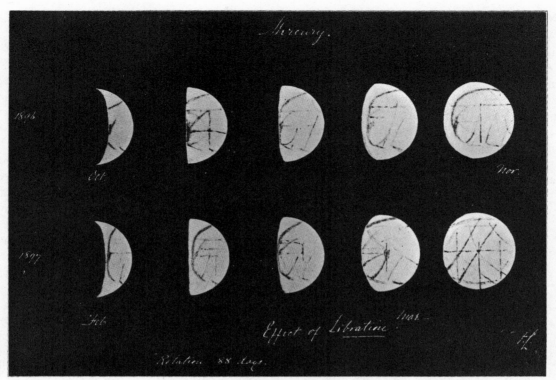

Rough drawings of markings on Mercury by Percival Lowell showing some rotation in what he thought was the effect of libration.
—PERCIVAL LOWELL, *The Evolution of Worlds*

that a great deal of work or drawing skill went into it. Yet, when compared with Schiaparelli's map and with those of later observers, Lowell's effort does show certain resemblances to the others.

One of the most careful Mercurian maps was made up by E. M. Antoniadi, a French astronomer, who used the 32-inch refractor at Meudon Observatory in France in daytime observations of Mercury from 1924 to 1929. Rather quaint Latin names are shown on this map for some of the features of Mercury. The "S" stands for the Latin word *solitudo*, or wilderness. "Prom." is an abbreviation of the Latin *promontorium*, a high point of land or rock, or a headland projecting into a sea.

Although Antoniadi underestimated Mercury's temperature by at least half, he was well aware that the surface must be very hot, and thought it likely to be a desert. This may be why he called the gray or dark areas "wilder-

nesses." Also, he believed that he observed changes in some of the markings. He commented that, "The dark area named Solitudo Atlantis, to the right of my map, is certainly much larger now than when it was first drawn 51 years ago. Yet the reality of this apparent change must be considered with the greatest diffidence, as vegetation seems impossible on a world where the temperature rises at least 200° above the zero of the centigrade scale [390° F.]."

The suggestion has been made that the dark markings on Mercury may represent large-scale maria, or outpourings of lava, similar to the maria on the moon and possibly produced by meteoritic impacts. Ralph B. Baldwin, who developed the impact theory of the lunar maria and craters in detail, wrote that, "through the telescope Mercury appears very much like a blurred version of the moon seen with the naked eye. The dark Mercurian areas look much like lunar maria. On both Mercury and the moon the dark markings are most prominent at the full phase. Within the limits of observation Mercury seems to be a slightly enlarged version of the moon."

In 1953 Audouin Dollfus published drawings and photographs of Mercury that he, together with Bernard Lyot and Henri Camichel, had made from observations of the planet be-

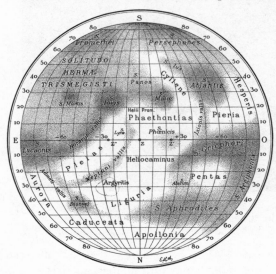

E. M. Antoniadi's map of the markings on Mercury with the names he assigned to some of the more prominent light and dark areas, based on his observations from 1924 to 1929, with the 32-inch refractor at Meudon Observatory, France.
—E. M. ANTONIADI, *La Planète Mercure et la Rotation des Satellites*

tween 1942 and 1950. Markings quite similar to those of the earlier maps appeared on these. Dollfus remarked that "the maps of Schiaparelli are in good accord with ours if they are rotated about 15° in a counterclockwise direction. These visual maps were based on evening observations, when the planet was close to aphelion, while . . . [ours] is based on morning observations when the planet was close to perihelion. The observations are brought into coincidence by assuming an obliquity of about 7° for the axis of rotation."

Dollfus did not find the kind of evidence supporting the existence of the atmospheric haze or veils, which, Antoniadi had written, often interfered with his view of the markings. It is possible but unlikely that Antoniadi misinterpreted turbulence in the earth's atmosphere as veils over Mercury. It is also possible that, observing between 1927 and 1929, over the maximum of a sunspot cycle in the middle of 1928, he was seeing the strong effects of solar disturbances and flares on the surface of Mercury, which is relatively close to the sun. The extreme radiation in such flares might pro-

Drawings of Mercury made at the 24-inch refractor of Pic-du-Midi by Audouin Dollfus, each based on a number of observations, made on October 6, 12, and 19, 1950. Some effects of rotation, then conceived as libration, are apparent.
—A. DOLLFUS, OBSERVATOIRE DE PARIS, MEUDON, FRANCE

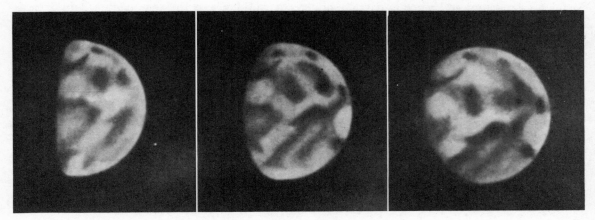

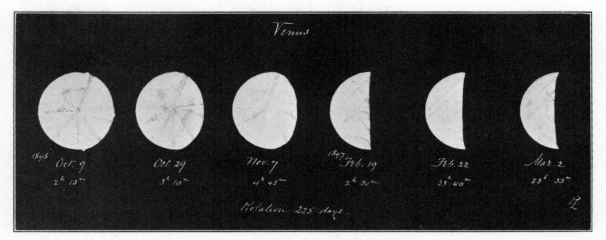

The radial markings of Venus as drawn in 1896 and 1897 by Percival Lowell. Below the series of drawings he noted "Rotation 225 days."
—PERCIVAL LOWELL, *The Evolution of Worlds*

duce fluorescence or luminescence in particles on Mercury's surface, though whether this would tend to obscure the dark markings or could cause anything to be viewed as a haze or veil is another question. Fluorescing free radicals (bits of broken molecules) and ions, or even dust particles electrostatically charged and repelled from the surface, might produce a kind of haze. And in a period of sunspot activity, a large number of solar flares and sub-flares are created on the sun, perhaps enough to account for the extensive variations noted by Antoniadi. The later French observations, during 1942 and again in 1944 and 1950 (the sunspot maxima fell in 1937 and 1947), were made during periods of a fairly quiet sun, so the same effects might not have been noted.

Venus' Unseen Surface

Just before the turn of the century, Percival Lowell's report and drawings of what he claimed were fairly permanent markings on Venus created an uproar. In some of his drawings, the darker markings look like the spokes of a wheel radiating out from a hub in the center of Venus, but without a rim around the edge of its disk. In the midst of all the hubbub about the existence of artificially created canals on Mars, a smaller rhubarb developed about whether the spokes tended to prove the exist-

ence of intelligent life on Venus. Perhaps, some speculated, this almost geometrical design on the face of Venus was another rational system of roadways or canals or some other planetary construction program.

To Lowell's credit, however, he himself did not interpret the Venusian wagon wheel as evidence of intelligence. He described these radial markings as akin to crow's feet. "In addition to some of more ordinary character," he wrote, "were a set of spokelike streaks which started with the planet's periphery and ran inwards to a point not very distant from the center. The spokes started well-defined and broad at the edge, dwindling and growing fainter as they proceeded, requiring the best of definition for their following to their central hub." While he believed the markings to be on Venus' surface, not in its atmosphere, he noted a number of times that he did not consider these streaks on Venus to be anything but natural, like those on Mercury. Other astronomers did not rush to confirm Lowell's charts of Venus.

French astronomers studying Venus in the 1940's and 1950's did find what they believed were fairly permanent markings radiating from the center of Venus' disk in visible light. A planisphere of Venus based on their composite photographs showed a rather radial pattern, though the contrast had been strongly in-

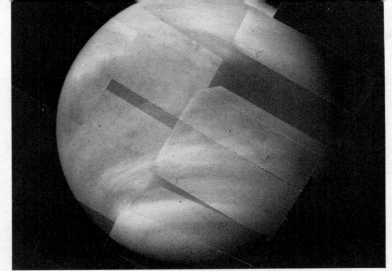

Cloud patterns, visible only in ultraviolet light, reveal the general circulation of Venus' upper atmosphere in this mosaic of TV pictures, taken by Mariner 10 in February 1974 from 440,000 miles away, a day after the probe's closest approach to the planet.

—NATIONAL AERONAUTICS AND SPACE ADMINISTRATION

creased in making these up. These observations make it quite likely that Lowell actually saw what he claimed to have seen, though in his drawing the pattern is straighter and more precise than in the more recent composite photographs and drawings.

This radial pattern may be related to currents in Venus' atmosphere, particularly if the planet only rotates very slowly in a retrograde direction, as now believed. Then the subsolar point in the center of the disk might be the point through which the heating of Venus takes place, with winds radiating out from it across the planet and passing around to the cooler, nighttime side, forming the principal heat sink opposite the subsolar point. Detailed models of such a movement of Venus' atmosphere have not been devised, but it appears to agree fairly well with the greenhouse or the aeolosphere models of the planet's atmosphere. Also, such a radial pattern in the atmosphere might only occur now and then, rather than being permanent—which would account for its not being always observable or being replaced by a different pattern. However, little is known about wave movements in the terrestrial atmosphere and still less about their nature on a planet which may rotate only very slowly.

The radial markings drawn by Percival Lowell are now generally accepted as occurring on

occasion. Even more remarkable is the fact that he explained them with detailed wind-convection diagrams. In this instance, too much of Lowell's work, showing both accurate observation and remarkable insight, had been thrown out like the baby with the bath.

While faint radial markings on Venus are found in visible light, observers claim to have seen straight, bright bands across Venus that are most distinct in ultraviolet light. Two different atmospheric levels or wind systems might be viewed in the ultraviolet and the visible light.

Recent radar data obtained at the Haystack Observatory, Westford, Mass., have indicated that Venus' surface temperature may run as high as 950° F. The same study showed a mountainous region that may extend some 300 miles north and south of Venus' equator, with altitudes of from 1.5 to 2.0 miles, and run for a long way along the equator. Averaged over its whole surface, however, the disk of Venus seems to be as smooth, if not smoother than, the moon's disk. These data, plus the yellowish tint of Venus' clouds, which may indicate some dust or the presence of complex molecules in the atmosphere, are about all there is to work on in inferring the nature of its surface.

The indication of a high surface temperature implies intense metamorphism of surface rocks and allows certain deductions about their nature. At such a temperature, calcium carbonate, iron oxides, certain hydrous silicates, and certain micas might be present in surface rocks, but not metallic iron, iron silicates, free carbon, magnesium carbonates, and hydrocarbons. The dream of vast petroleum oceans on Venus is vanishing. A partial equilibrium between the surface rocks and the atmosphere of Venus has been suggested, with quantities of fine rock dust in the lower atmosphere, helping to distribute the heat uniformly around the planet by convection and serving as a highly erosive force on the surface, literally sandblasting it down to the kind of low relief that radar studies have implied. This would correspond with the greenhouse, or possibly the aeolosphere, model of Venus' surface, indicating a hot desert, with veils of dust driven by howling winds.

Martian Kaleidoscope

When people study photographs of Mars for the first time, or even view it through a sizable telescope, their reaction is often a mixture of frustration and disappointment. Can this smudged little orb, slightly ochre-hued, a white polar cap its only outstanding feature, with the fuzzy edge of its dim disk almost merging with the blackness of space around it, be the much vaunted red planet? How can anything at all have been learned about such a vaguely outlined, dusky wad of stuff so far away in space?

As the planet is viewed intensely for a long time, and, with concentration, a few details are picked out, admiration grows for the observers who have produced maps of Mars like those shown.

The chart showing the major light and dark markings on Mars was developed between the years 1941 and 1952 by G. de Mottoni of the French Pic-du-Midi Observatory, on a Mercator projection, based on composites derived from hundreds of photographs of Mars. It was officially adopted by the International Astronomical Union at its 1958 meeting. The simi-

lar chart with the table gives the official names of the main features of Mars, many of which were identified and named by such renowned astronomers as Schiaparelli, Fournier, and Antoniadi.

These maps show a surprising amount of detail on Mars, perhaps more than could be distinguished on the earth, because of our more clouded atmosphere, if similar-sized Martian telescopes were aimed in this direction. Astronomers who concentrate on Mars are more familiar with its markings than most of us are with the shapes and positions of the continents and major islands on globes of the earth.

The bright and dark areas on the map are actually there, although they change mysteriously in size and shape over the years. With telescopic viewing, the bright areas were called "deserts" and the dark areas "vegetation," but Mariner-Mars flights have changed these ideas. Mariner 9 has shown that Mars is more rugged for its size than the earth, with overall elevation differences of 8 miles and peaks and plateaus that rise from 3 to 5 miles above their surroundings.

One of the intriguing features of Mars is that from opposition to opposition bits of its face change in a way the earth's certainly would not. In addition, Mars has four seasons, and these cause alterations in many of its observable features. Variability seems to be one of Mars's outstanding characteristics.

The inclination of the axis of Mars's rotation to the plane of its orbit of the sun is very close to that of the earth, and its full day is only 41 minutes longer than ours. Rough thermocouple measurements have been made of its temperature, which indicate that it is consistently colder on Mars than on the earth. Temperatures at the subsolar point on Mars have been estimated to average 55° F. at the planet's mean distance from the sun, 32° when it is at aphelion, and 81° at perihelion. On the Martian south polar cap, however, temperatures are very cold. They may average —80°

NOMENCLATURE MARTIENNE SELON L'U.A.I. - I.A.U. NAMED MARTIAN MARKINGS

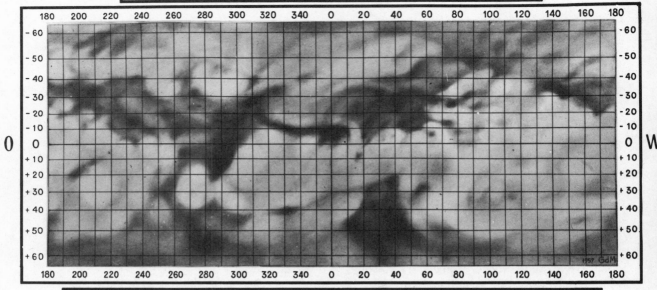

Longitude de l'équinoxe de printemps boréal =	84°,0	= Beginning of northern spring.	
Longitude du solstice d'été boréal =	174°,0	= Beginning of northern summer.	
Longitude de l'équinoxe d'automne boréal =	264°,0	= Beginning of northern autumn.	
Longitude du solstice d'hiver boréal =	354°,0	= Beginning of northern winter.	
Distance moyenne au Soleil	= 227,7.10⁶ km = 1,5237 U.A.	= Mean distance from the Sun.	
Excentricité de l'orbite	= 0,0933	= Excentricity of the orbit.	
Longitude du périhélie	= 334°35'	= Longitude of perihelion.	
Inclinaison du plan de l'orbite	= 1°51'	= Inclination of the orbit.	
Longitude du nœud ascendant	= 48°56',5	= Longitude of ascending node.	
Durée de révolution	= 686 j 23 h 30 m 41 s	= Martian year.	
Coordonnées célestes de l'axe de rotation	α = 316°,8 δ = + 53°,0	= Celestial coordinates of the axis of rotation.	
Inclinaison de l'axe sur la normale à l'orbite =	24°,8	= Inclination of the axis of rotation.	
Durée de rotation du globe	= 24 h 37 m 22 s 6	= Martian day.	
Diamètre équatorial du globe	= 0,530 = 6760 km	= Equatorial diameter.	
Valeur de 1° planétocentrique sur le disque	= 60 km	= 1° Planetocentric at the surface.	

Les petits détails sont désignés par leurs coordonnées planétographiques.
Small features are designated by their planetographic coordinates.

Les grandes régions sont désignées par un nom dont voici la liste et les coordonnées.
Main markings are designated by names according to the following record and coordinates.

Acidalium M. (30°, + 45°)
Aeolis (215°, — 5°)
Aeria (310°, + 10°)
Aetheria (230°, + 40°)
Aethiopis (230°, + 10°)
Amazonis (140°, 0°)
Amenthes (250°, + 5°)
Aonius S. (105°, — 45°)
Arabia (330°, + 20°)
Araxes (115°, — 25°)
Arcadia (100°, + 45°)
Argyre (25°, — 45°)
Arnon (335°, + 48°)
Aurorae S. (50°, — 15°)
Ausonia (250°, — 40°)
Australe M. (40°, — 60°)
Baltia (50°, + 60°)
Boreum M. (90°, + 50°)
Boreosyrtis (290°, + 55°)
Candor (75°, + 3°)
Casius (260°, + 40°)
Cebrenia (210°, + 50°)
Cecropia (320°, + 60°)
Ceraunius (95°, + 20°)
Cerberus (205°, + 15°)
Chalce (0°, — 50°)
Chersonesus (260°, — 50°)
Chronium M. (210°, — 58°)
Chryse (30°, + 10°)
Chrysokeras (110°, — 50°)
Cimmerium M. (220°, — 20°)
Claritas (110°, — 35°)

Copaïs Palus (280, + 55°)
Coprates (65°, — 15°)
Cyclopia (230°, — 5°)
Cydonia (0°, + 40°)
Deltoton S. (305°, — 4°)
Deucalionis R. (340°, — 15°)
Deuteronilus (0°, + 35°)
Diacria (180°. + 50°)
Dioscuria (320°, + 50°)
Edom (345°, 0°)
Electris (190°, — 45°)
Elysium (210°, + 25°)
Eridania (220°, — 45°)
Erythraeum M. (40°, — 25°)
Eunostos (220°, + 22°)
Euphrates (335°, + 20°)
Gehon (0°, + 15°)
Hadriacum M. (270°, — 40°)
Hellas (290°, — 40°)
Hellespontica Depressio (340° — 6°)
Hellespontus (325°, — 50°)
Hesperia (240°, — 20°)
Hiddekel (345°, + 15°)
Hyperboreus L. (60°, + 75°)
Iapigia (295°, — 20°)
Icaria (130°, — 40°)
Isidis R. (275°, + 20°)
Ismenius L. (330°, + 40°)
Jamuna (40°, + 10°)
Juventae Fons (63°, — 5°)
Laestrigon (200°, 0°)
Lemuria (200°, + 70°)

Libya (270°, 0°)
Lunae Palus (65°, + 15°)
Margaritifer S. (25°, — 10°)
Memnonia (150°, — 20°)
Meroe (285°, + 35°)
Meridianii S. (0°, — 5°)
Moab (350°, + 20°)
Moeris L. (270°, + 8°)
Nectar (72°, — 28°)
Neith R. (270°, + 35°)
Nepenthes (260°, + 20°)
Nereidum Fr. (55°, — 45°)
Niliacus L. (30°, + 30°)
Nilokeras (55°, + 30°)
Nilosyrtis (290°, + 42°)
Nix Olympica (130°, + 20°)
Noachis (330°, — 45°)
Ogygis R. (65°, — 45°)
Olympia (200°, + 80°)
Ophir (65°, — 10°)
Ortygia (0°, + 60°)
Oxia Palus (18°, + 8°)
Oxus (10°, + 20°)
Panchaia (200°, + 60°)
Pendorae Fretum (340°, — 25°)
Phaethontis (155°, — 50°)
Phison (320°, + 20°)
Phlegra (190°, + 30°)
Phoenicis L. (110°, — 12°)
Phrixi R. (70°, — 40°)
Promethei S. (280°, — 65°)
Propontis (185°, + 45°)

Protei R. (50°, — 23°)
Protonilus (315°, + 42°)
Pyrrhae R. (38°, — 15°)
Sabaeus S. (340°, — 8°)
Scandia (150°, + 60°)
Serpentis M. (320°, — 30°)
Sinaï (70°, — 20°)
Sirenum M. (155°, — 30°)
Sithonius L. (245°, + 45°)
Solis L. (90°, — 28°)
Styx (200°, + 30°)
Syria (100°, — 20°)
Syrtis Major (290°, + 10)
Tanaïs (70°, + 50°)
Tempe (70°, + 40°)
Thaumasia (85°, — 35°)
Thoth (255°, + 30°)
Thyle I (180°, — 70°)
Thyle II (230°, — 70°)
Thymiamata (10°, + 10°)
Tithonius L. (85°, — 5°)
Tractus Albus (80°, + 30°)
Trinacria (268°, — 25°)
Trivium Charontis (198, + 20)
Tyrrhenum M. (255°, — 20°)
Uchronia (260°, + 70°)
Umbra (290°, + 50°)
Utopia (250°, + 50°)
Vulcani Pelagus (15, — 35)
Xanthe (50°, + 10°)
Yaonis R. (320°, — 40°)
Zephyria (195°, 0°)

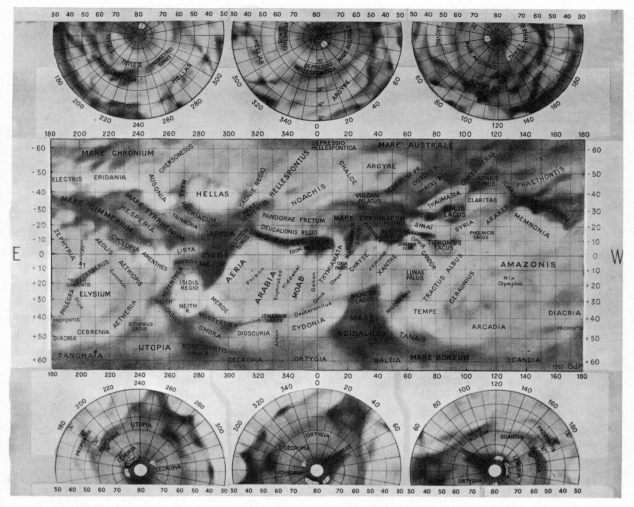

Map of Mars with the official names of the principal formations, with north polar detail (bottom) and south polar detail (top). The names "Iapygia" and "Pandorae Fretum" are misspelled.

—A. DOLLFUS, PRESIDENT, COMMISSION FOR PHYSICAL STUDY OF THE PLANETS AND SATELLITES, INTERNATIONAL ASTRONOMICAL UNION

Map of the principal formations of Mars for the years 1941 to 1952, drawn by G. de Mottoni on the Mercator projection, with a system of coordinates based on position measurements by Henri Camichel. (North at bottom, west at right, according to astronomical convention.) Data on the planet, its orbit, rotation, and the seasons are given, as well as the coordinates of the named features adopted at the 1958 meetings of the International Astronomical Union. This map is shown, rather than the MM '71 Mars Planning Chart for the Mariner 9 flight to Mars, because this map shows Mars as viewed with a telescope and traces the major features somewhat more strongly.

—A. DOLLFUS, PRESIDENT, COMMISSION FOR PHYSICAL STUDY OF THE PLANETS AND SATELLITES, INTERNATIONAL ASTRONOMICAL UNION

F. at mean solar distance, —97° at aphelion, and —62° at perihelion. The temperature of the equatorial regions in the summer do run above the freezing point and probably reach above freezing in other regions during the midsummer season. Under these conditions, ice would melt, or perhaps immediately vaporize or sublimate in Mars's thin atmosphere.

The polar caps of Mars expand and shrink, disappearing or almost disappearing during the hemispheric summer and appearing and in-increasing in size during the fall and winter. The south polar cap can vanish completely, while the northern cap reaches a minimum diameter of about 175 miles. During early winter, the caps appear bright white but often become murky and streaked with gray later in the season. Quantities of fine dust or other particles may be carried by currents in Mars's atmosphere to change the tint of the polar caps.

As the polar caps diminish in size during spring and summer, they develop gouges and rifts and sometimes leave islands of white beyond their line of retreat. Both infrared and polarization studies have indicated that the caps contain water, perhaps in the form of hoar frost no more than a fraction of an inch thick. During the fall they appear to be covered with white clouds or fog as they form, but in the spring the clouds vanish and, as the caps diminish, dark rings form around them in the area where they have receded. Mars is so dry it is difficult to see how these rings could be water, which would quickly be absorbed underground or sublimate into the atmosphere.

The differential forming and receding of the caps may indicate plateaus or higher hilly regions around the poles. Polarized light has suggested that bright spots around the edge of the caps may be ice-crystal clouds, located where the caps last longest, perhaps over plateaus. Similarly, small localized clouds, remaining for some days in about the same spot, as in the area of Meridianii Sinus or around Aurorae Sinus, may indicate hills, mountains, or rougher areas in these regions. Although polarimetric and early radar-echo data show that the surface of Mars is as smooth as, if not smoother than, that of the moon, the variability of the radar data suggests that there may be both smooth and rough areas.

The bright areas on Mars are colored light buff or orange-ochre; these so-called deserts cover about two-thirds of the surface. They do not vary much, though they may be encroached upon by dark markings. Occasionally, as in the Hellas region, these areas appear to become whiter than other bright areas. Opposition after opposition shows consistent polarization curves for these bright areas, which are very similar to those for limonite, a hydrated iron oxide ($Fe_2O_3 \cdot 3H_2O$), in a finely pulverized condition here on earth. Color photographs demonstrate that these desert areas give Mars its somewhat reddish or buff-orange hue, although this may also derive from its hazes.

As demonstrated in the chart of Mars, the renowned dark areas on Mars are concentrated mainly in its southern hemisphere. Although some appear in the northern hemisphere, they are not as complex or as widespread there. They have been called maria (seas), sinuses (bays), or lacus (lakes) in the naming of Mars's regions, although it is known that they are not bodies of water. They cover about a third of the Martian surface; most of them have been permanent, but they vary in size, shape, and density of shading, and from time to time entirely new areas appear, darken or brighten, or old areas disappear. The color of these dark markings is basically gray, varying from light to dark shades, though many observers have also seen blues or greens in them.

The dark areas vary with the seasons. As spring and summer come on, the seasonal color changes, becoming darker, at a rate of about 25 miles a day from the pole down toward the equator and sometimes beyond it. The markings tend to be faint during the winter, then darken in this odd progressive manner in spring and summer, shifting, some observers say, from gray or blue-gray to brown or violet. The seasonal changes of the dark markings seem to be related to the changes in the water-vapor content of the Martian atmosphere during the spring and summer after the polar caps have diminished, but so many other features may be related to the seasonal changes that it would be difficult to prove or disprove such an hypothesis. Water vapor may make up no more than $\frac{1}{1000}$ per cent of the Martian atmosphere, compared with perhaps $\frac{1}{10}$ per cent on the earth, and there may not be enough on Mars to cause such effects. It might merely have a triggering action.

Beyond the seasonal variations, irregular changes in the dark markings on Mars have been noted on many occasions. In some areas, like the Solis Lacus, the strengthening and fading of the seasonal variations tends to occur differently, so that such areas vary from year to year. Also, new dark markings occur, such as the spot that appeared in the region of Diacria in 1951 and was very definite in 1956. Remarkable changes were noted in the Solis Lacus and the Thaumasia regions in 1956 during a period of only one month. Some of the changes persist through opposition after opposition; others disappear after one opposition. The northern edge of Mare Cimmerium has shown a large new indentation since 1939.

The French astronomer Audouin Dollfus, who has devoted much attention to Mars, once wrote that, "It is tempting to assume that the changes in texture are due to vegetation. Many phanerogams [seed plants] and cryptogams [lower plants] with thalli have been measured; they show polarizations always quite different from the Martian surface. This is not true for microcryptogams. Scattered over the ground like a powder, these plants modify the polarization of the surface only slightly, but in a varying manner. On the other hand, these microorganisms have great pow-

ers of adaptation. Some protect themselves against extremes of climate by means of a brightly colored superficial pigment." But note that Dollfus only made a suggestion here; he did not say that polarization studies have identified such microorganisms with curves similar to those given by the dark Martian markings.

More recently, Dollfus reported that the yellow veil sometimes covering parts or nearly all of the Martian surface appears from polarization data to be made up of submicroscopic particles 2 to 5 microns in size, not like ordinary dust particles, but more the size of particles in cigarette smoke. Furthermore, he reported that the polarimetric data for the yellow veils and for the dark markings are about the same—that their structure may be similar, consisting of very tiny opaque granules which change in shape or size during the Martian seasons. He indicated that no known minerals have this property and was inclined to believe that, "therefore the most probable explanation is that they are a type of microscopic plant life." Again, this is in the nature of a supposition, not a positive identification or substantiated hypothesis.

Many hypotheses have been formulated to explain these seasonal and other variations in Mars's dark markings. They could be due to the presence of hygroscopic salts that absorb water from the atmosphere, although the exact composition of such salts on the Martian surface has not been specified, nor have they been identified by observations. Recent polarization data have been inconsistent with hygroscopic salts. Winds varying seasonally like the terrestrial trade winds might perhaps explain the yearly variations. They might be volcanic dust, dark by nature, scattered in shifting patterns across Mars' predominantly flat and dry surface. Mars was covered with one of its great yellow clouds when Mariner 9 went into orbit of it late in 1971. The storm lasted for over a month and then began to clear. Spectroscopic instruments on Mariner 9 proved this was a dust cloud, not pollen or microorganisms. It was silica dust, raised by high winds around Mars. It could come from igneous rocks or dry volcanic dust scattered across Mars' surface by the ancient volcanoes seen. Vast quantities of such dust could explain the Martian dark markings, the changes in the markings, and the yellow clouds, without assuming organic causes.

But where are the renowned canals of Mars? They do not seem to appear on these maps at all, although there are a few thin, stretched-out dusky or dark markings. Some observers of Mars have never seen these "canali," or "channels," as they were named by Schiaparelli in 1877; others spot them quite regularly. Much more heat than light has been generated over the years since 1877 as to whether the Martian canals exist or not. The solution accepted by most astronomers is that the canals do appear, at least dimly, at a certain stage or level in the resolution of distinct objects or features of Mars, partly determined by the clarity of the Martian atmosphere, by the seeing conditions in the earth's atmosphere, and by the capacities of the optical system of the telescope being used. With greater resolution, it is believed, they turn out to be composed of smaller dots or spots.

The canals are variously reported to appear as lines, bands, or streaks between dark regions on Mars, often running across and contrasting with the buff or ochre-colored, brighter desert regions, but also identified to some extent within the dark-marked regions. Some of the larger canals have been photographed, though only with exceeding difficulty. With a resolving power of 0.2 second or better, many observers claim, the dark areas tend to break up into multitudes of little dusky spots, sometimes called leopard spots.

Only a few short fracture lines across Mars' surface were picked out in photos from the first Mariner flybys. Many prominent impact craters were seen, however, like those on the moon, and it was assumed that the curving arcs of large craters or lines formed by chains of craters could explain the "canals."

Mariner 9 photographs from much closer to the planet added other features that could have caused the lines of "canals." Some long rills were viewed and large "canyons" or great rifts whose deep shadows would make such lines, in addition to the spots of many craters that the eye would tend to see as lines. And the rims of huge craters, fully as large as any on earth, could make the circular dark spots, or "oases," which telescopic viewers had seen. "Canals" had been there, but they came from natural causes, not as the products of intelligent beings irrigating their Martian deserts.

Surfaces of the Outer Planets

The atmospheres of the outer planets probably contain large quantities of methane and ammonia gases, as well as hydrogen, helium, nitrogen, and neon in quantities proportional to those in the sun.

Radio emissions of Jupiter and Saturn have been under study since 1955. If these emissions are being correctly interpreted, Jupiter must have an ionosphere and a strong magnetic field, running to at least 6 gauss. It also seems to have well-developed Van Allen belts, containing radiation of much greater intensity than that banding the earth. Some radio emissions appear to originate deep below the visible atmosphere of Jupiter. The nature of its lower atmosphere and distinguishable surface, if any, have been the objects of great speculation. No complex models among those being developed of the nature of the solid globes of Jupiter and Saturn, and among the simpler ones of Uranus and Neptune beyond, have won general acceptance as yet.

Russian and American scientists reported in 1963 that faint radar echoes had been returned from Jupiter. The data from the Soviet experiment have not been announced. The American report from the Goldstone radar system of the Jet Propulsion Laboratory interpreted the radar signals received back from Jupiter as indicating that a part of Jupiter's surface (or

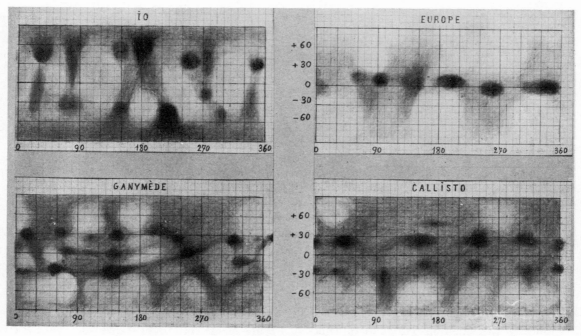

Planisphere drawings of Jupiter's satellites: Io (Jupiter I), the diameter of which is about that of the moon; Europa (Jupiter II), the markings of which concentrate along the equator, rather than well above and below it as on Io; Ganymede (Jupiter III), possibly a little larger than the planet Mercury and perhaps nearly as large as Mars; and Callisto (Jupiter IV), between the moon and Mercury in size. The drawings are after Bernard Lyot, with the contrast exaggerated, using the 24-inch refractor at Pic-du-Midi Observatory. The rotation of these satellites is believed to be synchronous with their revolution of Jupiter, like that of the moon.

—A. DOLLFUS, OBSERVATOIRE DE PARIS, MEUDON, FRANCE

whatever level the waves were reflected from) is smoother than those of Mars and Venus.

The larger (Galilean) satellites of Jupiter— Io, Europa, Ganymede, and Callisto—have quite distinct atmospheric or surface markings as has Saturn's satellite, Titan, the only satellite on which an atmosphere has been positively identified. Ranging from 250,000 to over 1 million miles from their primaries, these satellites should be beyond the intense radiation zones of Van Allen rings and should make good observation posts for the study of these great planets, and perhaps eventually for the transfer of exploratory techniques developed in the investigation of our own moon.

Jupiter III (Ganymede) is the size of Mercury, and except for Jupiter II (Europa) the rest of these satellites are larger than our moon. There have been indications that the smaller satellites may be only dirty balls of ice, possibly interspersed with rock. Nothing suggests that the larger satellites are to any great extent different in their constitution from small bodies like Mercury, the moon, or the larger asteroids, with rocky surfaces on which dim shadings or tracings can be noted from the distance of the earth. With the other smaller bodies in the solar system, they may in the long run offer the greatest possibilities for exploration and exploitation.

CHAPTER XII

"Dream not of other Worlds, what Creatures there
Live, in what state, condition or degree."

MILTON

"The human understanding," Francis Bacon wrote nearly 350 years ago, "supposes all other things to be . . . similar to those few things by which it is surrounded." It is this characteristic fallibility of the human mind that down through the ages has, as Bacon put it, "colored and infected" man's beliefs about the possibility of life on other planets. It was Bacon's own revolutionary proposals for the scientific method of inquiry that in time enabled men to approach this question free of myth, superstition, and "the authority of former conclusions."

What is this "life" that may or may not flourish on other planets in our system or in the universe at large? In simplest biological terms, life means the existence of organisms that absorb materials from the environment, process them to supply energy, and return the waste products to the environment. By these means organisms both sustain themselves and grow. Attaining greater complexity, they are able to disperse or move about, they reproduce themselves, and, finally, they react and adapt to their environment or remold it. Taken together, these characteristics roughly differentiate inanimate matter from living organisms ranging in size from the inconceivably tiny virus to the whale.

Living organisms are literally spread over the face of the earth. They crowd the oceans, teem in and on the soil of the continents, and permeate the air. Life's density and omnipresence have radically altered the earth's environment. A conspicuous example of this process is the availability of that necessary element for ani-

NATIVE LIFE ON OTHER PLANETS

mal respiration, oxygen, produced almost entirely by living plants.

Evidence of Life in Meteorites

Time and again in recent years certain light, crumbly meteorites called carbonaceous chondrites have challenged scientific thought about life in the universe. Scientists examining some fragments of these meteorites have found what they thought were living or fossilized organisms in the form of spores, viruses, or bacteria. Principally composed of silicates with lumps or globules of carbon and tiny spherical chondrules scattered through them, they constitute only 2 to 3 per cent of known meteoritic falls. Only 23 meteorites, all seen to fall, have been identified as carbonaceous chondrites.

In 1857 the German chemist Friedrich Wöhler extracted a small amount of organic material from a carbonaceous meteorite that had fallen that year near Kaba, Hungary. On analysis this proved to be a solid, high molecular hydrocarbon, a compound of carbon and hydrogen that was closely associated with life as we know it. Similar hydrocarbons were found in such other carbonaceous chondrites as the Orgueil meteorite.

It became a question of whether or not these hydrocarbons were in themselves a sure indication of life, because some complex hydrocarbons of the types found in these meteorites had been synthesized from inorganic materials. In 1828 Wöhler had converted an entirely inorganic compound, ammonium cyanate (NH_4NCO), which does not occur in nature, into urea (NH_4CONH_2), commonly a waste product of animal metabolism. This doomed the faith, held strongly until then—and the original basis for the distinction between organic and inorganic chemistry—that organic compounds could be produced only by living things. Many hydrocarbons are now known to have been formed naturally on the earth.

Two substantial claims have been made that dormant spores or tiny organisms have been extracted from meteorites. In 1932 C. B. Lipman reported that he had sterilized the surface of a number of stony meteorites and, taking every precaution not to contaminate them, had extracted particles from within them. These he pulverized and placed in a nutritive medium. Some of these samples produced bacteria colonies; these appeared to be bacilli or cocci, identical with or very similar to familiar terrestrial bacteria. However, careful as he was, Lipman himself was not altogether certain that contamination by terrestrial bacteria did not somehow take place—certainly it is implied by the similarity of the bacteria.

Orgueil carbonaceous chondrite meteorite, weighing 4.5 pounds, which fell on May 14, 1864, at Montauban, France. Containing much insoluble, black carbonaceous material (presumably a complex polymer of high molecular weight), this is one of the meteorites from which many organic molecules have been extracted. (Their ruler at left reads centimeters; 1 centimeter = 0.39 inch.)
—AMERICAN MUSEUM OF NATURAL HISTORY

Then in 1961 Frederick D. Sisler of the U.S. Geological Survey analyzed bits of the carbonaceous meteorite which fell near Murray, Kentucky, in 1950. He used a germ-free environment within which to obtain powdered samples from deep inside the meteorite. He found some strange particles, like organisms, which apparently multiplied rapidly, reappearing as portions of the samples were transferred from test tube to test tube. It was finally concluded that these were probably some terrestrial form of oxygen-dependent (aerobic) bacteria that somehow had contaminated the meteorite. The meteorite could have picked up the bacteria as it hurtled down through the atmosphere, or bacteria could have penetrated deep within the meteorite after it landed, possibly moving over wet surfaces by a kind of capillary attraction. But the possibility that life spores have reached, or perhaps returned to, the earth in a meteorite cannot be entirely excluded.

Could tiny organisms like bacteria exist over millions of years in an inert state and then return to active life? Heinz Dombrowski of the University of Geissen at Bad Nauheim, Germany, examined bacteria that were found buried, dehydrated and denatured, inside rock salt from a 180-million-year-old salt deposit in Germany, a Devonian salt deposit in Saskatchewan 320 million years old, a Silurian salt deposit 500 million years old near Myers, New York, and an Irkutsk, Siberia, salt deposit probably 650 million years old. When revived, the bacteria did not appear to employ the same carbohydrates as comparable bacteria in their metabolism today; when injected into mice they killed them, apparently by a generalized bacterial infection. If Dombrowski's results can be repeated by other scientists, and were not due simply to contamination, new records for the preservation of life in an inert state will have been set, and the possibility (not the probability) that bacteria survived for eons deep within meteorites would have to be admitted.

While most meteorites have been dated as 4.5 to 4.7 billion years old, some of their surfaces have been exposed to the radiations of space for only a few hundred thousand or million years. From this it is inferred that their parent body, which may have been fragmented in a collision, was as old as most bodies in the solar system, but the breakup producing some of the meteorites falling on the earth was much more recent.

An international debate about extraterrestrial life began in 1961 after Bartholomew Nagy of Fordham University and a number of colleagues reported that chemical analysis of the carbonaceous remains of four meteorites revealed the presence of chemical compounds normally only associated with life on earth. They had studied meteorites from Alais and Montauban in France, Tonk in India, and Tanganyika in Central Africa, using mass-spectrographic analysis, chromatographic methods of identification, and ultraviolet and infrared spectroscopy. They not only discovered evidence of very complex hydrocarbons in the carbonaceous material, such as cholesterol and the butyric acid found in butter, but also data indicating that liquid water may have been a part of a somewhat alkaline environment on the parent body of the meteorites; the evidence in their interiors was unaffected by high temperatures as they entered the earth's atmosphere. Perhaps even more significant was the conclusion that the meteorites contained many saturated hydrocarbons and aromatic compounds similar to those in sediments on earth normally associated with living remains. It was surmised that they might have come from the bed of a lake or small body of water on, or below, the surface of the parent body from which they derived.

This report was followed by another from Nagy and his co-workers that was even more startling. Certain microscopic "organized elements" had been found in some of these meteorites that could only be identified, they believed, as the fossil remains of once-living, one-celled organisms. Several such organized elements had walls with protuberances or spines on them. Another had a hexagonal shape with tubular appendages that could possibly have been used in movement. What may have been vacuoles appeared within the elements, they reported, but no cell nuclei—although the cells may have been dividing. A few terrestrial contaminations were identified in the meteorites, but they were either living or freshly killed and could be distinguished from the fossilized elements.

Other identifications of these organized elements have been urged. Some might be well-known types of pollen grains that have contaminated the pieces of meteorites as they landed on earth or afterward. Others might be starch grains, sulfur or hydrocarbon droplets, fossilized amino acids or other complex compounds of carbon produced naturally without living organisms, mineral grains, or jokes of nature resembling organisms by chance. But it is countered that some of the organized elements have been found deeply imbedded where pollen would not likely have settled or contaminants have found their way.

Robert Ross of the British Museum of Natural History extracted samples from the Orgueil meteorite and found elements that looked like collapsed spore membranes, as well as two microscopic bodies with an umbrella or mushroom form, which looked like fossils of living things. Harold C. Urey, who has been very much interested in these investigations, wrote, "If it can be shown that these hydrocarbons and the 'organized elements' are the residue of living organisms indigenous to the carbonaceous chondrites, this would be the most interesting and indeed astounding fact of all scientific study in recent years."

In 1963 Edward Anders and colleagues at the University of Chicago processed and stained various microscopic materials just as Nagy had, and reported the positive identification of one type of organized element as ragweed pollen, which must be contamination.

Notions of evidence of extraterrestrial life in meteorites received a blow when it was discovered that a part of the Orgueil meteorite showed definite signs of contamination with materials from living earth organisms, perhaps intentionally. Definite "organized elements," beyond contamination, could not be proved in other meteorites. The evidence seemed to have fallen apart.

With the discovery of complex molecules, precursors on earth in the formation of many biochemicals, in interstellar space, however, three of the carbonaceous chondrites were examined with more sensitive methods. All of these, including inner portions of the Orgueil meteorite, turned out to have hydrocarbon and amino-acid molecules in them, some identical with those found in living protein, some unknown in it. Also, while protein has amino acids of a so-called "D" structure, the acids in the meteorites showed an "L" structure in half the molecules, similar to those produced in laboratories under conditions like those of the primitive atmosphere.

The investigators concluded that these molecules seemed, then, to have been formed before the meteorites reached earth, occurring in meteorites as well as in interstellar space.

If life did thus come to earth in fossilized form, from where did it come? The asteroids mainly orbit in the space between Mars and Jupiter, although some of them, like Eros, Betulia, and Hidalgo, have very irregular orbits passing fairly near the earth. Originally, there may have been one, or perhaps two or more, fairly small, primitive planets, with a total volume less than that of the moon, which collided and broke into the thousands of asteroidal fragments in a number of subsequent collisions. Such a planet or planets could have served as the source of the rudimentary deposits and the pools of water indicated by the meteoritic material, and living forms may have developed. But would such small bodies so distant from the sun have received enough radiant energy from it to permit life to develop? And unless there was at least one sizable planet among them, could it have retained enough of a primitive atmosphere to nurture life?

Another possibility is that in striking the earth asteroids splashed huge fragments with their sedimentary deposits into escape velocity. Chunks blown up from the earth might have gone into solar orbits near the earth and then eventually have come back under the earth's gravitational influence and returned to it. The organized elements might then represent a much earlier stage in the terrestrial evolution of life. Perhaps similar fossil microstructures will sometime be found deep down within aged sedimentary layers of the earth. The study of meteorites might help to turn back the pages of the earth's own history.

It has been suggested that the carbonaceous chondrites may be fragments cast from the moon with the impact of a large meteorite or asteroid of the type that perhaps formed a great lunar sea like the Mare Imbrium. For that matter the carbonaceous chondrites may have been thrown from Venus or Mars by the impact of an asteroid hundreds of millions of years ago.

Origin and Conditions of Life

In the everyday world of the earth some 3 billion human beings are spread across every continent, thriving within hundreds of different complex social structures. Within, each one of us has extremely intricate, beautifully organized systems of specialized organs, tissues, and cells, all functioning together to preserve life, to adjust to environmental change, and to bring forth the generations to come. Around us are countless other species of living things, some adapted to land and some to air or water, blanketing the whole surface of the earth.

Life has penetrated to the utmost extremes of temperature, humidity, and pressure found on the earth. It burgeons most luxuriantly in the consistently warmer climates of the tropics, yet it can adapt to dry deserts, to the frigid Arctic and Antarctic, to mountaintops with eternal ice and snow and tenuous atmosphere, and to the beds of oceans, despite the massive pressure of water piled mile on mile above.

The total weight of living substance on the earth has been estimated at as much as a ten-thousandth of the weight of the earth's crust. The most striking impression from a broad view of life is that once it has come into existence it develops and spreads indomitably, in spite of extreme conditions around it. It is tough and resilient, not easy to destroy. Those responsible for the complete sterilization of spacecraft that might land on the moon or planets will testify to this.

Such is the great flowering of life over the earth on the levels immediately observable. When we turn to the eyepiece of a microscope, we enter a whole new world fully as complex, teeming with thousands of unicellular and multicellular organisms in every drop of water or particle of soil. The live weight of microscopic organisms in an acre of soil to the plow depth of 6 to 7 inches has been estimated at over 4,400 pounds. This includes 2,000 pounds of molds, 1,000 of bacteria, and another of branching unicellular organisms (actinomycetes). In addition, there are an estimated 200 pounds of protozoa, 100 of algae, and 100 of yeasts. Viruses are present in great numbers, but their weight is insignificant.

In view of the complexity of even the simplest organisms, the product of billions of years of evolution, and our ignorance of the details of most of the processes going on in them, to ask how life began seems highly presumptuous. But the accumulation of discoveries about a few organic processes and the trends revealed in evolution have brought us perhaps to that stage where tentative answers to such questions may be possible.

It is known, for example, that photosynthesis (the process by which plants with chlorophyll utilize the energy of the sun's rays with water and carbon dioxide to manufacture more complex compounds and give off oxygen) appeared relatively late in the evolution of life. With photosynthesis, free oxygen became available for animal respiration systems, providing vast effective energies previously lacking in organic substances. Earlier life must have developed without oxygen, utilizing energy from the transformation of simpler compounds. Many non-oxygen-using (anaerobic) forms of life exist today, as well as other forms that use oxygen when available but fall back on anaerobic metabolic processes when it is not.

During the last decade such investigators as the Americans Melvin Calvin, Stanley L. Miller, and Harold C. Urey, and others around the world, have obtained many organic compounds by synthesis from inorganic materials. By passing electrical discharges through gaseous mixtures of methane, hydrogen, ammonia, and water vapor, or by irradiating them with ultraviolet light, they have produced compounds as complex as amino acids, the building blocks of protein molecules and eventually of all living things. Hydrogen, methane, ammonia, and water, and their interaction products, such as hydrocarbons and aldehydes, may well have composed the primitive atmosphere of the earth. Under primitive earth conditions there probably was lightning, which would have provided the electrical discharges, and (without ozone) plenty of ultraviolet light from the sun, so this may have been the way in which the basic constituents of life on our planet were formed.

Some very complex organic molecules, components of nucleic acids, have already been produced by further combination of these amino acids. Adenosine triphosphate (ATP), a very complex molecule, used for the chemical storage and release of energy in cells and made by animals in the metabolic process and by plants by photosynthesis, has been produced in a California laboratory by directing ultraviolet light on a solution of compounds thought to resemble the composition of the earth's oceans about 4 billion years ago.

It is surmised that after such complex materials had developed in the nutrient broth of the oceans the next stage in the origin of life was the pulling together of very simple molecules analogous to coacervate droplets or of more complex proteinlike molecules, which form simple precellular models somewhat resembling bacterial cells. In either case, a selective relationship between the droplets and their environment was set up, and metabolic processes adaptive for the continuance of the organized droplets begun. But it must be emphasized that investigations of this type have just started and that knowledge of these processes is extremely limited and speculation rampant.

What conditions are necessary for the formation of the substances from which life might derive and grow? A good supply of simple hydrocarbons and water are considered basic. For chemical transformations to go on, the water would have to be, at least on occasion, in a liquid state, neither ice nor vapor all the time. Some electrical discharges, ultraviolet light, or heat would have to be present to produce high free-energy compounds like the amino acids. And there would have to be sources of the many other elements that play a role in all living things, such as nitrogen, phosphorous, calcium, and sodium.

At first thought, these may not appear to be very limiting conditions for the origin of life, but upon examination they prove to be quite restrictive. For one thing, they seem to imply that life could only begin in water, at or near the surface of a body large enough to retain a considerable enveloping atmosphere. While pockets of water might exist under the surface of the body, electrical or ultraviolet energy sources would not normally be present there, and it is difficult to think of energy sources which might take their place. It has recently been proved, however, that the amino acids and other organic compounds can be produced by heating nutrient solutions at from 300° to 400° F., temperatures consistent with the thermal history of the earth at or beneath the surface.

The occasional need for water in a liquid

Conditions conducive to life at New Quebec (Chubb) Meteor Crater, 2 miles in diameter and filled with a lake 800 feet deep, situated on the Ungava Peninsula, 130 miles south of Hudson Strait, Quebec. The bodies of water in the lake and in the distance and the atmosphere with its clouds reflected in the lake and stretching to the horizon represent environmental conditions on earth favorable to the development of life as we know it.

—ROYAL CANADIAN AIR FORCE

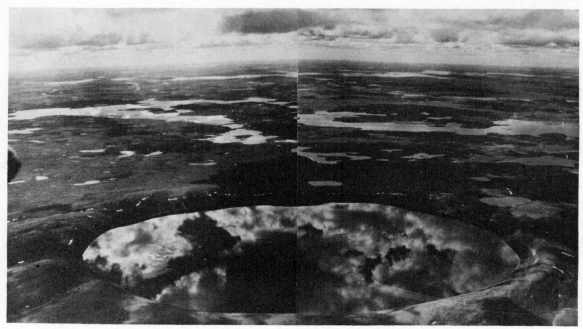

state is another very restrictive condition. While it has not been proved that thin films of water on the surface of rocks or particles of soil would be enough (nor to the contrary that water must be present in great quantities in seas and oceans), the water must enter into many of the chemical reactions and form a component in many of the compounds involved in living substance, as a liquid, not as a vapor or ice. And here the temperature range, though dependent on pressure and the concentration of solutions, is quite limited and seems to require a rather substantial atmosphere, as does the requirement for fairly permanent bodies of water. While the necessity for an atmosphere is a moot point, it is hard to conceive of adequate substitutes for it, such as vast caverns or nearly covered areas containing pools of water, which might not be too transient.

With water and hydrocarbons, it is possible to conceive how a variety of the kinds of life with which we are familiar could develop; with some other medium and basic building compounds, we might find it difficult to recognize the result as life.

Lunar Life

Many features of the moon have been seized upon in the past as indicators of life or even of intelligence there. William of Occam (1285?–1349) warned against this kind of speculation. "Beings ought not to be multiplied," he said, "except out of necessity" in the attempt to account for phenomena. This dictum became known as "Occam's Razor," a handy instrument for cutting theories down to the bone—that is, to the simplest explanation that covers the facts.

Geometrical forms of all sorts, thought to be representative of intelligence, were discovered on the moon in the early days of observation. One such formation, known as Mädler's Square, was described in the 1830's by the famous selenographers Johann Heinrich Mädler and William Beer, who used a 3.75-inch refractor in their mapping of the moon. North of the Mare Frigoris, near the north pole of the moon, they drew a strongly marked square that looked artificial.

Before long all of these early beliefs were discounted. Telescopes with greater resolution proved that the forms were natural and irregular, not strictly geometrical. Larger instruments show that Mädler's formation consists of very irregular ridges, walls, craterlets, rings, and hills, which do not constitute a precise square. Many of the walled plains or ring mountains on the moon appear to have a roughly polygonal (hexagonal) shape, but this is recognized as a natural geological phenomenon; sun-dried and caked mud flats on the earth crack into such forms.

But what of all the radial bands, spots, and linear markings that some observers, even in recent years, have taken as signs of vegetation or some other form of life growing and spreading during the lunar day? Here we should also wield Occam's Razor. Some physical markings undoubtedly extend or broaden as the angle of the sun's rays varies with the lunar dawn or evening. Others may be gases escaping infrequently from beneath the surface as it is warmed by the sun, or as the tugging tidal forces of the earth affect the moon. Still others may be the rare volcanic disturbances and emissions that are reportedly seen in particular locations. And some might be selective effects on certain materials of the solar radiation beating down on the moon's surface.

What actual possibility is there that some form of life may have originated and developed on the moon? It has remarkably little, if any, lasting atmosphere. Even if gases are puffed occasionally from crater cones, holes, or cracks in its surface, most of them would be constantly and quickly lost into space. Volcanoes might blow out methane, hydrogen sulfide, ammonia, and water, at the surface, but even the heavier gases would not remain there for long. They would be dissociated by the solar radiation and their ions swept into space. Small pockets of heavier gases like argon or xenon might form, as well as tiny amounts of lighter gases, and they might freeze and build up crusts or strata in locations sheltered from sunlight. In the main, though, these are not constituents of life. Oxygen, carbon dioxide, or nitrogen would be quickly dissipated into space. No one claims that water as a liquid stands anywhere on the surface of the moon, although it has been suggested that under the upper insulating surface of the moon there may be a heavy layer of ice, or pockets of ice or water, particularly near the moon's poles where the temperatures would not rise as high during the lunar day as around its equatorial zone. But the moon is hot underground.

The first Ranger and Surveyor vehicles that impacted or soft-landed on the moon were sterilized very thoroughly so as not to contaminate the lunar surface. If spacecraft were not sterilized and earth-like organisms turned up on the moon or another planet, who could say whether it was extraterrestrial life or not?

The lunar rocks and soil returned to earth by Apollo astronauts were kept under sterile conditions as far as possible. So, too, were the astronauts until it was quite certain that exposure to the lunar environment had given them no unearthly organisms that might contaminate the earth. But neither astronauts nor soil have shown any evidence of such organisms. The last of the Apollo astronauts were not even quarantined for a period, as were the early astronaut crews.

Samples of lunar soil and rock particles were sent to many crews of scientists to study for any evidence of lunar life. No evidence appeared under microscopes of any fossil life long since vanished. No living organisms grew from lunar dust placed in three hundred life-nurturing cultures. Except for possible contamination, no complex molecules associated with protein, like amino acids, were found. There was surprisingly little carbon on the moon, to say nothing of even its simple compounds, hydrocarbons. Evidence that the rocket engine jets helped form such hydrocarbons in the lunar soil when the Lunar Module landed was the best

that could be obtained from these explorations.

A typical report concluded that "the principal components of the various spectra were contaminant gases and traces of low-molecular-weight hydrocarbons which had adsorbed onto the surface of the lunar sample during handling and exposure to the [earth's] atmosphere."

In the sampling done so far, then, no primitive life forms have been found on the moon, and the lunar conditions do not offer any hope that they will be found in the future. Water seems almost nonexistent on the moon, although traces can be baked from its rocks at very high temperatures. The lunar atmosphere is only local and transient. Meteorites blasting into the moon would more likely carry life in them than would the lunar rocks themselves.

Life on Mercury or Venus

What of life on little Mercury, circling so close to the flaring sun? Direct evidence on this score is most unlikely, since Mercury is so far away and so difficult to observe in even a favorable position that no fine markings can be seen on its surface. While it appears to have real shadings and some observers have claimed to have seen a shifting haze condition, no one has come out with any direct evidence of life on Mercury, although in the past Mercury was populated in the customary way—by imagination working overtime.

Theoretically and by inference from the conditions, Mercury's atmosphere must be almost nonexistent. Radicals fluorescing under the impact of solar radiation could hardly support life. Such extreme heat scorches across Mercury's surface that it is difficult to conceive of liquid water on or anywhere near it. It would be steam before it reached the surface and would be immediately whisked away. Mercury rotates very slowly on its axis with a period of 59 days, two-thirds of its period of revolution of the sun. Thereby, the dark hemisphere away from the sun should radiate some of its scorching heat out into space, but again it would be scorched when it turned back into the blasting heat of the nearby sun once again.

An unmanned or manned spacecraft could match the speed of Mercury of about 30 miles a second (compared with the earth's 19) as the innermost planet circles the sun. If such a craft, sheltered in the shadow of Mercury at a distance, would come out to study the sun on occasion, it might make an ideal solar observatory. From Mercury's vantage point, the "doings" on our great sun would be awesome and the data obtained most precise.

It is difficult to think of any better use to which this little overbaked and rocky planet could be put. It makes much more sense to explore out beyond the earth, toward Mars and the moons of Jupiter, than in near the sun.

Venus has often been called our "sister" planet, because of its many similarities to earth. It is only a little smaller and less dense than the earth and its cloud cover has been obvious since telescopes were turned toward it. Many imaginative ideas have been formed about the conditions on Venus' surface under its clouds. The occasional "ashen light" of its dark side has been interpreted as dim glimpses of fires set by primitive Venusians as they burn the dense jungle for farming! And when at one time it was suggested that Venus' clouds might be hydrocarbons, it was imagined that a petroleum ocean covered Venus. But all these imaginings and any hopes of eventually discovering native life on Venus have gone glimmering in recent years.

It turned out that Venus rotates only very slowly, over a period of 245 days, in a direction backward to that of the earth and other planets—from east to west. While it sounds very strange, this gives Venus a combined day and night longer than its year! While this means that the dark side should be somewhat cooler than the bright side, the temperature differences observed seem small. It depends on how effectively Venus' heavy atmosphere transfers the heat around the planet from the side facing the sun. Also, the extreme temperatures on Venus' surface, roughly measured with

radar echoes, are far too high for life to exist.

The circulation of Venus' atmosphere may be what is called a symmetrical regime; that is, the major heat source for the atmosphere is thought to be the central, subsolar point. The heat sink, toward which the winds move with the heat and away from which the winds blow back the cooler air below, is the antisolar or midnight point diametrically opposite the solar point on Venus' sphere. This would help to explain the radial cloud pattern, like spokes radiating from near the subsolar center, which some observers have identified on Venus in visible light. On occasion such a pattern might switch to a similar symmetrical regime about the poles at the higher levels, and this may account for the parallel cloud bands seen on Venus in ultraviolet light and presumably at higher levels in the atmosphere.

A dense atmosphere probably filled with dust and circulating rapidly around Venus and a surface at temperatures ranging from 600° to 800° F. make it difficult to conceive of the development of life on the planet. Certainly terrestrial forms are not likely under these conditions, even if the dark side of Venus should turn out to be somewhat cooler than estimated. Temperatures above 600° F. also preclude manned exploration of Venus' surface. Unless the data have been misinterpreted, Venus can be written off for anything but more extensive investigations by unmanned space probes or satellites to work out its unusual atmospheric and surface conditions.

Martian Life?

Mars maintains its mysteries, accentuated rather than diminished by the partial visibility of its surface. While the planet is often wreathed in white clouds, yellow clouds or veils, and the unusual blue clouds or blue haze, astronomers can sometimes pick out numerous details on its surface, particularly when the planet is in close oppositions. Consequently, they have been able to map the relations

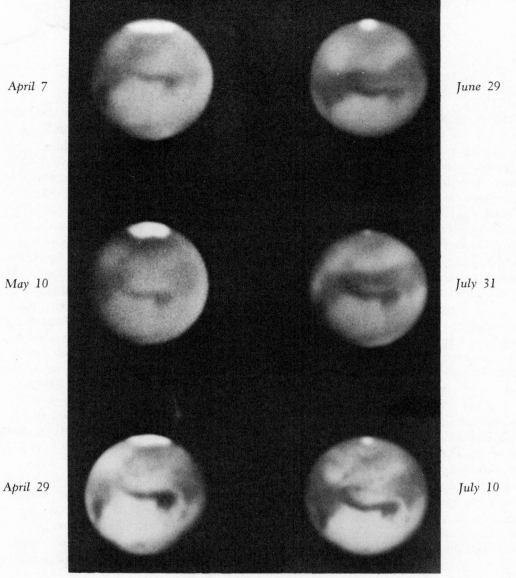

April 7

May 10

April 29

June 29

July 31

July 10

Mars photographed in its spring and summer, showing melting of the south (top) snow cap and striking seasonal development of dark markings in the tropics. Note the doubling of dark bands across the center. The Martian dates given correspond to our calendar dates in the Northern Hemisphere.

—E. C. SLIPHER, LOWELL OBSERVATORY, FLAGSTAFF, ARIZ.

between the bright regions and dark markings and to determine changes in the intensity and extent of both.

There is action on Mars. Its tawny bright areas, usually known as "desert" regions, sometimes become lighter or new bright areas develop. Some dark areas or markings, often called "vegetation," appear and disappear over the years or increase or decrease in size. The "canals" appear to extend out from the dark areas and sometimes pass through them. They manifest the same kind of coloring, ranging from shades of light gray to dark gray, and often cross bright areas to meet in blotches or "oases" of the same dark markings. Although with greater resolution the linear character of the canals appears to break down into arrangements of fine mottling or small, irregular black or dark markings, the canals are probably not entirely illusory. Asteroidal meteorites plunging into or raking Mars's surface, shrinkage, or other internal processes may have produced long cracks or crevasses radiating from central craters, and the visible markings of the "canals" may have been created when vegetation grew in and along them or tended to gather in clumps near them.

Even more mysteriously, the shade of the dark markings turns deeper progressively from the poles with the round of Martian spring and summer, the darkening passing at a rate of up to 25 to 30 miles a day down to and below the equator. This change in the surface appearance may be related to minute changes in the water-vapor content of the Martian atmosphere during the spring and summer with the evaporation of the polar caps, or to some concomitant changes of some kind.

Reports from the early Mariner spacecraft that flew by Mars and the preliminary returns from Mariner 9 in orbit of Mars indicate that the planet has an atmosphere consisting largely of carbon dioxide, but with small amounts of water vapor. The Martian poles have temperatures cold enough to freeze the carbon dioxide so that the polar caps are mostly "dry ice." However, in the Martian spring and summer the caps diminish rapidly in size. Mariner 9 recorded more water vapor over the poles in these seasons. Its photographs showed glacial features on the caps that must have come from water. It is therefore possible that there may be more water on Mars, either in the solid or vapor state, than has been supposed.

In identifying the dust clouds that enshrouded Mars when it first approached as silicates, Mariner 9 showed that Mars' soil must be largely silicates, or perhaps more generally basaltic in nature. With so much fine dust in the soil, pulverized perhaps by the constant impact of many meteorites from which its atmosphere does not protect it, Mars' physical surface may be much like that of the moon. Such silicate or basaltic soil can adsorb a great deal of water. Scientists infer that the quick release of this water when the surface is warmed may explain the brightening of Mars' surface.

With a larger supply of water than has been assumed, both in the polar caps and spread throughout the soil, the darkening waves toward the Martian equator may be explained now more adequately than before.

A favored explanation of the Martian dark markings is that in large measure they are caused by living organisms. Mars is large enough to have held its water and lighter gases for some time. The resulting reducing atmosphere and small surface bodies of water could have been conducive to the development of life during a primitive period through which Mars passed more quickly than the earth. Now Mars may have reached a state of equilibrium in which the water vapor emitted from its interior equals that being dissociated in its atmosphere and escaping as hydrogen from its exosphere, or it may simply be in the final stage of losing the last of its original supply of water. If, in addition, the planet was once warmer than it is now, so that its water was not frozen most of the time, the primary conditions for the development of simple forms of life could have been fulfilled.

The first good look at Mars in the full range of the infrared sector of the spectrum was made in 1963 with a 36-inch telescope carried 15 miles up over Texas above most of the earth's atmosphere in a Stratoscope balloon. The analysis of the infrared data obtained in five scans of Mars indicated that any water vapor and carbon dioxide in Mars's atmosphere were in very small quantities, marginal for supporting the existence of any but the most primitive forms of life as it is known on the earth. In the same year, another study of the near-infrared spectra of Mars made with the 100-inch reflector at Mount Wilson Observatory led to similar conclusions. Only very weak lines of water vapor were observed, and they seemed strongest over the Martian poles.

Life on Mars could never have been on any such scale as that on earth. For one thing, Mars never evolved the quantities of free oxygen in its atmosphere which photosynthesis in living things produced on earth. The respiratory processes of Martian organisms probably would not involve oxygen directly, so they

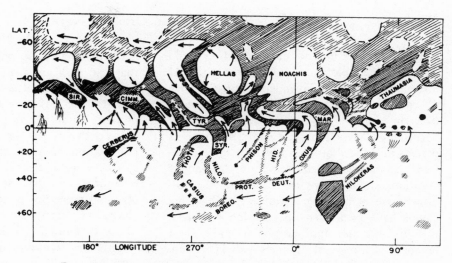

Dust carried and drifted by prevailing winds may create and change the dark markings on Mars. Wind arrows indicate the circulation during the southern summer on Mars (north at bottom), with bands running from southeast to northwest south of the equator, but curving to the north and northeast in the northern hemisphere and ending in bays (sinus), resembling the circulation of trade winds in the earth's atmosphere.

—DEAN B. MCLAUGHLIN, *Journal of the Royal Astronomical Society of Canada*

would be quite different from those on earth.

Are there other ways to explain the ever-changing dark markings on Mars? At the present time, no incontrovertible evidence has been accumulated for any particular theory, but some of the more sensible explanations put forward should be considered.

One view is that the Martian dark markings consist of water absorbing, or hygroscopic, salts or minerals that turn a darker color as they blot up water vapor. This theory was proposed by Arrhenius in his book *The Destinies of the Stars* (1918). A form of mineral known as Tikhvin bauxite, for example, turns noticeably blue when merely moistened. While the precise salts or minerals showing color changes at the varying Martian humidities have not been identified, this does not entirely demolish this theory. Although spectroscopic evidence casts doubt on this theory, we are certain of none of the minerals composing the Martian surface. But why is this material not completely covered up in the frequent Martian dust storms, or why, after being covered, does it keep reappearing?

A theory proposed by Dean B. McLaughlin of the University of Michigan is that the dark markings represent volcanic ash blown out of still occasionally active volcanic areas and carried about by the trade winds. The "canals" would be explained on the same basis as lines of ejection of the ash from volcanoes in definite directions, somewhat like the rays ejected from impact craters on the moon or explosion craters on the earth. The changes in color and darkness might be due to the action of carbon dioxide in the Martian atmosphere, with small amounts of water vapor on the fresh volcanic ash forming secondary minerals which might have a dark-green color.

From the reports of Mariner 9, it was learned that enough water and water vapor may be active in combination with the carbon dioxide in the air on Mars to bring about the color changes. In support of this theory are the many photographs of branching stream systems and valleys that look as if they had been worn by water at some time in the past and the many large and small volcanic features on Mars not recognized before. All or most of these volcanoes may be extinct now, although they could have provided plenty of ash in the past. At present there is no solid independent evidence of volcanic activities on a fairly continuous basis on Mars.

Although Mars is observed intensively at its periods of opposition, it is not kept under constant surveillance. Still some of the largest eruptions should have been seen, if the volcanoes are still active. Gray or yellow-gray clouds have been seen on occasion on Mars, two in 1950 and another two in 1952, by a Japanese astronomer, Tsuneo Saheki. These suggest volcanic activity, but such reports are infrequent.

The volcanic hypothesis does explain nicely why the dark areas are generally warmer than the buff-colored desert areas, indicating that the former consist of dark ash which would absorb and hold more radiation from the sun.

McLaughlin himself recognizes the possibility that the volcanic and vegetation theories are not mutually exclusive, "If the forms of the dark areas are due to fall-out of ash and if the dark colour is that of the ash, then what becomes of the Martian 'Vegetation'? It is farthest from my intention to deny categorically that vegetation exists on Mars. There is no observed phenomenon that fully convinces me of its presence, but its absence is not proven. . . . The seasonal changes may or may not be due in part to vegetation, but if they are, perhaps the forms of the areas where it can grow are determined by the fall-out pattern of the ash and the precious moisture that accompanies it."

The polarization of the dark markings on Mars has been studied by French astronomers and found to vary like the color with the Martian seasons and latitudes. As the contrast in the dark markings increases from winter to summer, the polarization curves change as well. It is asserted that crystalline deposits or surfaces on Mars would result in completely different polarization curves than those of the dark markings. The polarization data suggest that the dark areas may consist of very tiny opaque granules that change in shape and size with the seasons. Seasonal variation is not a property of any known minerals.

Recently it has been suggested that the polarization data for the dark markings and for the yellow veils are nearly the same; their structures, too, may be very similar. The yellow veils appear to be made up of small particles ranging from 2 to 5 microns in size, more like very tenuous smoke than ordinary dust.

This very tentative identification may tie in with the results of an American astronomer, William M. Sinton, who observed what he called absorption bands in infrared scans of the dark markings on Mars obtained with the 200-inch Mount Palomar telescope. These have been interpreted as evidence of carbon-hydrogen compounds in the markings, perhaps aldehydes. This is the most direct bit of evidence that some form of life may exist on Mars. Even if Sinton's identification is correct, however, it is still possible that such compounds were produced by chemical means as they have been on earth, without the mediation of living organisms.

The strength of Sinton's evidence has been diminished recently by the report of a group of scientists at the University of California, who checked the spectra of many substances, including varied samples of pure organic chemicals, polymers, biological specimens, pure inorganic compounds, and samples of minerals to see whether they could identify what had caused the three notable bands in Sinton's infrared spectra. They found the resolution of the spectra taken by Sinton quite low and pointed out that they may well have included considerable Martian bright areas as well as dark markings. They found reason to doubt that these were purely absorption spectra, but thought they might combine absorption and reflection spectra in a manner resulting in difficulties of interpretation. Their study practically eliminated the interpretation of the spectra as carbohydrates, such as sugars and starches, or even as cellulose, a major substance in plants, but it left open the possibility of free aldehydes, possibly even acetaldehyde, which might be the result of anaerobic metabolic processes.

Lichens were found to give a band similar to one of Sinton's bands, while the smooth, waxy leaf of the lily, *Agapanthus*, and that of the prickly-pear cactus, gave spectra somewhat like Sinton's. Thus a relationship to organic materials was found. It is of interest that these leaves have protective casings which can prevent such substances as water from passing into or out of them; this might be a significant capacity in Mars's water-poor environment. But a number of inorganic carbonates also gave spectra similar to two of Sinton's bands, though the third band might only have been produced inorganically by lead carbonate, not likely to be widespread on the Martian surface. No other inorganic samples showed much promise.

All of these possibilities were so inherently dubious that the group concluded that, "At present we know of no satisfactory explanation of the Martian bands. Observations of the planet with improved spectral and spatial resolution, in conjunction with radiometric temperature measurements, could possibly define the problem sufficiently to enable a solution to be found."

Surely, the final interpretation of the photographs of Mars from Mariner 9, as well as the data from its infrared and ultraviolet spectrometers, will greatly increase the understanding of what is going on in the atmosphere and on the surface of Mars. Whether the same observations will give solid evidence of the existence of living organisms on Mars is still an open question.

The data from Mariner 9 already reported do increase the probability that life may have started and developed on the planet at some time in the past, perhaps when conditions were more favorable. Whether it still thrives there probably will only be settled with the landing of instruments on Mars to test for its presence. A Soviet vehicle has already landed there, in 1972, but its transmissions ceased after 20 seconds. But other such vehicles are planned later in this decade. The odds for the existence of life on Mars in the past, if not at present as well, seem higher than the odds against it.

Although it has been suggested that under its cloud layers Jupiter may harbor fairly normal temperatures, most observers believe that it and the rest of the gaseous giants (Saturn, Uranus, and Neptune) with their great distances from the sun have very low temperatures. Also, their probable hydrogen, methane, and ammonia atmospheres are not conducive to life as we know it, although they might furnish appropriate conditions for the first stirrings of life. The most distant planet, Pluto, orbits the sun at 39.3 a.u., where the sun appears little brighter than any other stars. While it is thought now that it may approach the size of the earth, any gases on Pluto must be frozen.

The chances for native life on any of these outer planets are thus extremely slim. It is more probable on larger satellites such as Saturn's Titan, or Jupiter's Ganymede (Jupiter III), although, here again, the extreme cold and the improbability of productive atmospheres in which life might be nurtured militate against any forms of native living things similar to those on earth.

Life in the Universe

It is nearly a certainty that the phenomenon called life is not unique to our solar system. According to even most conservative estimates, there are at least 100,000 stars in the Milky Way Galaxy alone that provide an environment favoring the germination and support of living organisms. In the universe at large, esti-

Two photographs of Barnard's star (arrow), taken 11 months apart (July 31, 1938, and June 24, 1939), with the 24-inch Sproul Observatory refractor. In making the combined print, the second plate was shifted slightly with respect to the first. It is quite apparent that Barnard's star has moved against the background of other, virtually fixed stars; its large rate of motion (10.3 seconds of arc per year) can actually be detected on photographs taken one night apart.
—Peter van de Kamp, Sproul Observatory, Swarthmore College

mates of the probabilities that stars will have life-supporting planetary systems vary all the way from 6 per cent of all stars to 1 in a million.

What tests must a star pass if it is to qualify for this life-supporting role? Binary and multiple star systems can be eliminated at the outset. The unlikely planets in such systems would have orbits so complex as to introduce enormous variability in the amounts of radiation they would receive from their stars. Since dependable, fairly constant amounts of radiation are a must if life is to be maintained, the planets of this necessarily single star must have nearly circular, not too eccentric, orbits. Furthermore, the star must be neither "too young or too old," in the words of a popular song, nor too large or too small. If too young, it may be very large or not have given birth to planets; if too old, it may be expanding and intensely hot, burning its progeny to a crisp, or too small and cold to provide enough heat. Finally, the masses of its planets must be within certain limits—not as great, for example, as those of our own "gas giants," Jupiter and Saturn— if they are to have atmospheres hospitable to

life, and not be likely to become stars in their own right.

Lest this question of life elsewhere in the universe seem to be speculation of the wildest sort, some recent actual astronomical observations lend concrete support to the theory that there are planetary systems other than our own. Astronomers at Swarthmore College's Sproul Observatory have found a wobble in the motion of Barnard's star (Table 1), a star only 6 light-years, or about 36 trillion miles, from our solar system. The star itself shows a great motion in comparison with the relatively fixed stars in our sky nearby. The star's wobble can be most simply explained as a perturbation caused by an unseen companion. This companion could be a small, dark neighboring star. More probably, it can be classed as a planet, some 1.5 times as massive as Jupiter, or 500 times more massive than the earth. The Sproul Observatory astronomers have deduced from their observations that this invisible body moves in an elliptical, highly eccentric (0.6) orbit about Barnard's star, like that diagrammed. They estimate that its reflected light must make it about the 30th magnitude, much too dim to be seen in a telescope on earth or even on an orbiting earth satellite, and that it is probably too far from its primary star for the nurture of life. Further analysis of the wobble has implied that at least two dark bodies may orbit the star, rather than one. This begins to look more like a planetary system!

Earlier, Sproul Observatory astronomers had directed their attention to two other stars, the motion of which showed evidence of the presence of invisible companions. These were Lalande 21185, at a distance of 8.1 light-years, and 61 Cygni, 11.1 light-years away (Table 1). The companions of both stars may have masses about 1/100 that of the sun, placing them on the borderline between extremely large planets and very small stars.

It will be a long, long time before space probes can be launched toward stars near our own. Meanwhile, the best approach to the question of life elsewhere in the universe is radio or optical surveillance of possible emissions from other stars that could represent attempts to communicate. A practical trial surveillance has already been made to work out procedures and solve the problems entailed.

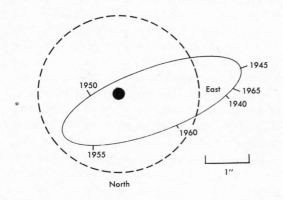

The inferred orbit of the invisible companion relative to the visible component of Barnard's star. The dashed circle indicates the greatest size of the latter's image on the plates taken with the 24-inch refractor at Sproul Observatory.
—PETER VAN DE KAMP, SPROUL OBSERVATORY, SWARTHMORE COLLEGE

In 1960–61, the 85-foot radio telescope of the National Radio Astronomy Observatory at Green Bank, West Virginia, followed the radio emissions from two nearby stars of the same type as our sun, Tau Ceti, 11.8 light-years away, and Epsilon Eridani, 10.8 light-years distant (Table 1). The trial was called Project Ozma, after the queen of the fictional land of Oz, a very distant and inaccessible place inhabited by strange beings. For receiving signals the 21-centimeter wavelength (1,420-megacycle frequency) was used. This is the wavelength of hydrogen emissions, the most commonplace of radio radiations in space, and thus likely to be chosen as the best medium for communications. It also has the virtue of passing freely through atmospheres like that of the earth.

No indications of any attempts to communicate were discovered. Despite this initial failure, it is expected that the surveillance of radio sources from nearby stars will continue as better equipment is developed. It may be carried out on the hydrogen frequency, twice this frequency, or possibly on other frequencies; and perhaps radiation sources other than stars will be investigated, as well as other types of radiation, such as the intense beams of light which could be emitted by lasers. Asteroids discovered in odd orbits in or near the solar system would be good sources to check, since intelligent beings might use them for interstellar space travel. Consideration may be given also to directing radio or laser signals in varying codes toward likely nearby stars to indicate our presence to those who could respond within a generation or two.

A respectable number of scientists are convinced of the strong possibility that messages from other worlds are constantly impinging on the earth and that all that remains to be done is to solve the riddle of how and where to look for the signals. Should signals of this kind be detected, the significance of such a discovery would be inestimable.

CHAPTER XIII

"And here an engine fit for my proceeding."

SHAKESPEARE

With the advent in the heavens of tiny, beeping Sputnik 1, the old earth's first artificial satellite placed in orbit by the Soviet Union on October 4, 1957, man finally achieved what had been theoretically possible since Isaac Newton formulated the laws of celestial mechanics in the seventeenth century.

In 1961, President John F. Kennedy announced ambitious plans to land men on the moon by 1970. By December 31, 1969, four American astronauts had landed on the moon via the Apollo 11 and 12 missions. A total of 1,052 spacecraft had been placed in orbit, of which 37 had been lunar missions and 4 planetary. Soviet spacecraft had totaled 407, American 616, and eight other nations 29. Thirty-seven manned space flights had been launched. American astronauts had spent 5,800 hours in space and Soviet cosmonauts 1,700 hours. Knowledge had flooded in about the sun, the moon, other planets, and most of all the earth itself from an entirely new environment, space.

All of this has been made possible by a piece of equipment called a rocket. What are rockets? What gives them such awesome power?

Types of Rockets

Rockets are vehicles carrying a load propelled by a reaction to the gases produced by the combustion of their self-contained fuel and oxidant systems. Already many families of rockets, and generations within these families, have been designed, built, and tested for different space transport missions. Perhaps the simplest type are the sounding rockets, which "sound" or probe the atmosphere; they usually consist of a series of stages or sections and

ROCKETS AND SPACE VEHICLES

carry an instrument package as a payload. They are designed to shoot straight up as high as 2,000 miles or more above the earth, conduct their observations, and return under the pull of gravity, often ejecting their instrument package to drop by parachute. But these are only midgets compared with their younger brothers.

The missile, a rocket structure designed for military use, is sent into a trajectory like a shell; it carries a warhead armed with a nuclear or other device to a target. Missiles may be shot from the surface of the earth to impact on the surface, or from air to air, or air to surface, and so on; they vary greatly in complexity, range, and destructiveness of the warhead they can carry. The IRBM's are intermediate range ballistic missiles; ICBM's are intercontinental ballistic missiles. The space launch vehicle or space carrier is another kind of rocket designed to go out into space with its complement of engines, other stages, and payload. Often the engines used in missiles and in space carriers are interchangeable.

Rocket engines or motors are called rockets in a more restricted sense, though they are simply one section—the propulsion stage—of the whole rocket. The rocket engine consists basically of fuel compartments and a combustion and thrust chamber with an escape nozzle and much auxiliary equipment. The rocket engine that lifts a vehicle off the ground, that is, the first-stage engine, is the rocket booster, constituting one section of the complete rocket. Sustainer engines are intermediate vehicle stages or are used to assist the booster engine; the terminal-stage engine is the final powered stage, designed to give an accurate final velocity and direction to the remainder of the rocket, placing the payload truly in orbit or on its course out into space. Booster, sustainer, and final-stage engines must be differently designed for their specialized functions. The Atlas-Centaur stages are shown.

Other stages or sections of rockets may consist of intermediate auxiliary instrument and equipment compartments, guidance, control, and service sections, and the payload. The payload carries the items for accomplishing the mission or objectives of the rocket. It may consist of a warhead or some other type of package in a missile and of some kind of satellite, space vehicle, or spacecraft in a space-carrier rocket. Usually the terms space vehicle and spacecraft, instead of referring to the entire rocket system, are used in this way for only the unmanned and manned payloads, respectively.

An artificial satellite is a payload injected into orbit around the earth or some other large body like the moon or planets. A space probe is a payload injected into an orbit with velocity sufficient to escape the earth's gravity, instrumented for observations and experiments in space. An artificial planet or planetary satellite is a payload that has been placed in orbit of the sun, aping the natural planets. A spaceship is a sizable spacecraft designed for carrying a number of human beings on lengthy trips into space. A space station or platform is a large, manned satellite of the earth or another body providing a base for scientific observations, the final release of space probes, and the refueling of spaceships, or a platform for assembling and launching satellites or missiles.

Components of the Atlas-Centaur launch vehicle or space carrier: the two booster engines and the sustainer engine (MA-5's) with their fuel tanks make up 66 feet of the 108-foot height of the rocket. The Centaur liquid hydrogen-oxygen terminal stage with two RL-10 engines is 30 feet tall.

—National Aeronautics and Space Administration

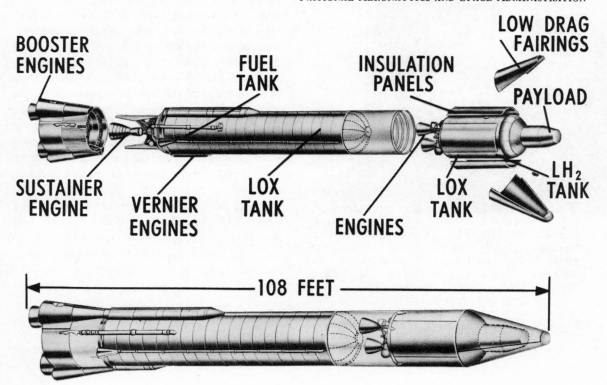

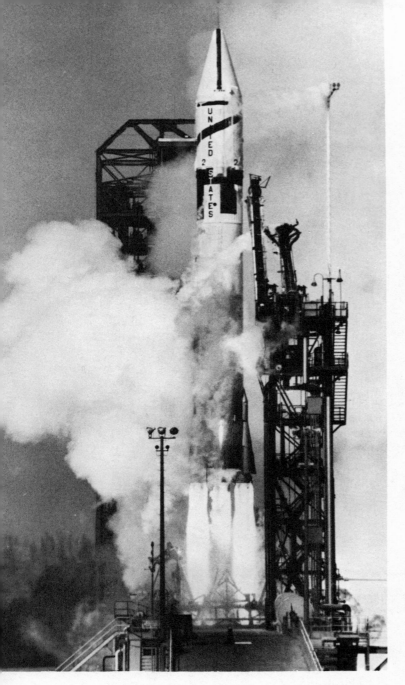

Atlas-Centaur lifts off Pad 36 at Cape Kennedy on November 27, 1963, on a flight which placed the whole 10,200-pound Centaur stage (top) into earth orbit after separation from the Atlas stage (below "United States" on the rocket). This rocket is designed to launch such payloads as Surveyors, to soft-land on the moon, and Mariners, to fly by the planet Mars.
—NATIONAL AERONAUTICS AND SPACE ADMINISTRATION

Rocket Principles

The whole rocket assembly of two to four towering stages topped by a payload lifts ponderously as it is finally unleashed from its launching pad, quivers up slowly foot after foot, and then slides swiftly up and away, disappearing in a twinkling into the blue and beyond. The launching of an Atlas-Centaur rocket from Cape Kennedy, Florida, is shown. The sound effects accompanying such a magnificent spectacle are just as striking—a vast, rolling, booming noise. All this apparent straining power comes from the rocket-booster engine, which has been billowing masses of flame on its pad. These lengthen into a diamond-brilliant shaft of extremely hot gases trailing behind the rocket as it shoots up toward space. How does all this happen?

In burning or combustion, oxygen combines with the material or fuel being burnt, whatever it may be, forming oxides and gases in combination with the material. Combustion goes on at varying rates. In a rocket engine, very rapid, but not quite explosive, combustion occurs. Rocket fuel is injected into the combustion chamber of a liquid-propellant engine, as is the oxidant or oxidizer, which consists of some material very rich in oxygen, or of pure gaseous or liquid oxygen. The fuel and oxidizer are ignited, or they spontaneously ignite in what are called hypergolic mixtures, and a jet of hot and expanding gases resulting from their combination roars out of the exhaust nozzles.

What makes the rocket engine drive the whole, heavy mass of the ship into space? If the gases produced in the combustion chamber were contained within its strong casing, and not released, they would simply push, or apply their force, outward in all directions equally, and nothing would happen. According to Newton's first law of motion, a body at rest remains at rest, and a body in motion remains in uniform motion, unless it is acted on by some external force. But Newton's third law of motion states that there is an equal and opposite reaction for every action of a force; thus, if there is a hole in the chamber, and the gases are allowed to escape through that hole, as the air does from the neck of the balloon, they push out forcefully, producing a great equal force in the opposite direction, thrusting back against the rocket. Similarly, if a youngster kneels on one knee in his wagon and pushes backward with one foot against the ground, the wagon will move forward. And, finally, following the second law of motion, any body acted on by a force is accelerated in the direction of the force in direct proportion to the force's strength and in inverse proportion to the mass of the body. In the case of the rocket, the force is the counterthrust provided by the escaping gases and the acceleration is determined by the total mass of all stages of the rocket.

Rocket Systems

Rocket launch vehicles are wracked by tremendous strains of expansion, contraction, and extreme or zero g's. In view of the conflicting forces within and without during a launching, it is remarkable that most rockets somehow hold together.

Rocket engines may burn either solid or liquid fuels. The basic, simple elements of solid-propellant rocket engines are diagrammed; they are similar to the powder-packed rockets used for patriotic celebrations. After ignition, the solid propellant is burned directly in the tube that contains it and forces its way, at high temperatures and pressures, out through the convergent-divergent exhaust nozzle.

For certain functions, solid-propellant rockets possess great advantages. The solid propellants are much easier and safer to store than liquid propellants. Solid-propellant engines can be ready to ignite at any moment; long countdowns are needed for liquid-propellant engines, for filling their tanks with the highly dangerous and corrosive fluids, often at exceedingly low temperatures.

Liquid-propellant rocket engines require large tanks to store the oxidizer and the fuel, usually separately, and a source of power to run the turbine pumps that force the oxidizer and fuel into the combustion chamber under pressure. Valves must be designed to carry extremely cold liquids, such as liquid oxygen (boiling point, $-361°$ F.) or liquid hydrogen (boiling point, $-422.9°$ F.), without locking up, and to open and shut almost instantaneously. So much piping is required that liquid-propellant systems have been called a plumber's nightmare.

The simplified diagrams show two different oxidizer and fuel flow systems for a liquid-propellant engine using liquid oxygen and liquid hydrogen. In the pump-fed system (left), the liquid hydrogen absorbs heat as it cools the combustion chamber by circulating through its casing. In what is called a topping cycle,

Major features of a solid-propellant rocket engine.
—Raymond E. Wiech, Jr. and Robert F. Strauss,
Fundamentals of Rocket Propulsion

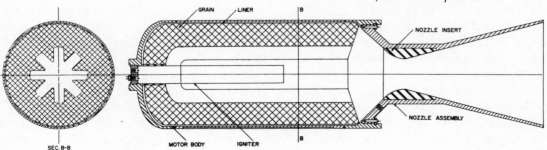

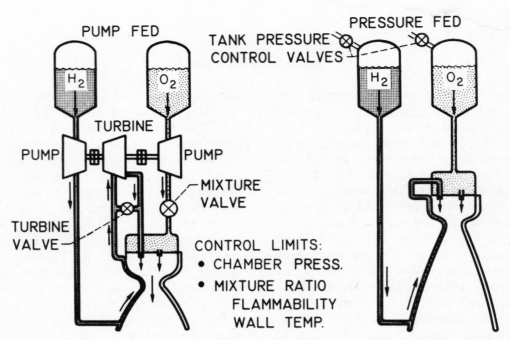

Pump-fed (left) and pressure-fed (right) flow systems for a liquid-propellant rocket engine using liquid oxygen and liquid hydrogen.
—NATIONAL AERONAUTICS AND SPACE ADMINISTRATION

the heated hydrogen is then injected into the combustion chamber by the turbine; control is obtained by the turning of the turbine valve to regulate the thrust in the chamber and the turning of the mixture valve to maintain the proper mixture ratio of oxygen and hydrogen. In the simpler, but less flexible pressure-fed system (right), the propellant-tank pressures force the hydrogen and oxygen into the combustion chamber, the hydrogen cooling the chamber walls by circulating in its casing on the way. The regulation of pressure in the supply tanks controls the combustion-chamber pressure and mixture ratio.

Such solid fuels as synthetic rubber and other complex organic molecules called polymers may be used, while liquid fuels often employed include ammonia (NH_3), hydrazine (NH_2NH_2), aniline, hydrogen (H_2), and the fuels, gasoline or kerosene. Ammonium perchlorate (NH_4ClO_4) is one of the most important solid oxidants, while the major liquid oxidizers are hydrogen peroxide (H_2O_2), nitric acid (HNO_3), and oxygen (O_2). Any of these provide pure oxygen directly, or loosely combined oxygen, which then rapidly combines with and oxidizes the fuel. Fluorine is also being used as an oxidant, since, like oxygen, it combines very actively with many substances. Its mixture with other oxidants may increase the efficiency of the propellants by as much as 30 to 60 per cent without raising the combustion temperature too high.

Liquid oxygen-hydrogen rocket engines provide many advantages in terms of high specific impulse and exhaust velocities. After extensive development, they have become operational in the Centaur and Saturn launch vehicles. The fluorine-liquid hydrogen combination is also receiving attention. Other so-called exotic mixtures are under development, such as diborane-oxygen difluoride and beryllium-oxygen-hydrogen combinations, which might provide very high energies if they prove to be manageable. Hybrid propellants, crosses between liquid and solid propellants, are being developed. Such a propellant might have an unoxidized solid fuel core and a liquid oxidizer, or liquid fuel and a solid oxidizer.

Somewhat like an umbrella on the end of your finger, rockets must balance delicately as they rise slowly from their launching pads; they encounter gusts of wind at low speeds, and are carried through a variety of atmospheric conditions and movements into what must be very precise trajectories. The accurate and split-second guidance and control of rockets has become a field of specialization in itself, dominated by transistors, computers, and a variety of special sensing instruments. Rockets can be steered in several ways: by rudders or jet vanes of such substances as graphite or tungsten, in, or at the perimeter of, the engine exhaust gases; by the gimbaling of the whole combustion chamber of the rocket or the exhaust nozzle, that is, swiveling it around a circle of small arc, directing its thrust in various directions to the center of gravity of the vehicle; by the firing of small auxiliary (vernier) rockets at the sides of the engines to apply angular thrusts in the desired directions; or by the injection of gases into the side of the combustion chambers or exhaust nozzles to deflect the main flow of exhaust gases and thus redirect the counterthrust. But who is to control these steering methods and how?

Remote-control guidance by radio from control stations on earth is one answer to this question. The rocket or missile transmits data to the control station on its position and flight, and such data may be received from tracking stations; these are coordinated by computer to compare the actual course with the programmed one. Then the necessary corrections are calculated in a split second by computer and transmitted by radio to the rocket to keep it on its precise course. In the instrumentation and control sections of the rockets, accelerometers indicate their positions very accurately, while fast-spinning gyroscopes maintain a constant position to which reference can be made in reporting and correcting the flight of the rockets.

If rockets are to be guided by radio, they must be within radio contact of some control station. This can lead to difficulties, since rockets pitch over after the immediate blast-off and may quickly be far distant from the control station; or they must be injected into escape trajectories from parking orbits at pre-

cise times, when they may not be conveniently near a control station. So self-contained inertial guidance systems have entered the rocket picture more and more. The rocket has its optimum course programmed into its own computer. In a split second, a check is made through data from sensors affected by accelerometers which show how the rocket is heading and accelerating. In the computer in the instrument section of the rocket this is compared with its program, the computer determines the necessary corrections, and these are sent out to control fuel flow and exhaust direction in the rocket engine and to bring the whole vehicle squarely back into its predetermined trajectory.

Space Payloads

Both the United States and the Soviet Union made remarkable leaps ahead during the 1960's in their rocket launch-vehicle programs. Thrust doubled and tripled and went on doubling and tripling. The high-energy propellant composed of liquid oxygen and liquid hydrogen came into use—hardly imaginable a few years before, since these gases were thought too dangerous to handle together. The payoff came with the much larger payloads carried into space and placed in orbit of the earth or sent to the moon, Venus, Mars, and beyond.

On May 15, 1958, the U.S.S.R. placed Sputnik 3 in earth orbit; it weighed, with a cylindrical case, an estimated total of 7,000 pounds, 2,925 of which were payload. On January 2, 1959, they sent off Lunik 1, now in solar orbit; its reported weight, given escape velocity, was over a ton and a half (3,245 pounds). They announced another vast step ahead with Sputnik 4, launched into earth orbit on May 15, 1960, weighing 5 tons (10,008 pounds). Their Venus probe, launched on February 12, 1961, weighed 1,419 pounds, and was injected into escape from Sputnik 8, which had been placed in a parking orbit and weighed 14,292 pounds (7 tons), perhaps including the weight of the probe. The Vostok space vehicles, carrying Soviet cosmonauts in low earth orbits, have all weighed in at about 10,000 pounds (5 tons). Later, Soyuz-manned space vehicles have weighed about 7 tons. The Soviet unmanned lunar space flights have consisted of vehicle payloads of from 3 to 7 tons. On the other hand, some Soviet research satellites, like the Proton craft, which was placed in orbit of the earth, have weighed from 17 to 20 tons. The Soviet unmanned lunar landers weigh about 2 tons, scooping up lunar soil and returning it to earth or carrying crawling lunar rovers to study the moon's surface.

With its Atlas-Centaur and then its Titan rockets, American launch vehicles forged ahead very rapidly. When the Saturn I was launched into orbit on January 29, 1964, weighing over 20,000 pounds (37,700 pounds including the second stage placed in orbit), the United States gained a lead over the Soviet launch-vehicle power that it never relinquished.

As the Saturn V launch vehicle was developed and tested, the United States was ready to meet the requirements for sending astronauts to the moon. The five huge engines of Saturn V's first stage produced a total thrust of 7.65 million pounds. The five second-stage rocket engines yielded up to 1.1 million pounds of thrust. The single-engine third stage generated up to 200,000 pounds of thrust at first burn and the same on a second burn. The total height of this great rocket with spacecraft atop was 362 feet. Suddenly, 110 tons could be placed by Saturn V in a low earth orbit and 45 to 55 tons sent into an escape orbit from earth, carrying men for the first time toward the moon!

The vibrations from the roar of one of the giant Saturn V's rockets were detected in New Jersey. With its thrust, three astronauts in a command module, a lunar landing module, and a service module could be shot to the moon. On July 20, 1969, Neil Armstrong was able to take ". . . one small step for a man, one giant leap for mankind" onto the lunar surface.

Astronaut Armstrong, first man on the moon, photographed Aldrin, his companion, setting up the experiment to test the composition of the solar wind, with the Lunar Expeditionary Module in the background.
—NATIONAL AERONAUTICS AND SPACE ADMINISTRATION

Nuclear and Electric Propulsion

Since the inception of intensive development of rocket propulsion in the United States, atomic nuclear reactions have been investigated as a source for power. Terrific punch is packed into small space with nuclear energy.

Very sizable specific impulses (indicating efficiency) of 750 to 1,000 seconds, and perhaps eventually 1,500 seconds, may be derived from the use of nuclear-power propulsion systems, as compared with a probable top of 400 to 500 seconds of specific impulse derivable from chemical propellants, whether solid or liquid. By this index, nuclear power offers two to three times as efficient a performance as chemical propulsion. Then, too, the long periods of time over which nuclear power can be available or stored without great deterioration make it a requisite for lengthy space journeys. Beyond the direct application of nuclear power lies its development for the so-called electric-propulsion engines, which may yield such extreme specific impulses as 3,000 to 10,000 seconds.

Nuclear reactions can perform two functions in space: first, the generation of electricity for auxiliary uses on rockets or spacecraft, providing a lightweight, long-term nuclear source of electric power; second, the nuclear production of power for actual propulsion. Great progress has been made in its first space application, using fissioning radioisotopes to generate electricity by means of thermocouples, as in what is called the Snap program, with a generator that employs polonium 210 as a heat source. A circuit is composed of two different metals in a thermoelectric converter or a thermocouple; this circuit develops a voltage when one metal is heated, directly converting the heat into electrical energy. The radioactive isotopes produce the heat for this reaction, but they are not useful for the generation of large amounts of power since they are heavy, weighing 1,000 pounds or more per kilowatt.

Another promising source for small supplies of electricity in space has been through gathering solar power. Earth satellites have guaranteed sunlight half of the time, when they are above the sunlit hemisphere of the earth, while space probes receive sunlight all the time on that side facing the sun. The ordinary satellite generates some electricity for "housekeeping" purposes and for charging its batteries through its silicon (or the more radiation-resistant cadmium-sulfide) cells exposed in blocks on broad surfaces to the sun's radiation. The solar cell is a form of photovoltaic cell, absorbing photons of light from the sun on its boron-treated surface. The photons release free electrons, leaving electron vacancies or holes. As the electron density builds up, some of the free electrons can move through the boron film into an external circuit, thus producing electric power.

Several new solar-powered systems are designed to use the direct heat from the sun, concentrate it, and produce power in turbo-

A design for a solid-core, heat-transfer nuclear propulsion unit (left) and some of the control problem areas and operational limits with such propulsion engines (right).
—NATIONAL AERONAUTICS AND SPACE ADMINISTRATION

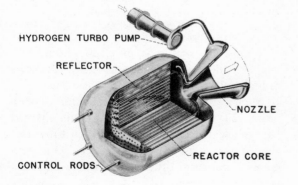

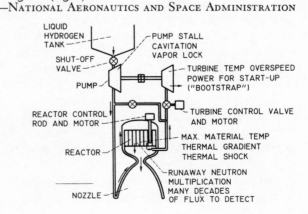

generator systems, rather than through solar cells or a nuclear reactor. In another system, the sunlight is focused by a mirror on a cavity which contains a hydrogen heat exchanger. Here liquid hydrogen is heated and then directly exhausted for propulsion.

A great deal of research and development has been devoted to the use of nuclear power to propel spacecraft themselves. A nuclear-reactor core is used to heat a light propellant gas, such as liquid hydrogen, to high temperatures within a pressure chamber; then the hydrogen is ejected at very high velocities through an exhaust nozzle. A series of such reactors and propulsion units are under development, with such project names for ground-test models as Kiwi, and Rover and Nerva for space-flight systems. Radiation protection without excessive weight, the prevention of uncontrolled explosions, and deterioration in the reactor core due to great heat and vibration are among the problems which make design difficult.

In a solid-core, heat-transfer nuclear propulsion engine the reactor is located forward of the exhaust nozzle. Nuclear-fission energy is produced within the solid materials of which the reactor core is composed. The propellant gas, hydrogen, is stored in the liquid state in a tank, pumped through the nozzle walls for cooling purposes, and also passed through the reflector regions around the reactor and any other parts of the engine that require cooling. On leaving the reflector, the already gasified hydrogen flows through the passages of the reactor core, where it is heated to as high a temperature as the core materials will stand, perhaps up to 6000° F. The high-temperature, high-velocity hydrogen blows down and is exhausted through a convergent-divergent nozzle to produce the thrust. In comparison with chemical-propellant engines, such nuclear engines offer a tenfold weight reduction most valuable for space flight.

Beyond such nuclear engines lies another family of electric-propulsion rocket engines being investigated as the eventual answer for efficient propulsion for periods of months or years in space. All of these will use nuclear reactors or radioactive fission in some form as a source of power. When controlled nuclear-fusion reactors (fusing hydrogen atoms to produce helium with the release of tremendous quantities of energy) are eventually constructed, they should have an immediate application in space to furnish this power, and they might in time be used as propulsion engines.

Arcjet systems, in which the propellant gas is heated to high temperatures in an electric arc and then ejected, appear to have the capacity for 50 to 60 per cent power efficiency and may eventually have specific impulses of from about 1,100 seconds up to 1,500 or 2,000 seconds. Such systems offer ready applications to satellite-orbit adjustment, attitude control, orbit-to-orbit transfer, and perhaps lunar-ferry functions, or to power communications satellite or spacecraft operations.

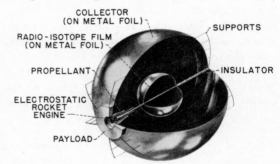

Theoretical design of a 500-kilowatt electrostatic propulsion system with a direct nuclear electro-generator using a radioisotope film.
—NATIONAL AERONAUTICS AND SPACE ADMINISTRATION

A most unusual type of electric-propulsion engine, called electrostatic, produces electricity of high voltages by electrostatic means in space. In such a system, a radioisotope film (polonium 120) on the inner sphere emits charged particles which build up a charge on the metal foil of the large collector sphere. This generates a potential difference between the concentric emitter and collector spheres, producing the electric current. This power is then used either in a contact-ionization engine, in which a stream of neutral cesium ions furnishes the jet driving out from the exhaust nozzles, or in a bombardment type of engine producing mercury-ion driving jets. The cesium-ion engine promises to be very efficient and long-lived, furnishing 5,000 seconds of specific impulse, if not more. In the contact-ionization type the liquefied cesium (melting point, 83° F.) is passed through a porous disk of tungsten, tantalum, or niobium metal, becoming a jet or beam of cesium ions in the electric field. These ions are accelerated to tremendous velocities, neutralized to become cesium atoms, and then shot out of the exhaust nozzle to provide the thrust.

Heavy-particle or colloid engines produce various kinds of colloidal particles, as small as a hundred-millionth of an inch in diameter, which are electrically charged, accelerated, and passed out in a fast-moving jet stream, as in the cesium engine. Experiments have been performed with droplets of aluminum chloride, glycerol, low-melting metal alloys, and oils.

Electromagnetic propulsion engines couple a strong magnetic field to currents flowing in a gas plasma (a hot, ionized gas), in such a way as to accelerate the plasma, using it then as the driving jet. One type is called a photon engine, driven with photons or light-energy particles. In another device, a pulsed, plasma-pinch accelerator causes a ring of hot plasma to contract and at the same time to be de-flected by curved electrodes so that the plasma is ejected in a jet-beam drive along the axis of the spacecraft. In still another design, plasma pulses are ejected by pulse currents through heavy single-turn coils. These electric engines may prove to be most efficient in the intermediate range of specific impulse.

One feasibility study has shown that about 50 per cent of the major planned scientific unmanned space missions of a planetary or interplanetary nature could not be carried out simply with a large chemical propulsion system such as the Nova; in addition, 40 per cent could not be completed even if nuclear heat-exchanger systems were also available. But most of the planned missions could be accomplished with an electric-powered rocket system launched into space by a Saturn 1 booster.

All these electric-propulsion engines involve thrusts which are limited to a few pounds. This restricts them to operations in space, since they do not have the great power needed to boost loads into orbit. Their small thrusts acting over extended periods can produce high velocities, and they can be engineered for extremely high specific impulses over very long operating periods, with great flexibility in starting, stopping, and applying thrust. Before the best designs for various functions in long-range spacecraft can be determined, much more research and development is required. There is probably no particular hurry, for other serious difficulties must be overcome before men venture into space beyond the moon.

Earth waxing like the moon from crescent (left) to half-earth (right), photographed from the moon by Surveyor 7.—J.P.L.

CHAPTER XIV

"I see a great round wonder rolling through space."

WHITMAN

In a few short years the science of launch vehicles has been mastered, and the earth has been "deflated" to a small globe that can be girdled in less than 90 minutes. Traditional standards of speed, size, and distance have been shattered—by the 59,000-pound thrust that rocketed the experimental X-15 up to 4,105 miles an hour, by the earth-orbiting manned flights that more than quadrupled this speed, and by the manned and unmanned space flights that have bettered 25,000 miles an hour.

Earth-Satellite Orbits

The terms that astronomers use for elements of the moon's orbit, such as apogee, perigee, and the argument of the perigee, are applied as well to the orbits of artificial earth satellites. These orbits must be circular or elliptical. If they were parabolic or hyperbolic, the space vehicle would not remain near the earth. One complete revolution of a satellite traces out a flat plane that can be conceived as cutting through the earth. When the satellite's orbit is circular, that is, at all points approximately the same distance from the earth's surface, then the center of the orbit coincides with the center of mass (or center of gravity) of the earth, and the radius of the orbit from this center is everywhere the same length. When the orbit is elliptical, one of the focuses of the ellipse must coincide with the earth's center of mass, or an orbit will not be established. A satellite in an equatorial orbit moves in the plane of the earth's equator, cuts the earth's center of mass with its plane, and has one focus there. In an inclined orbit, the plane is tilted from the equator. When this so-called angle

ARTIFICIAL SATELLITES AND SPACE PROBES

of inclination from the equator is 90°, the satellite is in a polar orbit and crosses both poles. An orbit of the earth cannot be established along a parallel at the latitude of, let us say, Buenos Aires or New York, since its focus would not coincide with the earth's center of gravity.

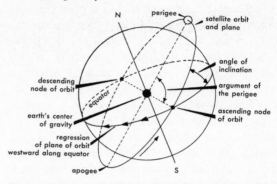

Features of artificial earth-satellite orbits. The angle of inclination is the angle that the plane of the satellite's orbit makes with the plane of the earth's equator. The regression of the plane of the satellite's orbit westward along the equator is caused by the earth's equatorial bulge. The apogee is the point where the satellite is most distant from the earth, the perigee is its point of closest approach, and the argument of the perigee is the angle which a line from the perigee to the earth's center of gravity makes with the point where the plane of the orbit cuts the plane of the earth's equator.

When a satellite is launched, the angle of inclination of its orbit to the equator is often announced, together with the apogee and perigee heights of the orbit. Although the perigee and apogee heights above the earth's center of mass are used in calculating earth orbits, news reports usually give apogee and perigee heights above the earth's surface, since these are the figures that interest most people. The argument of the perigee may be announced sometimes, particularly in technical publications, since it serves to determine the orientation of the orbit. Not usually reported is the line of nodes of the orbit, which is the line made by the plane of the orbit cutting the plane of the earth's equator. The ascending node is the point where the orbit crosses the equator moving northward; the descending node is the point on the opposite side of the earth where the orbit crosses the equator going southward.

To take advantage of the speed of the earth's rotation, satellites are normally launched in an easterly direction. When they are launched into polar orbits, their booster thrust must provide the whole orbital velocity. Polar orbits have certain advantages. In time, because the earth is rotating beneath it, the satellite will pass over its entire surface. In addition, as in an equatorial orbit, the satellite can be observed once in each revolution by a single tracking station, in this case at one of the poles. Though a satellite in an orbit with an inclination of 65° to the equator will in time cover most of the land masses of the earth except the polar areas, there must be a line of tracking stations from north to south to keep it under fairly constant observation.

Injection into Orbit

In the simplest case, to maintain a circular orbit at a given altitude above the earth's surface, a satellite must have a specific velocity if it is not to fall toward the earth or move away from it. A balance must be maintained between the gravitational pull of the earth and the centrifugal force imparted by the thrust. The satellite's height, velocity, and period of revolution all fit together for its particular orbit. Table 12 gives the velocities and periods of a satellite in a circular orbit at various heights above the earth's surface. A height of 1,044 miles determines a 2-hour orbit, a height of 22,300 miles puts the satellite in the 24-hour synchronous orbit, and so on. The speed required for a circular orbit diminishes with greater height. The closer to the earth a satellite is, the faster it must spin around the earth to maintain its orbit against the greater pull of gravity.

Satellites are lofted by their launching vehicles to their desired heights, where they must be accurately oriented in relation to the surface of the earth and then "injected" into orbit, that is, given the final exactly measured burst of speed by the last-stage rocket engine, which shoots them precisely into the planned orbit. If satellites are not injected in this manner and do not attain orbital velocity, they will behave as ballistic missiles do and fall

Effect of various injection angles on the orbits of earth satellites.

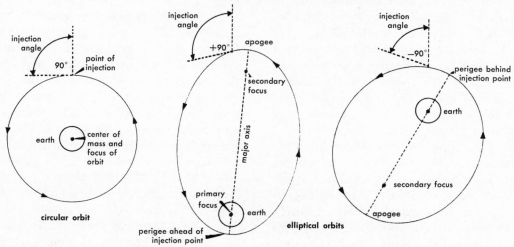

back to the earth. This is what happened to Alan Shepard and Virgil Grissom in their intentionally suborbital Mercury flights.

The two significant factors that establish a satellite in the desired orbit on injection, then, are the exact direction in which it is pointed and its exact velocity. When the velocity given to the satellite or spacecraft is correct for its altitude above the earth's surface, and when the angle at which that velocity is directed is precisely at right angles (90°) to the line straight down from it to the earth's center of mass, the satellite will be injected into a circular orbit. When the injection angle to the line between the earth's center of gravity and the satellite is greater than 90°, that center becomes the primary focus of an elliptical orbit, with the perigee of the satellite's orbit ahead of the point of injection, naturally, since the satellite will be closer to the earth at perigee. An elliptical orbit also results when the angle of injection is less than 90°, with the perigee behind the point of injection into the orbit.

If a satellite at 200 miles above the earth's surface is injected into orbit at 4.79 miles a second (Table 12), and the thrust is exerted at a precise 90° angle to the earth's center of mass, the satellite will go into a 200-mile circular orbit. If the velocity is increased by as much as 1 per cent, then the orbit becomes elliptical, with the same perigee height (200 miles) but with an apogee height of 370.8 miles. If the velocity is decreased by 1 per cent (about 0.05 mile or 264 feet a second), the perigee becomes the apogee, and the satellite will approach within 37.1 miles of the earth, right into the atmosphere, and will burn up as it plummets toward earth.

Thus the apogee and perigee of an elliptical orbit are radically affected by minute changes in velocity, and the satellite will not orbit at all if the velocity is 1.25 per cent below that required for a circular orbit. Similarly, with an error in the injection angle of 1°, for an injection elevation of 200 miles, the perigee

elevation would become about 170 miles and the apogee about 385 miles. Calculations made to correct orbits must take into account the fact that corrections produce new orbits, not simply shifts in directions or heights, with the new orbital injection point at the location where the correction was initiated. This is only one of the complications that make the rendezvous of two satellites in orbit such a tricky proposition.

Reentry

The orbit of an artificial earth satellite gradually decays or shortens. It is slowed down by atmospheric drag, the amount of which depends largely on the satellite's height above the earth. Revolution by revolution, it loses its original orbital velocity and begins to reenter the atmosphere. The apogee height is lessened and moves back closer and closer to the earth; the perigee height remains nearly constant until the last few revolutions. The satellite loses some velocity each time it passes through the lowest portion of its orbit about the perigee, so it cannot speed quite as far out to apogee as it did on the previous revolution, and the apogee approaches nearer to the earth. The perigee moves very slightly lower and the major axis of the ellipse of the orbit turns a little in each successive revolution.

The satellite tends to lose velocity faster and faster, because, as its apogee shortens, its ellipse loses more and more of its eccentricity, approaching a circular orbit and passing through more atmosphere each time it goes through perigee. When the ellipse finally degenerates into a circle, the atmosphere is dragging continuously on the satellite. The perigee height becomes the apogee height, and the two switch back and forth as the satellite spirals in through the atmosphere.

Without properly designed reentry configuration and materials, the satellite burns up like a great meteorite entering the earth's atmosphere. The recovery of unmanned-satellite instrument payloads without damage on reentry and the safe return of men in spacecraft through the earth's atmosphere have posed complex engineering problems. The great speed of the craft (17,500 miles an hour reentering from an earth orbit) is reduced to zero in short order. A controlled "skipping" approach would lengthen the time of reentry, that is, the spacecraft would duck into and out of the denser atmosphere a few times to reduce velocity slowly and keep the temperature down. This approach requires extensive rocket maneuverability.

The problem is that the craft must lose all of the energy of motion which has been imparted to it by the enormous energy of combustion of the rocket engines, which lifted it by reaction from its launching pad and injected it into orbit. In the drag of the atmosphere, this energy of motion is turned back into heat energy. Temperatures rapidly shoot up to 2000° to 3000° F. on the outer shell of the craft, as if it had been placed in an electric-arc furnace. The heating becomes even more intense at the more than 25,000-mile-an-hour velocities of craft which have escaped the earth's gravity and are returning to it, as from a trip to the moon. This speed could be reduced somewhat by using forward-firing retrorockets before the craft reenters the atmosphere, which would place the craft in a slower orbit around the earth.

A number of solutions of the reentry heating problem have been devised, and some combination of protective devices is customarily provided for reentry craft. A heat sink may be furnished by a nonmelting shielding material about the skin of the craft, with an absorbing thermal capacity great enough to protect the interior structure and equipment or astronauts. A cooling, circulating fluid may be used under the skin of the craft, absorbing the heat by a temperature rise or by a change in phase, such as vaporization, of the fluid. In transpiration cooling, gas or vapor is diffused through the porous skin of the craft or jetted through open-

ings to spread around the skin, carrying away the heat with it and helping to insulate the craft. One of the best methods is ablation, in which the heat is carried off by melting, vaporization, or sublimation of the surface material of the craft's skin or of a shield over it. Plastic ablation materials have been developed that have a great capacity for carrying off heat in this way.

Manned Orbital Flights

Two typical American satellite orbits are diagrammed against a representation of the earth as a globe, in what is called by map makers an oblique orthographic view. One satellite is represented (left) as launched into an inclined orbit from Cape Kennedy, Florida, into the Atlantic Missile Range (AMR), at an inclination of 34° south of the equator, while the other is represented as sent into a polar orbit from Point Arguello or Vandenberg Air Force Base, California, into the Pacific Missile Range (PMR), with an inclination of 90° to the equator. In the other diagram (right) the so-called flat Mercator projection map of the earth shows the satellite launched from Cape Kennedy in what is called a sine curve, which is like a stretched-out wave, because the vertical meridian lines in this projection have been given a constant parallel distance from each other, though actually, as shown in the left-hand diagram, the distance between them narrows to zero at the earth's poles. The plane of the satellite orbit actually cuts through the earth's center of gravity and has to be represented on the Mercator projection as a curve, but the satellite is not actually curving in orbit. It is moving in a great circle of the earth in the plane of its orbit as shown in the left-hand diagram. Similarly, the two vertical dashed lines indicate the polar orbit on the Mercator projection, but it is actually one great circle as shown in the oblique orthographic view.

A satellite cannot be launched at an angle

of inclination to the equator that is less than the latitude (north or south) of its point of injection into orbit, for then the plane of orbit would not pass through the earth's center of mass, which thus would not form one of the focuses of the orbit, as it must. However, a rocket can be launched from a launching site at a smaller angle of inclination and then turned into the correct angle to achieve an orbit (when it is close to orbital velocity or at the injection point). Because of its shape, this is called a "dog-leg" trajectory or "dog-legging" the satellite into orbit. It has not often been attempted because it takes extra thrust to change the direction of motion of

the satellite in this manner, and every pound of thrust has been at a premium in the early flights. On November 1, 1963, the Russians announced the launching of a craft called Polyot One (Flight One), which they claimed went through several dog-leg maneuvers, changing the inclination of its orbit before settling into its final path. On April 12, 1964, it was reported they had done it again—the Polyot maneuverable satellite became a series.

The periods of the manned orbits of the earth run from 88 to 89 minutes for a complete revolution. The United States craft used an inclination of 32.5° south of the equator, running from Cape Kennedy in a southeast direc-

A satellite orbit around the globe (left) at 34° inclination south of the equator from the Atlantic Missile Range (AMR) and another orbit at 90° inclination to the equator from the Pacific Missile Range (PMR). (Right) The same orbits against a Mercator projection of the earth, showing the effects of this projection on the flat, plane orbits. The wavy line from AMR (right) represents the orbit inclined 34°; the earth has rotated from west to east with each revolution of the satellite in orbit, so the orbital track against the earth moves toward the west with each orbit, though actually the orbital plane has not moved. The satellite orbit remains stationary and the earth rotates within it. The dashed line from PMR represents the polar orbit, with the track passing the South Pole, then coming up across the Indian Ocean and across the U.S.S.R. to pass over the North Pole and return down across Canada and the United States. The earth rotates inside this orbit in the same way, so with each revolution the orbital track would move westward, completing its swing around the earth in 24 hours.

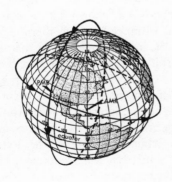

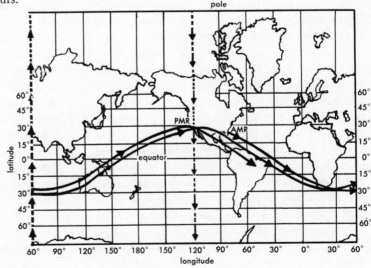

tion toward Africa, across the South Atlantic Ocean. On the other hand, Soviet spacecraft all orbited at an inclination of 65° north from the equator, which must be most convenient from their launch location, keeping their craft over Soviet territory as they move up into orbit toward the northeast. It is believed that the Soviet manned spacecraft launch site is located in southern Russia near Baikonur in the Kazakh S.S.R., some 220 miles northeast of the Aral Sea. Russian cosmonauts often land from orbit in this flat, desert country. The orbits of all the early spacecraft were elliptical, with perigees ranging from 99 to 111 miles and apogees from 141 to 188 miles above the earth's surface, well below the beginning of the intense radiation of the Van Allen belts. In retrospect, these early flights seem very rudimentary in contrast with the complex guidance, maneuvering, rendezvous, and changes in orbits now practiced by spacecraft.

Missile Ranges and Tracking Networks

The United States has three principal missile ranges from which missiles and space carriers can be launched: (1) the Atlantic Missile Range (AMR) at Cape Kennedy and Merritt Island, on the east coast of Florida about 50 miles directly east of Orlando and 60 miles south of Daytona Beach; (2) the Pacific Missile Range (PMR), which is headquartered at Point Mugu, California, about 25 miles northwest along the coast from the Los Angeles area, and which has launching complexes at Point Arguello and Vandenberg Air Force Base, about 100 miles farther up the coast; and (3) the Wallops Island (Virginia) Station, on the Atlantic side of the Delmarva Peninsula, about 40 miles south of Salisbury, Maryland.

The charts show the safe launching sectors, typical ground tracks of launchings, and the principal tracking stations supporting the Atlantic and the Pacific missile ranges. The American manned space program is carried on

from the Atlantic Range; military missiles and satellites are launched primarily at the Pacific Missile Range. With southeast launchings from Cape Kennedy, rockets can be tracked largely from island-based stations in the Caribbean Sea and the South Atlantic and need not pass over land (at the southern tip of Africa) until well into orbit. The Cape Kennedy launching center was expanded to Merritt Island to the north to allow plenty of room for the huge Titan and Saturn launch vehicles as America's space program burgeoned.

As demonstrated in the chart, the Pacific Range is ideal for polar-orbit launchings. It has the whole open stretch of the Southern Pacific Ocean, Antarctica, and the Indian Ocean for launchings. Missile launchings can also be made out over a series of tracking stations to the southwest in the Pacific, from which antimissile missiles can be shot up for test interception of the missiles. The landing areas for the Gemini and the Apollo manned vehicles were located in the wide expanses of the Pacific to give room for orbital shifts.

Complex facilities on the earth accomplish the tracking of missiles and space vehicles by radio and radar, eye, and telescope. Some facilities receive data from the vehicles by means of telemetry. Telemetry encompasses instrument data-sensing on the craft, coding of the data and their translation into radio signals sent back to earth, and guidance and control of the craft by transmitted radio signals. The Minitrack Network with over a dozen stations placed around the globe receives radio signals from unmanned satellites. The Manned Space Flight Tracking Network consists of another dozen or more complex ground stations placed along the band covered by manned orbits. These stations maintain contact with the spacecraft by means of radar, radio-listening devices, radio-telemetry equipment that can control the craft, and direct radio communication with astronauts.

The Deep Space Instrumentation Network follows and guides lunar and interplanetary

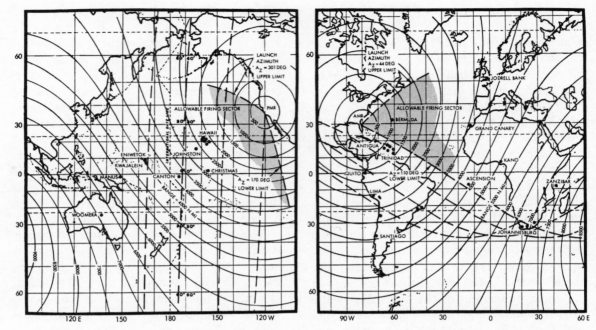

The Pacific (left) and Atlantic (right) missile ranges. Arcs indicate the distance in nautical miles (1 nautical mile = 1.15 statute miles) from the Cape Kennedy and California launching sites, the shaded area shows the possible launching sectors, and dots indicate the placement of the major tracking stations. Azimuth angles (A_Z) shown are measured as from the north point of the horizon (0°) eastward or clockwise through 360°. Dashed lines are typical ground tracks of satellite orbits, plotted against this Mercator projection map.
—Space Data, Thompson Ramo Wooldridge Space Technology Laboratories, Inc.

probes and spacecraft from four large radio antennas at Goldstone (near Mohave), California, and near Woomera and Canberra, Australia, and Johannesburg, South Africa. The stations are equipped to receive telemetry data from the spacecraft and to transmit commands to them. The Jodrell Bank radio telescope of the University of Manchester, near Macclesfield, England, has also assisted in tracking spacecraft and space probes.

An Optical Tracking Network at twelve locations around the world uses precision telescopic tracking cameras to accurately locate satellites when they are visible and determine their orbits; information on the rough positions to be used by this network is sometimes provided by organized groups of observers.

The American Space Surveillance System (Spasur) is a radar network of seven alternat-

ing transmitting and receiving stations along the 33rd north parallel in the United States from the east to the west coasts. They bounce radio signals from passing satellites, receive their echoes, accurately locate their positions, and predict their orbits.

Astronomical Satellites

The payloads boosted into orbit by powerful rocket-launching vehicles carry a wide range of instrumentation into space. Each sounding rocket payload, each satellite, and each space probe contains the instruments for one or more investigations and experiments. Perhaps the most fruitful observations so far have been the weather and communications studies and the studies of the earth itself; these have gathered more knowledge about its shape, gravitational and magnetic fields, and atmosphere. Other satellite series have already furnished data about interplanetary space, the planets, the sun, and the stars.

High-altitude (or stratospheric) balloons, sounding rockets, and satellites are releasing astronomers and astrophysicists from the exasperating limiting conditions of the earth's atmosphere. Telescope-carrying balloons like the Stratoscope series (1957, 1959, and 1963) are above 96 per cent of the earth's atmosphere at their observational height of about 80,000 feet. Sounding rockets penetrate much higher, but since they turn right around and come back, their observations of celestial phenomena are like snapshots as compared with the series of photographs or spectrograms that can be made from the balloons.

The Stratoscope balloons used 12-inch telescopes in their early flights and a 36-inch mirror in 1963. Stratoscope I made observations of the sun with a resolution of about 0.5 second of arc; later observations reached a resolution of about 0.3 second of arc. This is probably close to the limit because the sun heats the mirror and the thin atmosphere in and around the telescope structure, resulting in some turbulence. Valuable solar data can be obtained at these resolutions, however. On the nightside of the earth, without interference from the sun, the telescopes do better. The Stratoscope balloon telescopes are also being used to study the fine cloud structure of Venus (small "white dots" have been noted by some observers), analyze Mars's atmosphere and any detectable fine surface features, and investigate the atmosphere of Jupiter, the detailed structure of Saturn's rings, the optical diameter of Pluto, which is still in doubt, and the internal structure of galaxies and nebulas.

Astronomical observations are being carried out in another satellite series called Orbiting Solar Observatories (OSO), recording a broad spectrum of the radiation from the sun. The OSO I was launched on March 7, 1962, from Cape Canaveral by a three-stage Delta rocket

launch vehicle. The space observatory carried three solar x-ray experiments, a gamma-ray experiment, and a micrometeoritic dust-particle experiment. The majority of its instruments were concentrated on a whole series of observations of the sun, the causes of sunspots and solar flares, and solar radiations. The OSO 1 was highly successful. From March 7 until May 22, 1962, it transmitted nearly 1,000 hours of valuable data on the sun, opening up the solar ultraviolet and x-ray radiations to detailed study. The far ultraviolet spectrum was found to vary, its intensity increasing during solar flares. Some 75 flares and subflares were also measured. Flare information is sought to learn more about these storms on the sun and perhaps to be able to predict them, so that lunar or interplanetary space flights could be scheduled accordingly.

Another series of satellites in process of development and testing are called Orbiting Astronomical Observatories (OAO) to distinguish them from the solar observatories. The body of these satellites is octagonal, measuring some 7 feet wide and 10 feet high, with 4 large panels of solar cells mounted on the sides. A sunshade at the top of the satellite prevents sunlight from directly entering the experimental core, which bears telescopic mirrors ranging from 12 to 36 inches in diameter. Designed for more and more accurate astronomical observations in space, these satellites have a precision attitude-control system programmed from earth, and are instrumented to study the regions in the electromagnetic spectrum in space above and below the slot of the visible spectrum used by optical telescopes on earth.

Beyond this type of astronomical satellite the next stage might be the construction of a large observatory satellite or space station that could carry more varied and sizable equipment (a 100-inch telescope is being considered), or the establishment of a full-fledged astronomical observatory on the surface of the moon. These two possibilities have stirred a lively discussion; too many unknowns are involved for a decision as yet. The moon offers certain advantages as a site for large telescopes and other astronomical equipment, with a firm base for mounting them, slower motion, and protection under the lunar surface for observers from space radiation and meteorites. However, free dust on the moon's surface would be very harmful, the temperature variation with lunar day and night would be extreme, and lunar gravity, though slight, would require heavier structures to provide rigid equipment than a space station. The type of bearings on which the telescope would have to turn might give trouble in the near-vacuum of the moon.

On the other hand, an observatory orbiting the earth at a considerable distance would provide all the engineering advantages of weightlessness, easier accessibility by orbital rendezvous or from a nearby space station, less need for heavy bearings for mountings, and probably a lower cost. While personnel would not have the shelter from radiation offered by the moon and might not be able to stay long with the orbiting observatory, the ease of access to it could permit occasional adjustment, reloading, and repair between long periods of automatic operation of the instruments. Perhaps a small observatory can be established first on an artificial earth satellite. Later developments might justify a larger installation on our own natural satellite.

Lunar and Interplanetary Probes

Unmanned and manned space probes to the moon, as well as unmanned vehicles to Venus and Mars, and much later manned journeys for exploration of Mars, are certainly both the Russian and American space programs as far as the general public is concerned. The observations of a panel of space scientists (Pub. 944, National Research Council) indicate that the experts are similarly intrigued by the prospect: "There have been organisms on our planet for

about four thousand million years. By a remarkable stroke of fortune, it is in the next few decades that Man will first discover—rigorously, and in detail—what is happening on the neighboring worlds. The probability of our being alive at the present time, taken on a random basis, is therefore about a millionth of a per cent. We are immensely lucky to be living at the dawn of this era of planetary explortion and high scientific adventure."

Basic facts related to space flight to the moon and planets have been gathered together in Table 13. The surface gravities of these bodies are an all-important feature. The earth produces 1 g (1 gravity) on the average at its surface (a free-fall acceleration of 32 feet per second per second). The gravitational forces of the moon and other planets are given in the table in relation to the earth's gravity taken as 1, so they represent g's of gravity in relation to the earth's 1 g. The force of gravity on Mercury is only about $\frac{1}{3}$ that of the earth, Venus has nearly $\frac{9}{10}$ of our gravity, and Mars about $\frac{4}{10}$; the moon's gravity is only about $\frac{1}{6}$ that on earth. It will take little effort to walk or run on the surface of the moon.

The moon's low gravitational field makes it an ideal launching pad, and it may be a source of fuel and supplies for interplanetary missions. Only 0.1645 of the gravity of the earth must be overcome. If ready sources of rocket fuel are found near the moon's surface, it will be much cheaper and more efficient to launch rocket flights into space or toward the other planets from the moon rather than from the earth. Even though such launchings have already been made from so-called parking orbits of the earth, it is very costly to carry the space vehicles and all their fuel up from the earth. If quantities of ice or water were available not far under the surface of the moon, this would solve the problem of a water and an oxygen supply, as well as a supply of rocket fuel. But no ice or water has as yet been found there, and water would probably have to be baked from lunar rocks. Whether there are concentrated nuclear ores on the moon, containing uranium

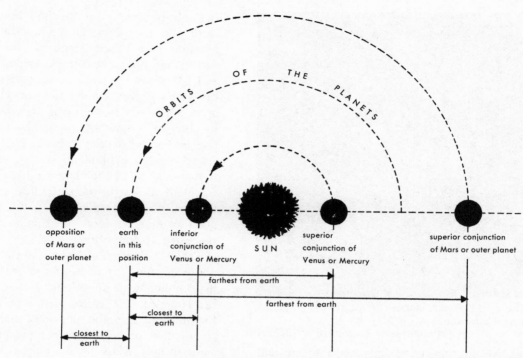

opposition
of Mars or
outer planet

earth
in this
position

inferior
conjunction of
Venus or Mercury

S U N

superior
conjunction of
Venus or Mercury

superior conjunction
of Mars or outer planet

farthest from earth

farthest from earth

closest to
earth

closest to
earth

Relative positions of the earth, the sun, and the other planets when they are in conjunction and opposition to the earth, looking down on the solar system from above. (Distances and sizes are not to scale.)

or other radioactive materials, is unknown. If there are, they would furnish the best initial source of energy on the moon except for solar radiation and provide fuel for nuclear-propulsion rocket systems.

A space probe or spacecraft must have a certain escape velocity if it is to pass beyond the predominant influence of the gravitational field of the body from which it is launched into the predominant influence of another gravitational field. Escape velocity from the earth's gravitational field in an hyperbolic orbit is about 7 miles a second (Table 13), which is 420 miles a minutes or 25,200 miles an hour. Escape velocity is substantially less from the other inner planets and only 1.5 miles a second from the moon. Because of the great masses of the outer planets (except for Pluto), their escape velocities are tremendous—a staggering 37.3 miles a second for Jupiter and 22.4 miles a second for Saturn. Aside from the extremely poisonous, perhaps cold, and very dense na-

tures of the atmospheres of the gas giants, their high escape velocities are the main reason why landings on only their satellites might be attempted.

The average speed at which the earth and other planets move in their orbits of the sun (Table 13) is a vital factor in planning flights to them. Venus and Mercury, closer to the sun, move faster than the earth's 18.6 miles a second and have shorter years; the moon's speed in earth orbit is slower, of course, as is that of Mars and the outer planets. A space probe directed toward the moon can disregard the velocity that the moon also has in tagging along with the earth around the sun, but the speed of the earth in relation to the other planets is a vital factor in choosing the most efficient orbits toward them. Fortunately, electronic computers are now available to make most of the necessary and formidable calculations.

Opposition and conjunction refer to certain

The 2,200 pound Mariner 9 spacecraft, launched from Cape Kennedy on May 30, 1971.
JET PROPULSION LABORATORY, NASA

relative positions of the planets, the sun, and other bodies in the solar system and are important in determining the shortest routes between planets. At inferior conjunction with the earth a body lies directly on a line between the earth and the sun (disregarding inclination to the ecliptic), while in superior conjunction

the sun is directly on the line between the earth and the other body. Only the planets Mercury and Venus, some of the asteroids with small orbits, like Adonis, Apollo, Hermes, and Icarus, and a few of the comets actually can reach inferior conjunction, in which they approach closest to the earth. In superior conjunction, which any of the planets can attain, they lie beyond the sun from the earth, at their greatest distance from our planet. At opposition, a position of the planets farther from the sun than the earth, the earth lies on the line between the outer planet and the sun. The outer planets, as does Mars, come closest to the earth at opposition.

Table 13 shows the closest and the remotest distances from the earth of the other planets in conjunction or opposition, as the case may be. However, the closest distances given are for the most favorable positions, and often at conjunction or opposition the planets do not come so near. Mars, for example, never comes closer than 34 million miles from the earth and on an average opposition is 48 million miles away. The outer planets lie at distances roughly ten times or more farther from the earth than Mars at their closest approaches.

The times of transit to the other planets by

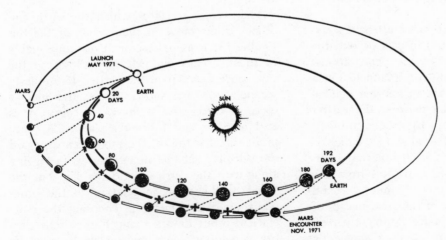

The flight path followed by Mariner 9 from earth to Mars from May 30 to November 14, 1971. One mid-course correction at about a million miles from earth on June 4 was the only maneuver required to hit the aiming zone next to Mars.
NATIONAL AERONAUTICS AND SPACE ADMINISTRATION

the most economical trajectory in terms of fuel use are also shown in Table 13. These routes are ellipses intersecting the paths of both planets and are called Hohmann transfer routes or transits for the scientist Walter Hohmann, who first worked them out in detail in the 1920's. Such periods of travel as 146 days for a flight to Venus and 237 days for one to Mars are necessary for the minimum energy expenditure, so that the greatest possible payload can be lifted into escape velocity. These periods can be shortened considerably if spacecraft are placed in parabolic or hyperbolic orbits at higher than escape velocities.

The earth and Mars moved toward a favorable opposition in 1971, within 34 million miles of each other. One Mariner craft failed to orbit when the upper stage of its Centaur launch vehicle malfunctioned because of a faulty circuit. Mariner 9 was off to a fine start, however, soon to be followed by two Soviet Mars craft. Mariner 9 functioned perfectly throughout its long trip to Mars, where it proved its correct aim by going into the desired orbit, the first artificial moon of another planet.

One Soviet craft released a soft-landing capsule slowed by a parachute. It landed in the

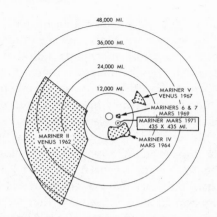

The aiming zones of the Mariner space probes became more and more precise from Mariner 2 toward Venus to Mariner 9 toward Mars (1971).
NATIONAL AERONAUTICS AND SPACE ADMINISTRATION

midst of great gales shrouding the planet in dust. Once settled on Mars, the capsule's signal lasted only 20 seconds—too short a time for transmission of even one photograph of the actual surface. A Viking mission to Mars, with a similar soft-landing vehicle is planned for a later opposition by the United States. The second Soviet vehicle entered an orbit of Mars like that of Mariner 9.

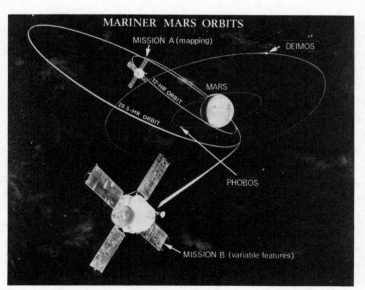

Slowed by the firing of its rocket engine, Mariner 9 swung into a 12-hour orbit of Mars and began to transmit information about the planet.

NATIONAL AERONAUTICS AND SPACE ADMINISTRATION

Mariner 9's TV cameras transmitted excellent close-up views of Mars once the dust storm had cleared. Preliminary reports indicated that Mars could now be mapped more precisely than before. Great long canyons were found with branches that may have been grooved by running water. Glacial layers in Mars' south polar cap implied water ice as well as frozen carbon dioxide in Mars' caps. Large volcanoes showed that Mars had been active, at least in the past, and must have a differentiated interior. Smoothed surfaces implied that weathering and erosion had occurred on Mars fairly recently.

Manned Space Flights

Before manned expeditions could be sent to land on the moon, many unmanned hard-landing, soft-landing, and lunar-orbiting vehicles had to make preliminary investigations of lunar conditions. These craft had to provide the basic information about the moon's surface before manned landing vehicles could be designed and engineered with full assurance. A start in this direction was made by Lunik 3, launched from the U.S.S.R. on October 4, 1959. In its initial orbit it approached the moon at a distance of about 3,800 miles from its surface and from the south. It swung past to about 40,000 miles from the moon, where it made the first photographs of the far side on October 7. The moon's gravitational attraction deflected Lunik 3 northward into another plane, and it came back around the earth in an elongated orbit with a period of 16.2 days, a perigee of 25,257 miles, and an apogee of 291,439 miles. This orbit decayed, and the vehicle burned up in the earth's atmosphere on April 20, 1960.

A total of eleven spacecraft hard-landed on the moon, among them the Ranger craft that sent back sharp photographs of the lunar surface as they fell faster and faster toward impact. Then came a group of soft-landing craft, like the American Surveyors and Soviet Lunas, which were slowed by their rockets as they approached the moon and landed gently in a cushion of

rocket flame. These research craft dug, tested, and photographed right on the lunar surface.

Finally, a series of Lunar Orbiters circled the moon at distances ranging from hundreds down to fifty miles from the surface, mapping both the moon's nearside and its farside. The amazingly clear, close-up photographs transmitted were studied for the selection of the landing sites which would offer the best samples of the lunar surface. Meantime unmanned and manned Apollo spacecraft were being tested and flown in earth orbits.

On December 21, 1968, three astronauts in Apollo 8 went into earth orbit and then became the first human beings to go into "escape orbit" of the earth as they blasted toward the moon. After ten lunar orbits they returned to the earth. Apollo 9 tested the manned lunar module in earth orbit in March 1969. With Apollo 10, in May 1969, when two astronauts came within 10 miles of the moon in the lunar module, preparations were completed for a manned lunar landing with Apollo 11.

The paths followed around the earth and moon in the Apollo 11 and later manned missions to the moon are shown in the chart. The numbered positions show fifteen key steps in Apollo expeditions: (1) Lift-off from Cape Kennedy; (2) earth-orbit checkout; (3) injection into orbit toward the moon; (4) turnaround of command, service, and lunar modules; (5) midcourse correction of orbit; (6) retrofiring to slow Apollo for lunar orbit; (7) elliptical lunar orbit; (8) separation of lunar module with 2 astronauts aboard; (9) landing on the moon; (10) LM ascending and docking; (11) ascent stage of LM left in orbit; (12) injection on homeward orbit; (13) course correction for precise reentry angle; (14) service module jettisoned; and (15) chutes lower command module to Pacific splashdown.

Seven manned Apollo missions to the moon were achieved. Each thoroughly studied another type of lunar surface and landscape, with the exception of Apollo 13, disabled by the blowout of an oxygen container in midflight.

All the key steps in an Apollo Manned Mission to the moon and return to earth, described by number in the text. The moon moves along in its orbit of the earth, from west to east, as the mission is accomplished.

—NATIONAL AERONAUTICS AND SPACE ADMINISTRATION

The astronauts on this mission rounded the moon and brought their ship back to earth, fortunately without mishap, despite the danger.

Each Apollo crew left instruments on the moon to report long-term conditions and changes back to earth, and brought back with them more and more precious lunar rocks and soils for study. Already a great deal has been learned by scientists about the moon today and its history back to its time of formation 4.6 billion years ago with the rest of the solar system. And only preliminary analyses have been made of all the observations gathered about our next-door neighbor in space.

Although the Apollo lunar exploration has now lapsed, it is small wonder that our attention is directed elsewhere, and scientists have enough data to try to fit together into an over-all picture of the earth-moon system for a long time to come. Undoubtedly, man will soon revisit the moon in order to begin to use it for purposes for which it is best fitted, as we come to understand it and learn much more about the great variety of human reactions under low-gravity and weightless conditions.

Meantime, space observations will be continued in manned space stations and observatories like the planned Skylabs, orbiting the earth. Human observers have a valid function in such space vehicles, as well as do the unmanned satellites. But both launch vehicles and their spacecraft will require extensive development in future before the manned exploration of another planet such as Mars will be feasible.

A Second Moon?

Some scientists have been peering even further into the future. They have studied proposals to establish a base on an asteroid or to change its orbit to make it a new and permanent satellite of the earth. The asteroids are such small bodies that their escape velocities are very low, they have no atmospheres, and rocket thrust or the thrust of the explosion of nuclear devices could change their courses.

Three small asteroids with diameters of about half a mile to a mile or two have been known to pass fairly close to the earth, but not close enough to be captured naturally by it. These asteroids are Apollo, passing at a distance of about 6.5 million miles in its most favorable position, Adonis passing within 1.5 million miles, and Hermes, which flashed by at only about 400,000 miles on at least one occasion. While their orbits have not been determined exactly or followed carefully, they, or others like them, might be located again (Table 9).

One possibility would be the launching of manned space ships to intersect the orbits of asteroids as they passed, land on them, and establish permanent installations beneath their surfaces for protection from solar radiations. In their natural orbits these asteroids would make fine space ships for the long-term investigation of the nearby reaches of the solar system. Eventually, with large rocket installations on their surfaces, using fuel developed

from their minerals, they might be nudged into changing orbits which would allow the close inspection of Venus and Mars, to say nothing of all the bodies in the asteroid belt past Mars, and the giant planets beyond.

A second possibility is to capture an asteroid. First it would have to be brought within a radius of about 230,000 miles of the earth, near the moon's orbit. Then its orbit would have to be changed slowly to bring it into a permanent orbit around the earth. The first modification of the orbit would guide the asteroid from a near-circular into an elliptical orbit which touches the orbit of the earth; it should require a change in the asteroid's velocity of only about 0.5 mile a second, it has been calculated. The second modification, bringing the asteroid as close as 22,300 miles in a circular, 24-hour, stationary orbit, might involve velocity changes under optimum conditions of only 0.5 to 1.5 miles a second, as compared with the 7 miles a second necessary to escape from the earth's gravitational field. Even the smallest asteroids would have tremendous weights compared with our puny rocket payloads, so the fuel to propel them and to change their velocity would have to be derived from sources on these bodies themselves. Rocket engines of low thrust applied over long periods or heavy nuclear detonations would have to be used.

The creation of the earth's own artificial satellite out of a natural body would be an astounding achievement for mankind. It might serve as a large space station and as a source for propellants, raw materials, and supplies for orbiting space stations, bases, and space vehicles, so that all these materials need not be lifted from the earth at great cost. It might furnish a site for new industries that would use the vacuum and other conditions of space, and raw materials might be shipped to earth at less expense than for shipping between points on earth. Finally, these captured asteroids, or others in their natural orbits, might become the great spaceships of the future, in which our descendants would set out on voyages beyond the solar system into interstellar space.

CHAPTER XV

"The long, long anchorage we leave."

WHITMAN

In May, 1927, Charles A. Lindbergh winged solo across the Atlantic from Roosevelt Field, Long Island, nonstop to Le Bourget Airport near Paris. In his stubby *Spirit of St. Louis,* which reflected the most advanced aeronautical design of the day, he flew at times as low as 10 feet above the waves and once climbed to an altitude of 2 miles to avoid clouds. His 3,610-mile record journey lasted 33 hours and 30 minutes, at an average speed of 110 miles an hour. A quarter of a century later, in May, 1963, it took Gordon Cooper 30 minutes longer to orbit the earth 22 times in his Mercury capsule, *Faith 7.* He covered a distance of some 600,000 miles at 17,550 miles an hour, at heights ranging from 100 to 166 miles above the earth's surface. In roughly the same time he had flown over 80 times higher and over 160 times faster and farther than Lindbergh. An astronaut blasting off for the moon must speed at over 25,000 miles an hour.

Whether or not man will be tough enough in body and spirit to endure lengthy space flights is an open question. The more we learn about space, the more man's chances for surviving it appear to shrink. The discovery of high-energy radiations in the earth's magnetosphere and of their increasingly great intensity at greater heights seemed to dash man's hopes of survival in space—until these radiations were found to be localized in the doughnut of the Van Allen belts that ring the earth. Gherman Titov's nausea after 6 hours of weightlessness in the second Soviet manned flight prompted gloomy talk of a natural period after which vertigo might occur in other astronauts. When other men flew weightless for more than 6 hours without difficulty, this hurdle toward

MAN IN SPACE

full space flight had been cleared. Another problem, still unsolved, is the deadly nature of the large solar flares or bursts. The exact time of the next maximum sunspot period in the 10- to 11-year cycle, to say nothing of the occurrence of individual solar flares, is as yet unpredictable.

The human body is made up of many systems that must function smoothly together to maintain an adequate adjustment to an alien environment. These are the musculoskeletal, cardiovascular, sensory, nervous, endocrine, respiratory, digestive, and excretory systems. The most important factors in space flight are acceleration, atmosphere, gravitational and magnetic fields, and particle and electromagnetic radiations. Numerous critical interactions between the human systems and these environmental factors will take place—the effect of acceleration on respiration and movement, the effect of weightlessness on the sensory and cardiovascular systems, and the bathing of the whole body in radiations, to name but a few.

Noise and Vibration

Enclosed in their snugly fitting space suits and lying on cushioned couches shaped to their own contours, astronauts are never safer than while they wait for the end of the countdown and the lift-off of the rocket. Even if the highly explosive liquefied gases being pumped into the fuel tanks of the booster blew up, they would be yanked out of danger by the small rocket in the escape tower above. But as soon as the rocket booster is ignited, rapidly builds up roaring thrust on the pad, and is released to hover and shoot upward bearing them with it, astronauts face many dangers, particularly in the first 5 minutes of flight. Among the most potent of these are noise and vibration.

A fearful, thunderous noise radiates around

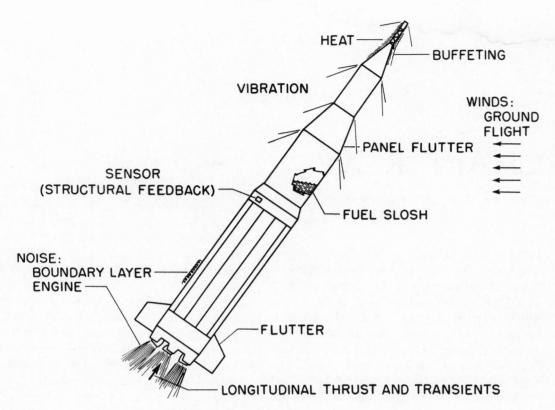

HEAT

BUFFETING

VIBRATION

WINDS:
GROUND
FLIGHT

PANEL FLUTTER

SENSOR
(STRUCTURAL FEEDBACK)

FUEL SLOSH

NOISE:
BOUNDARY LAYER
ENGINE

FLUTTER

LONGITUDINAL THRUST AND TRANSIENTS

Launch-vehicle problem areas.
—National Aeronautics and Space Administration

the rocket booster as it reaches full thrust. It is a snarling, growling, roaring monster. The rocket and spacecraft are enveloped in resounding waves of noise ranging from 100 to 180 decibels. For comparison, the noise intensity of heavy traffic is about 80 decibels and the roar of a jet engine can touch 140 decibels, the level that may cause ear pain. The noise levels reached by various booster systems range so high that breakage of equipment has resulted frequently at great distances from the launching site. The Saturn rocket booster makes a grumbling, roaring, shattering noise from near at hand. The National Aeronautics and Space Administration has bought up land far around Cape Kennedy, not only to provide more space for launching complexes but also to lessen the risk to people and things from the noise or explosions. As the rocket gains speed, the air begins to whistle past the nose cone or space cabin and build up acoustic pressures at the boundary layer, the area of greatest friction between the shell of the rocket and the air. This is added to the tremen-

dous noise produced by the rocket engines.

Men can tolerate noise in the range of audible frequencies at the 100 decibel level for up to 8 hours, at 120 decibels for about 5 minutes, and at 135 decibels for only about 10 seconds; 150 decibels and above may cause severe damage to the inner ear, pain, and nausea.

Special acoustic insulation may be required for space flights of the future. Space cabins might absorb up to 30 decibels of the noise and space suits and helmets about the same, but such absorption would be in the range of frequencies above 30 cycles per second. Below 20 cycles a second, sound is not audible but is felt as vibration, and in the larger rocket systems most of the sound may be subaudible, ranging from about 10 to 20 cycles per second. Below half a cycle per second, vibrations are felt simply as single jolts. It is believed that the chorus of noises in space cabins can be kept below the dangerous levels, but the problem could cause failure of manned flights within the first 2 minutes.

The many vibrations running up the length

of rocket systems as they are launched have a variety of sources. Among these are the shocks of engine ignition, pulsations in the engine thrust, the chugging of fast-spinning turbines, the effect of control forces and fuel sloshing in the tanks as the rocket veers, slightly unbalanced spinning components, and the shocks of engine burnout and the release of initial stages of the rocket. The human body can absorb or pass vibrations of high frequency without harm, but vibration below 100 to 200 cycles per second can be extremely damaging. Body systems, particularly in the abdominal area, may begin to vibrate in sympathetic resonance with the waves penetrating them, and this can shake and tear soft tissues apart, causing irreparable damage. After some 6 to 12 seconds of vibrations at 10 cycles per second, under an acceleration of 3 to 10 g, men experience severe chest pain, which may be followed by weakness and collapse if the vibration exposure continues.

Astronauts have said they get quite a jolting during the first minutes of their ride, partly from the buildup of high g's flattening them back against their couches and partly from the hammering up through the whole system of the rocket. Designers must provide adequate structural strength in the thin skins of rocket sections to absorb this hammering without damage and to prevent as much of the vibration as possible, without adding too much weight. The problem of eliminating vibration has sometimes slowed the development of larger rocket-engine systems.

As if the noise and jolting of launch were not enough, as the rocket accelerates up through the dense lower atmosphere it must pass through turbulent air currents. These buffet against the tall, slender shell of the rocket system and may cause fluttering of fins or thin panels. Rising through cross winds can produce severe shearing forces that tend to bend and twist the column this way and that, and the boundary-layer noise and vibration become more severe. Finally, the rocket spire is so long that it inevitably sways and bends

lengthwise as it is buffeted and the force of the rocket propulsion varies. Astronauts have reported that they can feel the rocket swaying behind them; somewhat like observers standing atop a tall building swaying in a wind, they feel their capsule weaving about at the peak of the rocket as it accelerates.

Too much of this swaying, buffeting, and longitudinal vibration has torn rockets and missiles apart. But the noise and buffeting decrease quickly as the rocket penetrates into the thinner upper atmosphere, and the vibrations and swaying are gone as soon as the capsule goes into orbit.

Acceleration and Deceleration

Extreme noise and vibration have not caused nearly as much trouble as have the high g's built up as rocket vehicles accelerate or decelerate to and from their earth-orbiting velocities of nearly 18,000 miles an hour. Greater velocity is required for orbits toward the moon or toward other planets—the escape velocity of 25,000 miles an hour or more away from the earth must be reached and later lost again when the vehicles reenter the earth's atmosphere. For earth orbits, the gravities jamming against the astronauts run up to 6 to 8 g's on launching and reentering; that is, the astronaut weighs about 7 times more than his normal weight at sea level as he accelerates up through the atmosphere. While the g's need not mount forever with launching into an orbit toward the moon or a planet, because lower acceleration developed for a slightly longer period will produce the same velocity as higher acceleration for a shorter time, or a space vehicle launched from parking orbit already has a good share of its required velocity, the astronauts must face and endure up to 10 g's on any direct flights from and direct returns to earth.

The giant test centrifuges in which astronauts are whirled around, at the end of long arms, in cabins simulating those designed for space, have proved that they can absorb such high accelerative and decelerative punishment for short periods of time. The human body is so resilient that it can take this mauling and drubbing without any permanent damage, though as the g's pile up the astronaut feels the effects strongly in his stomach and eyeballs and has difficulty in moving his arms and legs, and eventually in breathing.

Men totally immersed in water can absorb acceleration best, although space suits to restrain the body organs have been found adequate to meet the g's encountered, and the complexities of water immersion have not been necessary. It was soon discovered that the position of the astronaut's body is the key to the ability to absorb high g's. If the body is aligned head to foot with the direction of the acceleration, the blood is squeezed toward the head or feet as if by a hydraulic ram—this is the worst position. So the astronauts are seated across the rocket diameter. Extended across their capsules in this positon they face in the direction of flight on ascent ("eyeballs-in" is the graphic expression for the resultant pressure), and with their backs to the direction of flight on descent, again pressed back into their couches. High gravities are not as tolerable in the eyeballs-out position.

Practice makes astronauts more familiar with high-g conditions and enables them to continue to function fairly efficiently, using their fingers and even talking clearly under 6 to 8 g's, although the experience is far from pleasant. When astronaut Cooper was faced with the breakdown of his automatic attitude-control system for reentry into the atmosphere, he was able to fire his retrorockets at the end of his Mercury flight and then hold his bucking capsule in the proper position as it slowed down rapidly, operating so effectively that he landed within four miles of the aircraft carrier U.S.S. *Kearsarge*.

Weightlessness

We are thoroughly accustomed to the 1-g con-

dition here on earth—the gravitational force with which the massive sphere pulls at our bodies. This force acts in such a way that on a spring scale we have our given weight, and in free fall near the surface of the earth we would experience an acceleration of 32 feet per second per second. But strange things happen to the effects of gravity under the conditions of space flight.

When an express elevator accelerates upward very rapidly, its passengers feel that they weigh a little more, being thrust down with greater force against the floor of the elevator. When they accelerate downward very fast, the floor of the elevator seems to drop out from under them and they feel light and airy. Going up, they are experiencing hypergravity; going down, hypogravity, or gravity conditions greater and less than 1 g. At the zero g of a space vehicle in earth orbit, the astronaut feels no weight at all, the sensations of actual weight of limbs and muscles are lacking, and he seems simply to float or be suspended just where he is.

The human body is equipped with a number of little-understood sensory receptors acted on mainly by gravity, the so-called gravireceptors. The nonauditory labyrinths of the middle ear contain otoliths, tiny calcium-carbonate granules surrounded by nerve endings. Linear accelerations in any direction can be detected by the pressure of the otoliths against the nerves. The semicircular canals, located in the same labyrinths, register angular accelerations of the head, and the sense of balance derives from their functioning. Mechanoreceptors report localized pressures, tensions, and accelerations, such as Merkel's corpuscles in the skin and the nerve networks around the hair follicles. Within the body, proprioceptors immediately under the skin and in muscular and connective tissues supply the sensations of the position and posture of the body.

Weightlessness, it was predicted, would knock out the astronaut's sense of balance and position and sooner or later produce nausea, loss of body sense, disorientation, and inability to function. While this would be known as spacesickness, rather than seasickness or airsickness, it was thought it would involve the same symptoms and be just as incapacitating.

Most of the fears arising from the oddity of weightlessness vanished as the first astronauts, Russian and American, experienced the condition. While it took them a few seconds to get their bearings as they were injected into orbit and weightlessness, and to place their craft in relation to the earth, they were not disoriented. They found weightlessness pleasant, even exhilarating. They did not experience nystagmus, the involuntary eye movements associated with motion sickness or vertigo. They found that they maintained their postural sense and could write, take photographs through the ports, operate many types of delicate equipment and instruments, and control the attitude of their craft—in short, could carry on habitual activities without difficulty. When the body is properly restrained, the speed and accuracy of hand motions appear to be about the same during zero g as during normal g.

The Russians, perhaps because their space cabins were larger, experimented with releasing themselves from their restraining harnesses, floating about, and exercising as they went through long periods of weightlessness. In a very long flight of 119 hours (almost 5 full days), the Soviet astronaut Valery F. Bykovsky reported no serious problems in a total of 81 earth orbits. So fears vanished and astronauts have been weightless for over three weeks.

A few discordant symptoms have been detected, however. The Soviet cosmonaut Gherman Titov became dizzy and almost nauseated after about 6 hours of weightlessness. (Physicians have warned astronauts to avoid vomiting while weightless, since it might cause suffocation. The American astronauts, and perhaps also the Russian, have carried drugs to prevent such an eventuality.) While Titov's vertigo became worse when he turned his head or observed fast-moving objects, and may have involved the organs of balance of the middle ear, it was apparently controllable, and he continued on in his flight to a total of over 25 hours, able to sleep and carry out his assigned tasks. It is generally assumed that certain persons may be more subject to disorientation and nausea than others, and Titov may well have been one of these. In any case, his experience has been unique thus far.

When first released from the long hours of their confinement, astronauts have exhibited varying degrees of blood pooling in the major body cavities, such as the abdomen and legs, a reaction known as orthostatic hypotension. This depletes the brain circulation and is reflected in difficulty in standing upright and walking steadily. It appears to be caused by a combination of diminished blood volume, which many astronauts have shown, reduced blood pressure, and changes in the sympathetic nervous system that reduce the tone or elasticity of blood vessels. The condition lasts only a few hours, however, and regular vigorous exercise in the capsule during flight may well prevent it entirely.

No way has yet been found to overcome the diminished blood volume and an accompanying loss in sodium and potassium levels in the blood, however. Apparently, gravity has something to do with maintaining blood volume in the body. Under weightlessness, the kidneys excrete larger quantities of liquid and drinking a lot of water does not correct the situation. Loss of sodium and potassium led to abnormal heart rhythms of the astronauts on Apollo 15. Astronauts of Apollo 16 imbibed a great deal of potassium to prevent this condition, but still lost weight during their 11-day flight. The results of this and other tests are not yet in.

Under long weightlessness, calcium is lost from the bones and becomes more concentrated in the blood, which might damage the kidneys and weaken bones during long periods of weightlessness. All of these reactions will be investigated on Skylabs to check for any real danger. A regular program of exercise, working the body against springs, has been suggested as one preventive measure, and ample fluid intake and a low-calcium, high-phosphate diet as another.

Artificial gravity may yet have to be produced in space vessels if adverse and uncontrollable physiological reactions occur during longer periods of weightlessness. Weight could be created in a number of ways. A space station with a sizable diameter might be rotated on a central axis, a miniature earth, to reproduce part or all of normal gravity at its perimeter. Or a spacecraft might be made to revolve around its propulsion engine, to which it could be connected by a cable or pole, to produce artificial gravity. Systems like this would be very complex, however, and would require equipment that would add an enormous amount of weight. Preliminary experiments have uncovered many design problems in rotating spacecraft around a short radius.

Basic physiochemical reactions may unhappily prove to be limiting factors in space flight under weightless conditions, until ways can be discovered to counteract them directly in the body or with artificial gravity. Gravity on the moon is about one-sixth that on earth, on the planet Mars it is only a quarter, and in orbit it is zero. What effects these gradations in reduced gravities might have on human growth and the maintenance of health and energy cannot now be foretold, but they offer exciting experimental conditions. Space stations or lunar or Martian installations could offer laboratory environments most fruitful for comparison with the ways our bodies react here on earth under 1 g.

Life-Support Systems

In recent years, nuclear-powered submarines have provided an example of many men living together in a completely closed environment for long periods of time. While much has been learned about human requirements and reactions under these circumstances, a submarine is not a space capsule, unfortunately, and nothing comparable to it in size and weight could be launched into space. Weight, whether in equipment, stored provisions, or air, must be cut to an absolute minimum in space flight.

The creation of small, light, closed, life-support systems for spacecraft flights to the moon and beyond, and for short-term life support in space suits under the conditions of interplanetary space or the surface of the moon, has been the objective of much research. One of the most clear-cut results to date has been the conclusion that for periods of up to 2 weeks, and possibly a month, storage of the necessities of life and of its waste products will be adequate, and recycling from waste back to necessities will not be required. The more regeneration the better, of course—if body evaporation and wastes can be reprocessed in part to potable water, the spacecraft will have to carry less extra weight. At any rate, fully recycling life-support systems will not have to be developed before man can travel to the moon.

On earth, it has been customary to say, man's basic needs are food, clothing, and shelter. Those who climb or fly to great heights soon find that an atmosphere of a certain consistency is another basic necessity. Atmosphere is the main thing lacking in space. In addition, the equivalents of food, clothing, and shelter in space are very different from those on earth; when life must be maintained in a tiny, light container, the design of the life-support system becomes a veritable nightmare.

The atmosphere in space cabins, for instance, must be kept above a minimum pressure, below which oxygen would not be transferred properly to the blood in the lungs; but if the pressure is too high, carbon dioxide will not be adequately vented from the lungs in breathing. The air must be kept in motion to allow perspiration to evaporate from the skin. It must contain enough oxygen to fill the astronaut's needs without extra exertion; but experiments have indicated that too great a proportion of oxygen over weeks of time can have detrimental effects.

The American Mercury astronauts were provided with a pure oxygen atmosphere at a pressure of 5 pounds per square inch (about a third of that at sea level) and worked easily in it. The Russian Vostok cosmonauts were given an atmospheric mixture of oxygen and nitrogen, more like our normal terrestrial air. The oxygen in American space capsules was mixed with some helium during the critical first phase of flights, after the disastrous fire in a grounded Apollo capsule that killed three astronauts. Astronauts need a rich oxygen supply with as little effort in breathing as possible.

Oxygen can be carried as a compressed gas or as a liquid at low temperatures in bottles. It can also be generated from oxygen-rich chemicals such as potassium or sodium superoxides, sodium chlorate, or hydrogen peroxide. The superoxides have the advantage of removing carbon dioxide, odors, water vapor, and bacteria from the atmosphere. Oxygen can also be made by the electrolysis of water into its components or by the decomposition of carbon dioxide exhaled in breathing. Plant photosynthesis also creates oxygen. But producing oxygen by these methods demands complex, heavy systems or uses too much electricity to be feasible for short space trips.

While American astronauts have sometimes had trouble adjusting the temperatures within their space suits and cabins to the ideal range, this was because of mechanical malfunctioning of the air-conditioning systems, and not because there were any inherent environmental extremes in space that could not be overcome. The old fears about the near-absolute-zero chill of the vacuum in space have evaporated. The sun is a reliable source of heat, which the thinly dispersed particles in space cannot carry away rapidly from the space vessel. The heat exchange between the spacecraft and space can be adjusted by the absorptive, radiative, and reflective character of the skin of the craft so that a fairly normal temperature range can be

maintained, and temperature can be regulated within the craft by air-conditioning to suit the needs of the astronauts. A greater problem is radiation of the heat produced in the vehicle's systems into space fast enough to maintain a normal operating temperature without requiring too much radiative equipment.

Astronauts can eat and drink (and even be merry) in space and can perform all of their normal bodily functions; they have simply had to learn to do so under weightless conditions. Water and food are carried in squeeze tubes or canisters, because, if loosely carried, bits of food and droplets of water float helter-skelter about space cabins. Water poured into the air in a cabin simply forms a globule wherever it is released. Astronauts can eat cubed, dessicated foods or cookies with ease as long as they keep their mouths shut. Muscles of the mouth, throat, and esophagus deliver the food and drink efficiently to the stomach despite the weightlessness. A fairly normal, tasty diet can be predicted for the astronauts of the future, although they won't be grilling steaks in celestial barbecue pits.

The average daily requirements on a strict regimen for an astronaut in a space cabin total about 8 pounds. This includes about 4.8 pounds of water, 1.9 pounds of oxygen, and 1.2 pounds of solids—foods and minerals. The atmosphere can be carried in pressure bottles and the food and water in containers. The waste output of the astronaut is balanced with this at about 8 pounds—5.6 pounds of water a day, 2.2 pounds of carbon dioxide, and 0.1 of a pound of solid waste, urea, and minerals. The excess water vapor from breathing and perspiring can be wrung out of the capsule's atmosphere by the conditioning system. Exhaled carbon dioxide can be filtered from the air with an absorptive compound like lithium hydroxide, a pound of which can absorb about 0.75 of a pound of carbon dioxide before it is saturated. This can be stored, as can the body wastes, until the return of the spacecraft to earth. For some years and maybe decades, human needs in space flight can be taken care of quite adequately in this manner. As more power becomes available and greater weight can be carried, more recycling from human output to input items can be effected.

Completely closed and fully recycled life-support systems should not be needed until there are long-term space stations, or spacecraft fly beyond the moon toward the planet Mars, toward an asteroid, or past the planets Mars and Venus and back to earth. These self-sufficient systems might use plants or algae, which could recycle the wastes, produce fresh oxygen, use up carbon dioxide, and in the process create food. A similar system would be necessary for astronauts proposing to dig in and live on the moon for extended periods, although there the problem of the weight of the equipment would not be as serious as it would be for spacecraft. Unmanned craft could land equipment in separate loads on the moon; some input raw materials, such as ice for water, might be discovered on the moon itself, or water might be processed from its rocks.

Effective Human Action in Space

There is concern that essential physiological rhythms might be disturbed or destroyed by weightlessness or by the lack of the normal magnetic and gravitational fields experienced continually at the earth's surface. Can the effective energy for a normal day's work be maintained by a human being under space conditions? Will the lack of familiar day and night cycles tend to disrupt the astronaut's sleep, rest, and work cycles? Will the confinement and resultant need for tedious, energy-consuming exercise diminish the effectiveness of human beings in space?

On the short trips that have been made in near space, astronauts have tended to follow their natural day-night cycles on earth with little difficulty. Astronaut Gordon Cooper, who had a habit of catnapping, was able to refresh himself in the same way in his *Faith 7* Mercury capsule, sleeping on 13 different oc-

casions for a total of 4 hours and 9 minutes in his 34-hour journey. Astronauts on longer flights have slept as long as 8 hours at a stretch. Lengthier flights, of course, may have detrimental effects on physiological rhythms, but on the basis of experience to date, the prognosis is good.

The long-term psychological effects are another matter. Without frequent communication with others, aviators on lengthy missions at high altitudes have occasionally reported what is known as the "breakoff phenomenon." That is, they began to feel detached from the earth and from human beings, to lose sight of their objectives, and to indulge in unrealistic feelings and actions. Flight up in space might well give rise to more acute breakoff, even to a desire never to come back. For this reason among others, the long flights in prospect are planned for two or more astronauts together. In the relatively brief flights to date, astronauts traveling alone have not reported this psychological reaction.

Of greater concern is the so-called sensory deprivation. In space there promises to be much less input into the sensory receptors, partly because of the weightless condition and the limited environment in which the astronaut must remain. In sensory-deprivation experiments, subjects have become unrealistic, turned within themselves, and grown prone to hallucinations. The work load of the astronaut, however, is such that he can occupy every minute of his waking time fruitfully. The astronauts were wise to insist that the Mercury craft be equipped with a porthole and a periscope. To see the earth rolling beneath them, the awesome sunsets, and the brilliance of stars in the black sky above must surely have compensated for whatever slight sensory deprivation the capsule environment imposed. In long-term confinement in a capsule, however, the sensory factor might prove crucial.

Harsh and long-endured stress has its inevitable effects on the human system. The battle fatigue of war has often led to lasting mental deterioration, inability to face obstacles with aplomb, and sometimes complete breakdown. While individual tolerances range widely, the continuation of stress over long periods will eventually crush almost anyone's strength, efficiency, and spirit. Reactions to great stress are manifold: forgetfulness, failure to notice changes in one's surroundings, slowness or lack of ability to analyze problems and make decisions, lack of balanced coordination, erroneous estimates of speed and distance, carelessness, and improper operation of controls or reading of instruments. A single lapse of this kind could spell death to an astronaut.

With the wide individual differences in the capacity to withstand stress, astronauts can be selected who have proved their greater than normal endurance. Battle conditions have shown the remarkable ability of some men to live through ever-present threats to life for weeks at a time. And the more nearly the space environment is made to approach a normally dressed condition, the longer astronauts will be able to endure the strains they face.

Probably the best way to ward off the effects of continued stress is by sending groups of astronauts, two or three at the beginning, into space to allow adequate rest and sleep periods and to maintain normal communication with each other and with the earth. The inclusion of women to maintain a more normal atmosphere has been suggested for more extended flights and would probably be a must on the long flights to other planets in the solar system. The first space flight of a woman, Valentina Tereshkova, a 75-hour journey, indicates that the Russians may be planning that men and women will enter space together.

As soon as two or more human beings get together, to be sure, particularly in a confined space, complex social interactions come into play. Petty irritations and repeated frustrations develop. One of the group may try to dominate the others, openly or indirectly, or to vent hostility on them. One of the group may become a scapegoat for the aggressions of the others and may rebel at this role. The possibilities of trouble are endless. Nonetheless, with a com-

pletely planned schedule of operations, properly allocated functions, and a high morale deriving from the over-all objectives, there is little doubt that two or more astronauts will carry off space flights together much more efficiently than could one man alone.

In the flights to come unforeseen psychological reactions are bound to occur, but thus far physiological and psychological limitations, in theory at least, do not raise insuperable obstacles. Man is an adaptable and extremely tough creature. It appears that he can take it, for trips of at least a month in duration, which has permitted the long missions orbiting the earth and traveling to the moon and back. Longer missions will be tested in space stations.

But what of the external obstacles and hazards that space may have in store for us? What are the chances that meteors might blow out a spacecraft, or that penetrating radiations in space might sometimes be too strong for human beings to survive?

Meteors and Dust

Interplanetary space is almost entirely empty. Its emptiness surpasses by a good margin that of the emptiest vacuums obtainable in laboratories here on earth. In recent years, with liquid-hydrogen systems, vacuums of a ten-trillionth of a torr have been approached, a torr being the feather pressure necessary to support a column of mercury 0.04 of an inch high. The pressure from the thin gas in interplanetary space is estimated to be on the order of a ten-quadrillionth of a torr, on the other hand.

Despite this great emptiness, a variety of bodies and radiations criss-cross interplanetary space: the solar wind, consisting of high-velocity protons, electrons, and neutrons ejected from the sun; solar radiation and cosmic rays from interstellar space; interplanetary dust and gases; and asteroids, meteors and meteoric dust, and comets or their remains.

The solid bodies—the asteroids, large and small, the meteoroid streams, the cometary particles, and smaller meteoroid grains—hurtling at tremendous velocities through space could be terribly destructive to spacecraft. The larger comets and asteroids, the orbits of which are known with fair accuracy, and the identified meteoroid streams can be avoided, it is true, by launching at appropriate times or by simply steering clear of them. Before the Mars 1 spacecraft failed far out on its probe toward Mars, it had passed through a previously un-identified meteor stream about 23 million miles from the earth, according to Soviet reports. Given adequate instrumentation, the approxi-mate position and movement of such a stream could be mapped. As more and more un-manned probes are dispatched into cislunar and interplanetary space, many streams will be charted. When their trajectories have been worked out accurately, they can be avoided. Only meager bits of information are available on the dangers of encountering solid particles or bodies in space. Knowledge must be ac-quired swiftly and surely by means of ground-based observations and unmanned, heavily instrumented probes.

Various estimates have been made of the probability that spacecraft will be damaged or holed by meteoroids or solid grains in space. One authority has stated that a $\frac{1}{10}$-inch stain-less-steel hull of a space vehicle 1 yard in diam-eter is likely to be penetrated by a sporadic meteoroid only once in 1,450 years, while if the hull is $\frac{1}{2}$-inch thick, penetration would occur only once every 180,000 years. Another estimate indicates a penetration by sporadic or shower meteoroids every 110 days for a hull $\frac{1}{10}$-inch thick of a craft 50 yards in diameter, or every 300 years if the hull is 1 inch thick.

In a report to the National Aeronautics and Space Administration, another expert estimated that a spacecraft shielded with a double hull of aluminum, padded with glass wool between the layers, might be punctured by a meteoroid and seriously damaged on a trip to the moon only once every 57,000 days. On 14-day trips to the moon, only 1 in every 4,070 trips would suffer a puncture in the craft. However, assum-ing the variables at the most frequent end of

the range in each instance, he believed that a puncture might occur every 11 days on 14-day missions to the moon. The satellite Vanguard 3, in an orbit ranging from 320 miles above the earth at perigee to 2,300 miles at apogee, re-ported some 6,000 impacts from small particles over an 80-day period; however, 2,800 of these impacts, or nearly half, occurred during a 70-hour period from November 16–18, 1959, at the period of peak activity of the Leonid me-teor shower (Table 10). The size of the im-pacting particles was not indicated in this Van-guard investigation. How many of them were sizable grains and how many sheer fluff?

The rate of erosion of spacecraft shells by meteoroid dust and small solid grains does not seem to pose much of a problem for space flight. The erosion rate has been estimated, by studying meteorites that have landed on the earth, at an upper limit of 8 to 12 hundred-millionths of an inch per year or less. At this rate, even telescope mirrors exposed for several years in space should escape serious damage. But larger grains and bodies in space are an-other story, and ways must be found to pro-tect spacecraft against their impact.

In any investigation of the density and size of particles near the earth, it would be wise to study the regions of the Lagrangian points in the earth-moon system, where it is suspected that clouds of particles are located. It would be a sensible precaution to place manned spacecraft in trajectories that avoid these posi-tions until they have been thoroughly searched. It might also be wise, depending on future ex-perience with unmanned space probes, for manned ships to arch up over the plane of the ecliptic in passing from one planet to another, particularly in flights through the asteroid belt between Mars and Jupiter, or even in the vicin-ity of Mars, where there seems to be a great concentration of asteroids. A fast-moving chunk of asteroid or meteoroid weighing only half a pound or so could completely demolish a space-craft. The energy available in a meteoroid of first magnitude, for example, moving at about 25 miles a second, is the equivalent of the explosion of about ⅕ of a pound of TNT, roughly the amount contained in a hand grenade.

On explosive decompression of the interior of a spacecraft by the penetration of a body, experiments have shown that the crew would have only 30 to 60 seconds to take action before becoming unconscious, slightly longer if the atmosphere escaped more slowly from the craft. This would give the crew time to pull down the visors of their helmets and start their suit systems operating before unconsciousness overtook them; if their suits were intact, they could carry on the job of repair in their individual environments. But it would be touch and go, a grave emergency, and the spacecraft might not be repairable.

What is the best way to shield a spacecraft from solid particles? A very thick wall, which would really protect, adds too much weight. Meteor bumpers have been proposed, employing the device of making the wall of the spacecraft or an outer wall beyond it double, so that the outside skin would absorb most of the heat and blast of the impact of a particle, without holing the inner shell. Another device would be to locate the larger approaching bodies or streams by radar and to avoid those approaching too close by propulsion into a different course. This would not be feasible until spacecraft design and propulsive power have advanced far beyond their current state. A great deal of work remains to be done before the solid-particle hazards of space flight are fully explored, understood, and coped with. Still it is heartening that space probe and meteoritic studies to date suggest that encountering solid particles much larger than fine dust in space will be a rare event.

Space Radiations

Out beyond the earth's effectively shielding atmosphere, radiations are another serious threat to manned space flight. The discovery of the Van Allen belts, extending up to 40,000 miles above the earth's surface in the magneto-

sphere, for a time fostered the belief that they might prevent or greatly hinder manned exploration of space, since their radiation levels were intense, with unknown upper limits. Some people even theorized that the Van Allen radiation might be a manifestation of the sun's corona, which would become more and more intense nearer the sun. These Van Allen radiation levels were increased when the United States exploded a high-altitude nuclear device in July, 1962. Fortunately, this added radiation is gradually clearing, and it has been discovered that the Van Allen belts have an outer limit within the earth's magnetosphere since this particle radiation is trapped in the earth's magnetic field.

Earth satellites and spacecraft over 300 to 500 miles above the earth's surface are exposed to much more intense radiation than were the astronauts in orbits 100 to 200 miles high. The major types of dangerous radiation include primary cosmic rays, protons and electrons in the Van Allen belts, and the low- and high-energy protons and electrons from solar flares. Doses might run anywhere from a fraction of a rad a day up to thousands of rads an hour. A rad in living tissue is about equivalent to a dose of 1 roentgen. For comparison, in the United States a maximum permissible dose rate of 0.3 roentgen per 3 months has been set for radiation workers, or up to 12 roentgens a year. For industrial safety it is generally assumed that a person can take without serious effects a single exposure of about 25 rads of hard (short wavelength) x-rays or gamma rays, or several exposures adding up to about 15 rads a year. The normal radiation level from stray radiations from the earth and the atmosphere, which we all receive, is about 0.001 of a rad a day, or 0.38 of a rad a year.

The rem, another and more practical standard unit from the viewpoint of space flight, takes into account the type of radiation particle, its energy, the dose rate, and the size of the absorbing target, such as a person, in defining the "roentgen equivalent for mammal or man." This is the dose in rads multiplied by a factor of *relative biological effectiveness* (RBE), which depends on the variables named above. The RBE may be about 2 for the Van Allen belts, and about 10 for solar-flare radiation, based on the nature of the complex radiations in them. For comparison, the RBE is 1 for x- and gamma radiation, 2 to 5 for slow or thermal neutrons, 10 for fast neutrons, 8 to 10 for protons (hydrogen nuclei), and 15 to 20 for alpha particles (larger helium nuclei).

What are the effects of different levels of doses of radiation on man? A dose of about 20 rads is likely to cause nausea; 100 rads of hard x- or gamma rays may well cause vomiting, which might suffocate an astronaut in a weightless condition. Even though medication can be carried to control nausea and vomiting, doses of 150 rads of radiation would probably be about the limit of exposure, because of the increasing probability of serious disease or death beyond this limit.

The primary cosmic rays, despite their tremendous energies, are not concentrated enough in space to cause very serious effects. The Soviet cosmonauts reportedly received about 10 millirads (0.01 rad) a day during their orbital journeys. This totals something less than 4 rads a year, well below the so-called "permissible" annual dosage rate of 15 rads, which presumably can be absorbed without ill effects. While cosmic rays can produce genetic changes, so can many other types of radiation to which we are all exposed daily here on earth.

In or above the Van Allen zones, the radiation field inside the cabin of a spacecraft will consist of a mixture of many types of radiations—primary and secondary cosmic rays, gamma and x-rays, electrons, nucleons, fast and slow neutrons, high-energy protons, alpha particles, and mesons, as well as radio-frequency waves and microwaves. The composition and intensity of the radiation at any point will be the result of the radiation around the spacecraft, its shielding, and the total interior radiation from within the craft itself.

Relatively light shielding will probably be adequate to check the danger from the outer parts of the Van Allen belts, because it is relatively easy to stop the kind of electrons of which the belts are principally composed. Because of its protons, the inner Van Allen belt is a worse hazard and harder to shield against. With some heavy equipment around them and some fairly thick shielding, and if they follow a course angling north or south of the earth's equator, thus avoiding the greatest depth of the Van Allen belts above the equator, the astronauts should not receive too large a dose of radiation. The Van Allen belts no longer seem quite so formidable.

Solar Flares

Stormy weather in space consists of solar-flare outbursts of radiation, of the magnetic fields they sweep along with them, and of fast solar particles that find their way through these fields. The intensity of solar x-radiation, probably produced mainly in the sun's inner corona, runs in cycles related to the cycles of dark blotches on the sun, called sunspots, and a lot of x-radiation is released with solar flares.

Solar flares are the greatest known danger to men in space. They could inflict tremendous doses of radiation. The problem is complicated by the fact that while most major flares occur at times of sunspot maxima, some have taken place well down in the 11-year sunspot cycle. Flares emit particles of various energies in a range of intensities. It has been estimated that about 9 flares of low-energy particles at a low intensity occur per year, and that the probability of encountering one in a 1-week space flight is about 16 out of 100; only 3 flares of low-energy particles and very high intensity may occur in a year, with a probability of about 6 chances out of 100 of meeting one. Flares of high-energy particles of high intensity occur only about once in 4 years, so the chance of meeting one in a 1-week flight would be 3 in 1,000. In a week's flight to or around the moon and back to the earth (unmanned space probes have taken from 35 to 70 hours one way), there are, then, about 20 chances out of 100 that a hazardous solar flare might develop and spray its radiation through the spacecraft. The protons largely composing the flares are flung across space with such monumental energy that it would be much simpler to at least avoid the worst ones than to try to shield spacecraft against them.

Solar flares occur in or near sunspot fields on the sun, flare up very abruptly, and then gradually decay and subside, usually in less than 1 hour, though they may last as long as 10 to 20. They move out from the sun at velocities of about 700 miles a second or even faster, covering the distance to the earth in from half an hour to several hours, depending partly on the magnetic configuration then prevalent in space.

The solar flares are at a statistical minimum during the periods of a "quiet" sun, about midway in the 10- to 11-year periods of sunspot maxima. During quiet periods there is little likelihood of intense solar flares. Then their frequency builds up in three to five years to the next sunspot maxima. Still the danger of the occurrence of a great flare is not too great for short periods in space. Many Apollo astronauts accomplished trips to the moon and returned to earth without being overexposed to radiations during the 1969 to 1970 high period.

In the long run, it is believed that large solar flares will probably be predictable far enough ahead so that a 2-day trip between the earth and the moon can be made in comparative safety, particularly after the next maximum sunspot period. Once on the moon, adequate protection can be obtained by tunneling under its surface. But what of flares that may appear suddenly around the limb of the rotating sun? And what of flights to the inner planets, which will take at least 3 to 6 months? Some kind of shielding against the solar-flare radiation likely to be met on such a trip is a must.

Shielding

Passive shielding, simply a protective body which will absorb the radiation, and active shielding, setting up strong magnetic fields about a spacecraft, are both under study. Oddly enough, the lighter elements provide more adequate passive shielding per unit of mass than do the heavier elements. Carbon, nitrogen, and oxygen are quite effective; hydrogen would be best. Any of these might be incorporated into some metallic or plastic material to produce a passive shield.

The great weight is one objection, but not the only one, to passive shielding. It would cut down the primary radiation, but would produce secondary radiation of such particles as protons, gamma rays, and neutrons as it strikes the shielding material itself, and these would add to the dosage received within the spacecraft. The shield weights required for moderate solar flares have been put at several tons at least, and this on a spacecraft on which every ounce has to be justified. For extreme solar flares, tens or hundreds of tons would be needed for the primary radiation alone, forgetting about the secondary radiation that would be produced. Passive shielding thus involves many serious problems, and much research will have to be done to devise its most effective form.

Small, heavy inner capsules into which astronauts could crawl when solar flares were reported might reduce the requirements for outer shielding for the spacecraft as a whole, as well as the secondary radiation it would produce. It is also possible that selective shielding of the spleen, a vital, blood-forming organ greatly affected by radiation, might enable astronauts to live through heavy flares. Experimental animal studies of drugs that might reduce the effects of heavy radiation are under way, but as yet no effective compound has been found without dangerous side effects. So far it is a case of the cure being as bad as the disease.

In considering the use of active shielding,

which requires a magnetic field about a spacecraft strong enough to deflect most, if not all, of the incoming radiation, the effects of intense magnetic fields on human beings must be taken into account.

Reports of scientists who are often in strong magnetic fields in their laboratory work indicate in a preliminary way that man can tolerate fields up to 20,000 gauss without sensations for at least short periods of time, and that no cumulative effects are apparent from exposures to fields up to 5,000 gauss for total periods of 3 days a year. Taste sensations have been experienced in strong magnetic fields by some people with metallic fillings in their teeth. This kind of evidence, however, does not cover situations in which people might live in strong magnetic fields for many hours or days at a time. In all probability, active magnetic shielding would be used only when intense radiation threatened.

What would be the effect of magnetic fields lower than normal? Little or nothing is known in this area. The human reaction over long periods of time to living in very weak magnetic fields, or almost none at all, as in interplanetary space or on the surface of the moon, is an important possible hazard and must be investigated. One expert suggests that very weak magnetic fields might change biological rhythms or the perception of space and time. This is a variable in the space environment which will have to be studied closely as men venture into it. However, reports from persons working in conditions of very low magnetic fields, as in "degaussing" coils in ordnance work, do not indicate that there are specific detrimental effects on their health.

High-intensity magnetic fields, built up by superconducting coils, might provide active shielding for spacecraft, without the great weights necessary for passive shields. In entirely warding off the primary radiation, they would also prevent the formation of the secondary radiation produced in the passive shields. Hard, superconducting, solenoid coils, made of such materials as niobium tin or niobium-zirconium alloys, might be necessary to withstand the effects of the high magnetic fields they produced. Nonsuperconducting magnetic coils would not effect any great saving in weight over the passive shielding and would require great amounts of power and volumes of coolants. The magnetic fields produced would deflect both negatively charged particles like electrons and positively charged bits like protons.

Active protection might also be obtained by electrostatic shielding, that is, a positively charged outer spherical shell to deflect the positively charged particle radiation such as protons. This, however, has the weakness of actually accelerating negative electron radiation toward it, which would produce secondary x-radiation. The addition of a negatively charged sphere within the outer one would overcome this reaction, but it presents knotty design problems.

All these techniques—the passive, the active magnetic, and the electrostatic shielding—are under study. What will turn out to be the most effective space-radiation shielding is unpredictable. But some very strong shielding against solar-flare radiation is certainly in order, if men are to venture safely on lengthy space flights.

Men have successfully broken the bonds holding us to our planet earth on brief expeditions circling the earth and traveling to the moon and back. Long missions lasting for months or years are still untried, however. Can astronauts adjust to prolonged weightlessness or the weak gravitation of the moon, an asteroid, or Mars? Can adequate protection from space raditions be fashioned? How will extra strong or weak magnetic fields affect people? Undoubtedly the answers to these and many other questions will be worked out during the 1970's and following decades. Man's future in the solar system offers many challenging goals to be achieved.

TABLE 1. THE TWENTY-FIVE NEAREST STARS [a]

Name of star	Distance (light-yrs.)	Parallax (sec. of arc)	Visual apparent magnitude [b]	Absolute magnitude	Radial velocity (miles/sec.)	Transverse or tangential velocity (miles/sec.)
Sun	—	8.79 [c]	−26.73	4.7	—	—
Alpha Centauri [d]	4.3	0.762	−0.01	4.7	−14.3	14.3
Barnard's star (BD + 4° 3561)	6.0	0.545	9.54	13.2	−67.1	55.9
Wolf 359	7.7	0.425	13.66	16.6	+8.1	33.6
Lalande 21185 (BD + 36° 2147)	8.1	0.402	7.47	10.5	−53.4	35.4
Sirius [d]	8.7	0.375	−1.43	1.3	−5.0	9.9
Luyten 726-8 [d]	8.7	0.375	12.50	15.6	+18.0	23.6
Ross 154	9.6	0.340	10.60	13.3	−2.5	5.6
Ross 248	10.3	0.316	12.24	14.7	−50.3	14.3
Epsilon Eridani	10.8	0.303	3.73	6.2	+9.3	9.3
Ross 128	11.0	0.297	11.13	13.5	−8.1	13.7
Luyten 789-6	11.0	0.297	12.58	14.5	−37.3	32.9
61 Cygni [d]	11.1	0.294	5.19	7.9	−39.8	52.2
Procyon [d]	11.3	0.288	0.38	2.8	−1.9	12.4
Epsilon Indi	11.4	0.285	4.73	7.0	−24.9	47.8
Sigma 2398 (BD + 59° 1915) [d]	11.7	0.278	8.90	11.1	+5.0	23.6
Groombridge 34 (BD + 43° 44) [d]	11.7	0.278	8.07	10.3	+11.2	30.4
Tau Ceti	11.8	0.275	3.50	5.8	−9.9	20.5
Lacaille 9352 (CD − 36° 15693)	11.9	0.273	7.39	9.4	+6.2	73.3
BD + 5° 1668	12.2	0.266	9.82	12.2	+16.2	41.6
Lacaille 8760 (CD − 39° 14192)	12.8	0.255	6.72	8.6	+13.0	39.8
Kruger 60 (BD + 56° 2783) [d]	12.9	0.253	9.77	11.9	+16.2	9.9
Kapteyn's star (CD − 45° 1841)	13.0	0.251	8.81	11.2	+150.4	103.1
Ross 614 [d]	13.1	0.248	11.13	12.9	+15.0	11.2
BD − 12° 4523	13.4	0.244	10.13	11.9	−8.1	14.9
van Maanen's star	13.8	0.236	12.36	14.2	+18.6	36.7

[a] From G. H. Herbig and C. E. Worley, "Some Basic Astronomical Data," Astronomical Society of the Pacific Leaflet No. 325, rev. Dec., 1960, and other recent sources. [b] Photoelectric "visual" determinations when available; otherwise, photovisual or visual magnitude. [c] Approximate only. [d] A double or multiple star; data given represent the main star in the group.

TABLE 2. LOCAL GROUP OR CLUSTER OF GALAXIES OF WHICH THE MILKY WAY GALAXY IS A MEMBER, BY DISTANCE FROM THE MILKY WAY GALAXY [a]

Name of galaxy	Type	Distance (thousand light-years)	Diameter (thousand light-years)	Visual magnitude[b] (mv)	Absolute magnitude[c] (MV)	Radial velocity[d] (miles/ sec.)	Mass[e] (solar masses)
Milky Way Galaxy	Spiral		100		(-21)[f]		2×10^{11}
Large Magellanic Cloud	Irregular	160	30	+0.9	−17.7	+171	2.5×10^{10}
Small Magellanic Cloud	Irregular	180	25	+2.5	−16.5	+104	
Ursa Minor system	Dwarf elliptical	220	3		(-9)		
Sculptor system	Dwarf elliptical	270	7	+8.0	−11.8		$(2 \text{ to } 4 \times 10^{6})$
Draco system	Dwarf elliptical	330	4.5		(-10)		
Fornax system	Dwarf elliptical	600	22	+8.3	−13.3	+24	$(1.2 \text{ to } 2 \times 10^{7})$
Leo II system	Dwarf elliptical	750	5.2	+12.04	−10.0		(1.1×10^{6})
Leo I system	Dwarf elliptical	900	5	+12.0	−10.4		
NGC 6822[g]	Irregular	1,500	9	+8.9	−14.8	−20	
NGC 147	Elliptical	1,900	10	+9.73	−14.5		
NGC 185	Elliptical	1,900	8	+9.43	−14.8	−190	
NGC 205	Elliptical	2,200	16	+8.17	−16.5	−149	
NGC 221 (M32)[h]	Elliptical	2,200	8	+8.16	−16.5	−133	
IC 1613[i]	Irregular	2,200	16	+9.61	−14.7	−148	
NGC 224 (M31)[j]	Spiral	2,200	130	+3.47	−21.2	−165	4×10^{11}
NGC 598 (M33)[k]	Spiral	2,300	60	+5.79	−18.9	−117	8×10^{9}
Maffei 1	Elliptical	3,300			−20.5		2×10^{11}
Maffei 2	Spiral	3,300					

[a] The Local Group contains at least the seventeen galaxies listed, grouped in an ellipsoidal-shaped space about 3 million light-years in its longest diameter. Data adapted from George O. Abell, "Clusters of Galaxies," *The Griffith Observer*, August, 1963.   [b] Apparent magnitude when observed by the naked eye or the eye with the aid of a telescope.   [c] Magnitude as the object would appear if observed from a distance of 10 parsecs (32.6 light-years), at which its stellar parallax would be 0.1 of a second.   [d] Toward, +, or away, −, from the earth.   [e] Number of times greater than the mass of the sun.   [f] Merely educated guesses within parentheses.   [g] NGC stands for New General Catalogue.   [h] M stands for Messier system of identification.   [i] IC stands for International Catalogue.   [j] The Andromeda Galaxy. NGC 205 and 221 (M32) are small companions of NGC 224 (M31).   [k] The Triangulum Galaxy.

TABLE 3. SATELLITES OF THE SOLAR SYSTEM

Planet primary	Satellite	Year of discovery	Discoverer	Approximate mean distance from primary's center (miles)	Approximate diameter (miles)	Sidereal period[a] of revolution (days, hours, minutes)		
Mercury	None							
Venus	None							
Earth	Moon			238,866	2,160	27	7	43
Mars	Phobos	1877	A. Hall	5,825	5 ?	0	7	39
	Deimos	1877	A. Hall	14,580	3 ?	1	6	18
Jupiter	I Io	1610	Galileo	261,900	2,020	1	18	28
	II Europa	1610	Galileo	416,600	1,790	3	13	14
	III Ganymede	1610	Galileo	664,600	3,120	7	3	43
	IV Callisto	1610	Galileo	1,169,000	2,770	16	16	32
	V Amalthea	1892	E. E. Barnard	112,000	70 ?	0	11	57
	VI Hestia[b]	1904	C. D. Perrine	7,130,000	50 ?	251		
	VII Hera[b]	1905	C. D. Perrine	7,290,000	8 ?	260		
	VIII Poseidon[b]	1908	P. J. Melotte	14,600,000	9 ?	735	(Retrograde)	
	IX Hades[b]	1914	S. B. Nicholson	14,700,000	6 ?	758	(Retrograde)	
	X Demeter[b]	1938	S. B. Nicholson	7,200,000	5 ?	254		
	XI Pan[b]	1938	S. B. Nicholson	14,300,000	6 ?	714	(Retrograde)	
	XII (not named)	1951	S. B. Nicholson	13,000,000	4 ?	625	(Retrograde)	
Saturn	I Mimas	1789	F. W. Herschel	115,200	400 ?	0	22	37
	II Enceladus	1789	F. W. Herschel	147,700	500 ?	1	8	53
	III Tethys	1684	J. D. Cassini	182,900	630	1	21	18
	IV Dione	1684	J. D. Cassini	234,300	550	2	17	41
	V Rhea	1672	J. D. Cassini	327,100	810	4	12	25
	VI Titan	1655	C. Huyghens	758,400	2,990	15	22	41
	VII Hyperion	1848	G. P. & W. C. Bond	918,700	250 ?	21	6	38
	VIII Iapetus	1671	J. D. Cassini	2,210,000	?	79	7	55
	IX Phoebe	1898	W. H. Pickering	8,040,000	100 ?	550	9 (Retrograde)	
	X Janus	1966	A. Dollfus	135,000 ?	?	0	18	—
Uranus	I Ariel	1851	W. Lassell	119,100	450 ?	2	12	29
	II Umbriel	1851	W. Lassell	165,900	350 ?	4	3	28
	III Titania	1787	F. W. Herschel	272,100	700 ?	8	16	56
	IV Oberon	1787	F. W. Herschel	363,900	600 ?	13	11	7
	V Miranda	1948	G. P. Kuiper	80,700	200 ?	1	9	56
Neptune	I Triton	1846	W. Lassell	219,500	2,000 ?	5	21	3 (Retrograde)
	II Nereid	1949	G. P. Kuiper	3,450,000	200 ?	360		
Pluto	None							

[a] The sidereal period of revolution is the time required for one complete revolution around the primary, from alignment with a given star to alignment with the same star again as seen from the primary. [b] Names not yet officially accepted.

TABLE 4. FEATURES OF THE ORBITS, SIZE

Characteristics	Mercury	Venus	Earth	Moon	Mars
Mean distance from sun (miles)	35,983,000	67,235,000	92,956,000	238,866 [a]	141,637,000
Mean distance from sun (a.u.)	0.3871	0.7233	1.0000	0.00257 [a]	1.5237
Least distance from sun, at perihelion (miles)	28,584,000	66,778,000	91,404,000	221,463 [b]	128,412,000
Greatest distance from sun, at aphelion (miles)	43,382,000	67,692,000	94,511,000	252,710 [c]	154,862,000
Eccentricity of orbit (circle = 0)	0.20563	0.00679	0.01673	0.0549 [d]	0.09337
Mean orbital velocity (miles/sec.)	29.7	21.7	18.5	0.64 [d]	15.0
Length of year [e] (sidereal period of revolution in tropical earth years)	0.24085	0.61521	1.00004	27 days,[d] 7 hr., 43 min., 11.5 sec.	1.88089
Length of year (earth days)	87.969	224.70	365.24	(Same as sidereal period)	686.98
Diameter at equator (miles)	3,030	7,550	7,927	2,160	4,220
Mean radius (earth = 1)	0.38	0.961	1.00	0.273	0.523
Oblateness [f]	?	?	0.0034	[g]	0.0052?
Volume (earth = 1)	0.06	0.86	1.00	0.0202	0.150
Mass (earth = 1)	0.0543	0.81485	1.00	0.012304	0.1069
Density (earth = 1)	0.9	0.89	1.00	0.643	0.7609
Mean density (water = 1)	5.2	5.1	5.52	3.33	4.1
Inclination of equator to orbit (deg. min.)	?	32° ?	23° 27'	6° 41'	25° 0'
Sidereal rotation period [h]					
days	59	243		27	
hours			23	7	24
minutes			56	43	37
seconds			4.1	11.5	22.67

AND ROTATION OF THE PLANETS AND MOON

Characteristics	Jupiter	Saturn	Uranus	Neptune	Pluto
Mean distance from sun (miles)	483,715,000	890,602,000	1,777,021,000	2,799,435,000	3,654,407,000
Mean distance from sun (a.u.)	5.2037	9.5809	19.1168	30.1157	39.3133
Least distance from sun, at perihelion (miles)	460,260,000	844,629,000	1,698,281,000	2,778,887,000	2,747,712,000
Greatest distance from sun, at aphelion (miles)	507,170,000	936,575,000	1,855,761,000	2,819,983,000	4,561,102,000
Eccentricity of orbit (circle = 0)	0.04849	0.05162	0.04431	0.00734	0.24811
Mean orbital velocity (miles/sec.)	8.1	6.0	4.2	3.4	3.0
Length of year[e] (sidereal period of revolution in tropical earth years)	11.8622	29.4577	84.013	164.79	248.4
Length of year (earth days)	4,332	10,759	30,685	60,188	90,700
Diameter at equator (miles)	88,700	75,100	29,000	28,000	3,600 ?
Mean radius (earth = 1)	10.97	9.03	3.72	3.38	0.46 ?
Oblateness[f]	0.062	0.096	0.06	0.02	?
Volume (earth = 1)	1,317	762	50	42	0.09 ?
Mass (earth = 1)	318.45	95.22	14.54	17.23	0.03 ?
Density (earth = 1)	0.2409	0.13	0.23	0.29	0.37 ?
Mean density (water = 1)	1.34	0.68	1.75	2.2	?
Inclination of equator to orbit (deg. min.)	3° 7'	26° 45'	97° 59'	29°	?
Sidereal rotation period[h]					
days					6
hours	9	10	10	15	9
minutes	50	14	45	48	22
seconds	30				

[a] Moon's distance from earth. [b] Moon's least distance from earth at perigee. [c] Moon's greatest distance from earth at apogee. [d] Eccentricity of orbit, orbital velocity, and period of revolution of the moon around the earth in tropical earth days. [e] The sidereal period of revolution is the period calculated from a position of alignment of the body with the sun and any given star through the revolution and back to the same position again. [f] Oblateness, an index of the degree of flattening of the planetary disk, is the equatorial diameter minus the polar diameter of the planet, divided by the equatorial diameter. [g] Since the shape of the far side is not known, the moon's oblateness cannot be given. [h] The sidereal period of rotation is the period calculated from a position of alignment of the body with the sun and any given star through a rotation and back to the same position again. Venus' direction of rotation is from east to west, retrograde to that of the other planets, in a period greater than that of its revolution.

TABLE 5. SOLAR CONSTANTS, ALBEDOS, AND MAGNITUDES OF THE EARTH, MOON, AND OTHER PLANETS

Celestial body	Solar radiation available (solar constant)[a]	Proportion of light reflected (mean visual albedo)[b]	Brightness in magnitudes Unit visual magnitude[c]	Mean opposition magnitude[d]
Inner planets				
Mercury	6.7	0.056	−0.36	—
Venus	1.9	0.76	−4.29	—
Earth	1.00	0.36	−3.87	—
Moon	1.00	0.067	+0.21	−12.74
Mars	0.43	0.16	−1.52	−2.01
Outer planets				
Jupiter	0.04	0.73	−9.25	−2.55
Saturn	0.01	0.76	−8.88	+0.67
Uranus	0.0031	0.93 ?	−7.19	+5.52
Neptune	0.001	0.84	−6.87	+7.84
Pluto	0.0006	0.14	−1.01	+14.90

[a] The solar constant, the amount of solar radiation present immediately above the atmosphere of a planet in a unit area at right angles to the surface, is given in relation to that of the earth taken as 1.00; actually, the solar constant of the earth is about 0.13 kilowatt per square foot. [b] The average proportion of the light reflected by the planet in relation to that impinging on it. [c] Magnitude or brightness at identical unit distances from the earth and the sun. [d] Mean magnitude or brightness at opposition for the planets, when the earth is on a straight line between the sun and the planets; not applicable, therefore, to Mercury and Venus, always nearer the sun than the earth, nor to the earth. The apparent visual magnitude of the planets, their visible brightness as viewed from the earth, is given in Chapter 3.

TABLE 6. U.S. STANDARD ATMOSPHERE, 1962, AT 40° LATITUDE [a]

Geometric altitude (Z) Feet (thousands)	Miles (approx.)	Pressure (in. of mercury)	Density (lb./cu. ft.)	Kinetic temperature (deg. F.)	Speed of sound (ft./sec.)	Mean molecular weight	Temperature zones
2,320	440	3.2713×10^{-11}	8.895×10^{-15}	+2253.53		16.12	
2,000	380	9.1699×10^{-11}	2.596×10^{-14}	+2250.84		16.77	
1,500	285	5.5234×10^{-10}	1.761×10^{-13}	+2221.16		18.68	Thermosphere
1,000	190	5.1288×10^{-9}	2.675×10^{-11}	+2124.55		22.52	
500	95	1.3799×10^{-7}	1.020×10^{-10}	+1203.81		26.86	
300	57	3.7368×10^{-5}	1.488×10^{-7}	− 126.77	894.5	289.61	—
200	38	5.84575×10^{-3}	1.6957×10^{-5}	− 2.671	1047.98	289.64	Mesosphere
100	19	3.29046×10^{-1}	1.0676×10^{-3}	− 51.098	990.90	"	—
60	11	2.05528	6.9866×10^{-3}	− 69.700	968.08	"	Stratosphere
40	8	5.55844	1.8895×10^{-2}	− 69.700	968.08	"	—
36	7	6.73197	2.2853×10^{-2}	− 69.161	968.74	"	
30	6	8.90289	2.8657×10^{-2}	− 47.831	994.85	"	Troposphere
20	4	13.7612	4.0773×10^{-2}	− 12.255	1036.93	"	
10	2	20.5808	5.6483×10^{-2}	+ 23.355	1077.40	"	
0	0	29.9213	0.076474	+ 59.000	1116.45	"	
−10	−2	42.4456	0.10150	+ 94.679	1154.21	"	

[a] *U.S. Standard Atmosphere, 1962* (Superintendent of Documents, U.S. Government Printing Office, Washington, D.C.).

TABLE 7. BASIC UNITS AND EQUIVALENTS IN THE METRIC AND U.S. SYSTEMS AND SCALES FOR CONVERSION BETWEEN THE SYSTEMS [a]

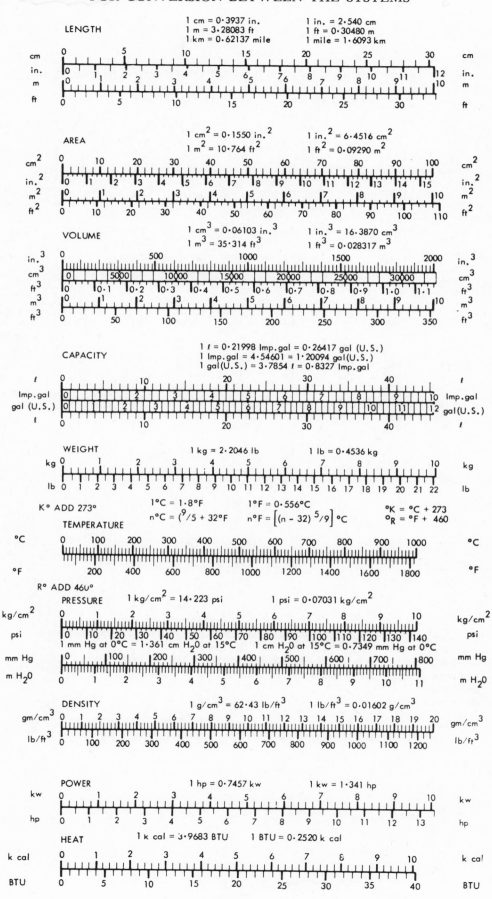

LENGTH

$$1 \text{ cm} = 0 \cdot 3937 \text{ in.} \qquad 1 \text{ in.} = 2 \cdot 540 \text{ cm}$$
$$1 \text{ m} = 3 \cdot 28083 \text{ ft} \qquad 1 \text{ ft} = 0 \cdot 30480 \text{ m}$$
$$1 \text{ km} = 0 \cdot 62137 \text{ mile} \qquad 1 \text{ mile} = 1 \cdot 6093 \text{ km}$$

AREA

$$1 \text{ cm}^2 = 0 \cdot 1550 \text{ in.}^2 \qquad 1 \text{ in.}^2 = 6 \cdot 4516 \text{ cm}^2$$
$$1 \text{ m}^2 = 10 \cdot 764 \text{ ft}^2 \qquad 1 \text{ ft}^2 = 0 \cdot 09290 \text{ m}^2$$

VOLUME

$$1 \text{ cm}^3 = 0 \cdot 06103 \text{ in.}^3 \qquad 1 \text{ in.}^3 = 16 \cdot 3870 \text{ cm}^3$$
$$1 \text{ m}^3 = 35 \cdot 314 \text{ ft}^3 \qquad 1 \text{ ft}^3 = 0 \cdot 028317 \text{ m}^3$$

CAPACITY

$$1 \ell = 0 \cdot 21998 \text{ Imp.gal} = 0 \cdot 26417 \text{ gal (U.S.)}$$
$$1 \text{ Imp.gal} = 4 \cdot 54601 = 1 \cdot 20094 \text{ gal(U.S.)}$$
$$1 \text{ gal(U.S.)} = 3 \cdot 7854 \ell = 0 \cdot 8327 \text{ Imp.gal}$$

WEIGHT

$$1 \text{ kg} = 2 \cdot 2046 \text{ lb} \qquad 1 \text{ lb} = 0 \cdot 4536 \text{ kg}$$

K° ADD 273°

TEMPERATURE

$$1°C = 1 \cdot 8°F \qquad 1°F = 0 \cdot 556°C \qquad °K = °C + 273$$
$$n°C = (^9/_5 + 32°F) \qquad n°F = \left[(n - 32) \, ^5/_9\right]°C \qquad °R = °F + 460$$

R° ADD 460°

PRESSURE

$$1 \text{ kg/cm}^2 = 14 \cdot 223 \text{ psi} \qquad 1 \text{ psi} = 0 \cdot 07031 \text{ kg/cm}^2$$

$$1 \text{ mm Hg at } 0°C = 1 \cdot 361 \text{ cm H}_2\text{0 at } 15°C \qquad 1 \text{ cm H}_2\text{0 at } 15°C = 0 \cdot 7349 \text{ mm Hg at } 0°C$$

DENSITY

$$1 \text{ g/cm}^3 = 62 \cdot 43 \text{ lb/ft}^3 \qquad 1 \text{ lb/ft}^3 = 0 \cdot 01602 \text{ g/cm}^3$$

POWER

$$1 \text{ hp} = 0 \cdot 7457 \text{ kw} \qquad 1 \text{ kw} = 1 \cdot 341 \text{ hp}$$

HEAT

$$1 \text{ k cal} = 3 \cdot 9683 \text{ BTU} \qquad 1 \text{ BTU} = 0 \cdot 2520 \text{ k cal}$$

[a] From *Space Data* (Thompson Ramo Wooldridge Space Technology Laboratories, Inc.).

TABLE 8. FEATURES OF SOME ASTEROIDS OR MINOR PLANETS[a]

Name	Number	Date of dis- covery	Diam- eter (miles)	Mean distance from sun (million miles)[c]	Period (earth- years)	Inclina- tion to ecliptic (deg.)	Eccen- tricity of orbit (index)	Magni- tude at mean opposi- tion[d]	Features
Ceres[b]	1	1801	480	257.0	4.604	10.6	0.076	6.8	Largest known as- teroid and first discovered
Pallas[b]	2	1802	304	257.4	4.615	34.8	0.234	7.9	Great inclination to ecliptic
Juno[b]	3	1804	120	247.8	4.359	13.0	0.258	8.8	Third asteroid dis- covered
Vesta[b]	4	1807	240	219.4	3.629	7.13	0.089	6.1	Only asteroid rarely visible to unaided eye
Astraea	5	1845		239.3	4.136	5.33	0.190	10.3	First asteroid dis- covered by an am- ateur
Hebe	6	1847	137	225.2	3.775	14.75	0.204	8.6	
Iris	7	1847	124	221.6	3.684	5.5	0.230	8.6	
Flora	8	1847		204.5	3.267	5.9	0.156	8.8	
Metis	9	1848		221.7	3.687	5.6	0.123	9.0	
Hygiea	10	1849	205	292.6	5.593	3.8	0.100	10.0	
Brucia	323	1891			3.68	24.1	0.301	12.1	First asteroid dis- covered by photog- raphy by Max Wolf
Eros	433	1898	15	135.0	1.761	10.8	0.223	10.7	Used in determin- ing a.u.
Achilles	588	1904	35		11.98	10.3	0.148	14.2	First of Jupiter's Trojan asteroids discovered
Hidalgo	944	1924	15		13.9	42.5	0.656	17.1	Greatest known aphelion distance —9.6 a.u.
Feodosia	1048	1924			4.52	53.8	0.180	12.6	Largest inclination to ecliptic

[a] Data on those asteroids occasionally approaching close to the earth are given in Table 9. [b] Called the "big four" of the as- teroids and diameter measurable directly. [c] Mean distance only approximate and varying, because the orbits of asteroids are frequently perturbed by the major planets. [d] Visual magnitude; magnitudes of most asteroids vary because of their axial rotations.

TABLE 9. ASTEROIDS WITH ORBITS WITHIN THAT OF MARS AND OCCASIONALLY APPROACHING NEAR THE EARTH'S ORBIT

Name	Number	Approximate diameter (miles)	Distance from sun (a.u.)			Closest approach to earth			Period (years)
			Mean	Perihelion	Aphelion	a.u.	Miles (million)	Year	
Betulia[a]	1580	1	2.1954	1.1135	3.2773	0.156	14.5	1963	3.253
Eros[b]	433	5	1.4581	1.1084	1.8078	0.15	13.9	1975	1.761
Amor[c]	1221	3	1.9190	1.0850	2.7530	0.115	10.7	1956	2.658
Geographos	1620	1	1.2441	0.8271	1.6611	0.073	6.8	1969	1.388
Apollo[c]	1932 HA	2	1.4861	0.6445	2.3277	0.0699	6.5	1932	1.812
Icarus	1566	1	1.0777	0.1869	1.9685	0.042	3.9	1968	1.119
Adonis[c]	1936 CA	1	1.9692	0.4348	3.5036	0.0161	1.5	1936	2.763
Hermes[c]	1937 UB	1	1.2904	0.6780	1.9028	0.0052	0.4	1937	1.466

[a] On May 21, 1963, Betulia made a very close approach to the earth, ranging in within 14.5 million miles. It was followed carefully by astronomers to determine its orbit more accurately. [b] Eros was the asteroid used for extensive observations in 1900–1 and again in 1930–31, when it passed near the earth, to arrive at a more exact estimate of the length of the a.u. by the so-called gravity method. [c] Not observed recently, but possibly recoverable on search; calculated orbits may be based on too short an arc to be accurate. These small bodies pass so close to the earth so quickly and their magnitude (15 to 20) is so low that they can be followed for only a brief time.

TABLE 10. MOST PROMINENT NIGHTTIME METEOR SHOWERS [a]

Name of shower	Period of detectable meteors	Date of peak activity	Visual hourly rates[b]	Duration of peak (days)	Radiant coordinates (deg.)[c]	
					Right ascension	Declination
Quadrantids[d]	Jan. 1–4	Jan. 3	35	0.5	231	+50
Corona Australids	March 14–18	March 16	(5)[e]	(5)[e]	245	−48
Virginids	March 5–April 2	March 20	(less than 5)	(20)	190	0
Lyrids	April 19–24	April 21	5	2	272	+32
Eta Aquarids[d]	April 21–May 12	May 4	12	10	336	0
Ophiuchids	June 17–26	June 20	(20)	(10)	260	−20
Capricornids	July 10–Aug. 5	July 25	(20)	(20)	315	−15
Southern Delta Aquarids[d]	July 21–Aug. 15	July 30	20	15	339	−17
Northern Delta Aquarids[d]	July 15–Aug. 18	July 29	10	20	339	0
Pisces Australids	July 15–Aug. 20	July 30	(20)	(20)	340	−30
Alpha Capricornids	July 15–Aug. 20	Aug. 1	5	(25)	309	−10
Southern Iota Aquarids	July 15–Aug. 25	Aug. 5	(10)	(25)	338	−15
Northern Iota Aquarids	July 15–Aug. 25	Aug. 5	(10)	(25)	331	− 6
Perseids[d]	July 25–Aug. 17	Aug. 11	50	5	46	+58
Kappa Cygnids	Aug. 18–22	Aug. 20	(5)	(3)	290	+55
Orionids	Oct. 18–26	Oct. 20	20	5	95	+15
Southern Taurids	Sept. 15–Dec. 15	Nov. 1	(5)	(45)	52	+14
Northern Taurids	Oct. 15–Dec. 1	Nov. 1	(less than 5)	(30)	54	+21
Leonids	Nov. 14–20	Nov. 16	(5)	4	152	+22
Phoenicids	(Dec. 5)[e]	Dec. 5	(50)	(0.5)	15	−55
Geminids[d]	Dec. 7–15	Dec. 13	50	6	113	+32
Ursids	Dec. 17–24	Dec. 22	15	2	217	+80

[a] Adapted from D. W. R. McKinley, *Meteor Science and Engineering* (New York: McGraw-Hill Book Company, Inc., 1961).   [b] Number of meteors observable visually by a single observer at maximum shower activity.   [c] +, North of the celestial equator, south of it, −.   [d] Among the stronger and more consistent meteor showers.   [e] Figures in parentheses less reliable than other figures for duration of the peak activity, visual hourly rates, and period of detectable meteors.

TABLE 11. MAJOR TERRESTRIAL METEORITIC CRATERS BY APPROXIMATE DATE OF DISCOVERY [a]

Name and/or location	Number of craters	Diameter[b] (ft.)	Date of discovery	Features
Barringer Meteor Crater, between Winslow and Flagstaff, Arizona	1	3,937	1891	Created possibly 25,000 years ago by a metallic (iron or stony-iron) meteorite, estimated the equivalent of a 5-megaton nuclear explosion; many meteoritic fragments around main crater; natural coesite first found here
Odessa, Ector County, Texas	2	530	1921	Metallic meteoritic crater; with many small fragments, thousands of years old, since fossil bones of extinct animals found in excavation of crater
Dalgaranga, Western Australia	1	230	1923	Metallic meteoritic crater
Oesel Island, Baltic Sea, Kaalijarv, Estonia, U.S.S.R.	6	327	1927	Small metallic meteoritic fragments found; largest crater filled with a lake
Tunguska, Central Siberia, U.S.S.R.	10+	164	1927 (fell 1908)	Possibly produced by a comet head or cometary materials; no meteorites recovered; trees leveled by fall
Henbury, Central Australia	13	660 × 360	1931	Metallic meteoritic crater; largest may be two overlapped smaller craters; irons excavated from smaller craters
Wabar, Al Hadija, Saudi Arabia	2	328	1932	Possibly metallic meteorite; coesite and large amounts of silica glass (impactite) found
Brenham, Haviland, Kansas	1	56	1933	Originally thought a buffalo wallow, then identified as a meteoritic crater; stony-iron meteorites scattered over a wide area
Campo del Cielo, Gran Chaco, Argentina	Many	184		Several metallic meteorites of over a ton in weight found
Box Hole, Central Australia	1	575	1937	Metallic meteoritic crater; number of iron meteorites found; crater probably very old
Sikhote-Alin, north of Vladivostok, Siberia, U.S.S.R.	106	92	1947	Metallic meteorite of an estimated 75 tons fragmented in passage, creating many craters; iron meteorites found
Wolf Creek, Northwestern Australia	1	2,756	1947	Stony-iron meteorites found
New Quebec (formerly Chubb) Crater, Ungava Peninsula, 130 miles south of Hudson Strait, Quebec	1	2 miles	1949	Crater contains lake, observed in aerial photography; may be fossil crater, 2 billion years old; no meteorites found
Brent Meteor Crater, Algonquin Provincial Park, Ontario	1	2 miles	1961 (verified)	A fossil meteor crater, perhaps 600–900 million years old; crater walls 3,000 ft. high
Clearwater Lake Craters, 50 miles east of Richmond Gulf, east coast of Hudson Bay, Canada	1 / 1	20 miles / 14 miles	1963 (verified)	May be fossil craters 2 billion years old; no meteorites found; two craters form Clearwater Lake; shattered rocks found
Holleford Crater, 16 miles north of Kingston, Ontario	1	1.46 miles	1963 (verified)	Fossil crater produced perhaps 500–600 million years ago; coesite identified, as well as shocked quartz

[a] Excludes some craterlike structures found which may be impact craters of great age, or "fossil craters," and sites not yet fully authenticated as meteoritic, even though shatter cones (Sierra Madera, Texas; Vredefort Ring, Transvaal, South Africa) or coesite (Ries Kessel Basin, Southern Germany; Ashanti, Lake Bosumtwi, Ghana, Africa) may have been found. [b] Diameter of largest crater given when more than one crater is involved.

TABLE 12. VELOCITIES OF SATELLITES AND THEIR PERIODS IN CIRCULAR ORBITS AT INCREASING ALTITUDES ABOVE THE EARTH

Height above earth's surface (miles)	Velocity		Period of revolution (min.)
	miles/sec.	miles/hr.	
100	4.85	17,500	87.7
200	4.79	17,244	91.0
300	4.73	17,028	94.3
500	4.63	16,668	101.0
1,000	4.39	15,804	118.4
1,044	4.37	15,732	120.0[a]
2,000	4.00	14,400	155.9
22,283	1.91	6,876	1,440.0[b]
230,000	0.64	2,304	38,880.0

[a] Orbit with a period of revolution of 2 hours. [b] A synchronous orbit with a period of revolution of 24 hours, like the period of rotation of the earth.

TABLE 13. FEATURES OF THE MOON AND PLANETS RELATED TO SPACE FLIGHT

Body	Surface gravity (g's) (earth=1)	Mean orbital velocity		Escape velocity (miles/ sec.)	Farthest distance from earth[a]	Closest approach to earth[a]	Time of transit by most economical routes[b] (days)	Inclination of orbit to ecliptic	
		(miles/ min.)	(miles/ sec.)					(deg.)	(min.)
Mercury	0.35	1,785	29.9	2.2	138	48	115	7°	0'
Venus	0.86	1,299	21.9	6.3	162	24	146	3	24
Earth	1.00	1,112	18.6	6.88	—	—	—	—	—
Moon	0.1645	38	0.64	1.5	0.253	0.222	116 hrs.[c]	5	9
Mars	0.408	902	15.1	3.1	249	34	237	1	51
Jupiter	2.815	486	8.2	37.3	602	366	937	1	19
Saturn	1.296	359	6.0	22.4	1,031	743	2,043	2	30
Uranus	0.975	254	4.3	13.1	1,964	1,606	5,466	0	46
Neptune	1.142	202	3.4	14.3	2,915	2,678	10,972	1	47
Pluto	0.40	177	3.0	1.9	4,680	2,650	10,972	17	9

[a] Distances in millions of miles. [b] Elliptical, Hohmann transfer routes. [c] Transit time to the moon might range anywhere from 10 to 116 hours depending on the orbit followed and the velocity given the craft; feasible orbits range from perhaps 30 to 85 hours.

BIBLIOGRAPHY

The Universe

Alter, Dinsmore, Cleminshaw, Clarence H., and Phillips, John G. *Pictorial Astronomy*, 3rd rev. ed. New York, Thomas Y. Crowell Company, 1969.

Baade, Walter. *Evolution of Stars and Galaxies*. Edited by Cecilia Payne-Gaposchkin. Cambridge, Mass., Harvard University Press, 1963.

Bailey, Kenneth Vye. *Telescopes and Observatories*. Toronto, S. J. Reginald Saunders & Company, Ltd., 1961.

Bok, Bart Jan, and Bok, Priscilla F. *The Milky Way*, 3d ed. Cambridge, Mass., Harvard University Press, 1957.

Bova, Ben. *The Milky Way Galaxy: Man's Exploration of the Stars*. New York, Holt, Rinehart and Winston, Inc., 1961.

Chisnall, George Allen, and Fielder, Gilbert. *Astronomy and Spaceflight*. Toronto, Clarke, Irwin & Company, Ltd., 1963.

Deutsch, A. J., and Klemperer, W. B., eds. *Space Age Astronomy*. New York, Academic Press, Inc., 1962.

Dodson, R. S., Jr., *Exploring the Heavens*. New York, Thomas Y. Crowell Company, 1964.

Hoyle, Fred. *The Nature of the Universe*, rev. ed. New York, Harper & Row, Publishers, 1960.

———. *Astronomy*. Garden City, N.Y., Doubleday & Company, Inc., 1962.

Kienle, Hans. *Modern Astronomy: An Introduction*. New York, Thomas Y. Crowell Company, 1969

King, Henry C. *Pictorial Guide to the Stars*. New York, Thomas Y. Crowell Company, 1967.

Kopal, Zdeněk. *Widening Horizons: Man's Quest to Understand the Structure of the Universe*. New York, Taplinger Publishing Company, 1970.

Kuiper, Gerard Peter, and Middlehurst, Barbara M., eds. *Telescopes*. (Vol. 1 of *Stars and Stellar Systems*.) Chicago, University of Chicago Press, 1960.

Lovell, Sir Bernard. *Our Present Knowledge of the Universe*. Cambridge, Mass., Harvard University Press, 1967.

Lyttleton, R. A. *Man's View of the Universe*. Boston, Little, Brown & Company, 1961.

Miczaika, Gerard Robert, and Sinton, William M. *Tools of the Astronomer*. Cambridge, Mass., Harvard University Press, 1961.

Muirden, Janes. *The Amateur Astronomer's Handbook*. New York, Thomas Y. Crowell Company, 1968.

Page, Thornton, and Page, Lou Williams, eds. *Beyond the Milky Way—Galaxies, Quasars, and the New Cosmology*. New York, The Macmillan Company, 1969.

Ronan, Colin A. *Changing Views of the Universe*. New York, The Macmillan Company, 1961.

Sciama, D. W. *The Unity of the Universe*. Garden City, N.Y., Doubleday & Company, Inc., 1959.

Shapley, Harlow. *Galaxies*, rev. ed., New York, Atheneum, 1967.

Struve, Otto, and Zebergs, Velta. *Astronomy of the 20th Century*. New York, The Macmillan Company, 1962.

The Solar System

Abetti, Giorgio. *Solar Research*. New York, The Macmillan Company, 1963.

Allen, Clabon Walter, *Astrophysical Quantities*, 2d ed. New York, Oxford University Press, 1963.

Berlage, H. P. *The Origin of the Solar Ssystem*. New York, Pergamon Press, 1968.

Blanco, V. M., and McCuskey, S. W. *Basic Physics of the Solar System*. Reading, Mass., Addison-Wesley Publishing Company, Inc., 1961.

Jastrow, Robert, and Cameron, A. G. W. *Origin of the Solar System; Proceedings of a Conference held at the Goddard Institute for Space Studies, New York*. New York, Academic Press, Inc., 1963.

Kuiper, Gerard Peter, and Middlehurst, Barbara M., eds. *Planets and Satellites*. (Vol. 3 of *The Solar System*.) Chicago, University of Chicago Press, 1961.

———. *The Sun*. (Vol. 1 of *The Solar System*.) Chicago, University of Chicago Press, 1953.

Kurth, R. *Introduction to the Mechanics of the Solar System*. New York, Pergamon Press, Inc., 1959.

Lyttleton, R. A. *Mysteries of the Solar System*. Oxford, Clarenden Press, 1968.

Menzel, Donald H. *Our Sun*, rev. ed. Cambridge, Mass., Harvard University Press, 1959.

Nautical Almanac Office, U.S. Naval Observatory. *The American Ephemeris and Nautical Almanac; also Explanatory Supplement to the Astronomical Ephemeris and the American Ephemeris and Nautical Almanac*. Washington, D.C., U.S. Government Printing Office, published annually.

Urey, Harold C. *The Planets: Their Origin and Development*. New Haven, Conn., Yale University Press, 1952.

Whipple, Fred Lawrence. *Earth, Moon, and Planets*. Third Ed. Cambridge, Mass., Harvard University Press, 1968.

Planet Earth

Asimov, Isaac. *The Double Planet*. New York, Abelard-Schuman, Ltd., 1960.

Bates, D. R., ed. *The Earth and Its Atmosphere*. New York, Basic Books, Inc., 1957.

Beiser, Germaine. *The Story of Gravity*. New York, E. P. Dutton & Company, Inc., 1968.

Dobson, G. M. B. *Exploring the Atmosphere*. New York, Oxford University Press, 1963.

Evans, C. *Geology and Astronomy*. New York, Cambridge University Press, 1962.

Gamow, George. A *Planet Called Earth*. New York, The Viking Press, Inc., 1963.

Geiss, J., and Goldberg, E. D., compilers. *Earth Science and Meteoritics*. New York, Interscience Publishers Inc., 1963.

Howell, Benjamin F., Jr. *Introduction to Geophysics*. New York, McGraw-Hill Book Company, Inc., 1959.

Johnson, Francis S., ed. *Satellite Environment Handbook*. Stanford, Calif., Stanford University Press, 1961.

Kuiper, Gerard Peter, and Middlehurst, Barbara M., eds. *The Earth as a Planet*. (Vol. 2 of *The Solar System*.) Chicago, University of Chicago Press, 1954.

Mason, Brian Harold. *Principles of Geochemistry*, 2d ed. New York, John Wiley & Sons, Inc., 1958.

Massey, Harrie Stewart Wilson, and Boyd, R. L. F. *The Upper Atmosphere*. New York, Philosophical Library, Inc., 1959.

Phillips. O. M. *The Heart of the Earth*. San Franciso, Freeman, Cooper & Company, 1968.

Ratcliffe, John Ashworth, ed. *Physics of the Upper Atmosphere*. New York, Academic Press, Inc., 1960.

Rumney, George R. *The Geosystem: Dynamic Integration of Land, Sea, and Air*. Dubuque, Iowa, William C. Brown Company, Publishers, 1970.

Spar, Jerome. *Earth, Sea, and Air; A Survey of the Geophysical Sciences*. Reading, Mass., Addison-Wesley Publishing Company, Inc., 1962.

Stumpff, Karl. *Planet Earth*. Ann Arbor, Mich., University of Michigan Press, 1959.

Tarling, Don, and Tarling, Maureen. *Continental Drift: A Study of the Earth's Moving Surface*. Garden City, New York, Doubleday & Company, Inc., 1971.

Interplanetary Space

Berkner, Lloyd V., and Odishaw, Hugh, eds. *Science in Space*. New York, McGraw-Hill Book Company, Inc., 1961.

Cranshaw, T. E. *Cosmic Rays*. New York, Oxford University Press, 1960.

Jastrow, Robert, ed. *The Exploration of Space; A Symposium on Space Physics*. New York, The Macmillan Company, 1960.

Le Galley, Donald P., ed. *Space Science*. New York, John Wiley & Sons, Inc., 1963.

Liller, William, ed. *Space Astrophysics*. New York, McGraw-Hill Book Company, Inc., 1961.

Merrill, Paul Willard. *Space Chemistry*. Ann Arbor, Mich., University of Michigan Press, 1963.

Smith, Henry J., and Smith, Elske V. P. *Solar Flares*. New York, The Macmillan Company, 1963.

Asteroids, Meteorites, and Comets

Hawkins, Gerald S. *The Physics and Astronomy of Meteors, Comets, and Meteorites*. New York, McGraw-Hill Book Company, 1964.

Krinov, Evgeny Leonidovich. *Principles of Meteoritics*. Translated from the Russian by Irene Vidziunas. New York, Pergamon Press, Inc., 1960.

Kuiper, Gerard Peter, and Middlehurst, Barbara M., eds. *The Moon, Meteorites, and Comets*. (Vol. 4 of *The Solar System*.) Chicago, University of Chicago Press, 1963.

Ley, Willy. *Visitors from Afar: The Comets*. New York, McGraw-Hill Book Company, 1969.

Lyttleton, Raymond Arthur. *The Comets and Their Origin*. New York, Cambridge University Press, 1953.

Mason, Brian Harold. *Meteorites*. New York, John Wiley & Sons, Inc., 1962

Nininger, Harvey Harlow. *Out of the Sky; An Introduction to Meteoritics*, 2d ed. New York, Dover Publications, Inc., 1959.

O'Keefe, John A., ed. *Tektites*. Chicago, University of Chicago Press, 1963.

Richter, Nikolaus B. *The Nature of Comets*. Trans. and rev. ed. by Arthur Beer. London, Methuen & Company, Ltd, 1963.

Roth, Günter D. *The System of Minor Planets*. Translated from the German by Alex Helm. Princeton, N.J., D. Van Nostrand Company, Inc., 1962.

Watson, Fletcher Guard. *Between the Planets*, rev. ed. Cambridge, Mass., Harvard University Press, 1956.

The Moon

Alter, Dinsmore. *Introduction to the Moon*. Los Angeles, Griffith Observatory, 1958.

———. *Pictorial Guide to the Moon*. Rev. and expanded ed. New York, Thomas Y. Crowell Company, 1973.

Baldwin, Ralph B. *The Measure of the Moon*. Chicago, University of Chicago Press, 1963.

———. *A New Photographic Atlas of the Moon*. New York, Taplinger Publishing Company, 1971.

Fielder, Gilbert. *Structure of the Moon's Surface*. New York, Pergamon Press, Inc., 1961.

Kopal, Zdeněk. *The Moon, Our Nearest Celestial Neighbor*. New York, Academic Press, Inc., 1961.

Kuiper, Gerard Peter, and Middlehurst, Barbara M., eds. *The Moon, Meteorites, and Comets*. (Vol. 4 of *The Solar System*.) Chicago, University of Chicago Press, 1963.

Mason, Brian, and Melson, William G. *The Lunar Rocks*. New York, John Wiley & Sons, Inc., 1970.

Markov, Aleksandr Vladimirovich, ed. *The Moon; A Russian View*. Translated from the Russian by Royer and Roger, Inc. Chicago, University of Chicago Press, 1962.

Moore, Patrick. *A Survey of the Moon.* New York, W. W. Norton & Company, Inc., 1963.

Mutch, Thomas A. *Geology of the Moon: A Stratigraphic View.* Princeton, New Jersey, Princeton University Press, 1970.

Wilkins, Hugh Percy. *Moon Maps; With a Chart Showing the Other Side of the Moon Based Upon the Soviet Photographs.* London, Faber & Faber, Ltd., 1960.

Wilkins, Hugh Percy, and Moore, Patrick. *The Moon; A Complete Description of the Surface of the Moon.* New York, The Macmillan Company, 1961.

Planets and Satellites

Alexander, Arthur Francis O'Donel. *The Planet Saturn; A History of Observation, Theory and Discovery.* New York, The Macmillan Company, 1962.

De Vaucouleurs, Gérard. *Physics of the Planet Mars; An Introduction to Aerophysics,* rev. ed. London, Faber & Faber, Ltd., 1954.

Glasstone, Samuel. *The Book of Mars.* Washington, D.C., National Aeronautics and Space Administration, 1968.

Grosser, Morton. *The Discovery of Neptune.* Cambridge, Mass., Harvard University Press, 1962.

Kellogg, William W., and Sagan, Carl. *The Atmospheres of Mars and Venus; A Report by the Ad Hoc Panel on Planetary Atmospheres.* Washington, D.C., National Research Council, Publication 944, 1962.

Kuiper, Gerard Peter, ed. *The Atmospheres of the Earth and Planets,* rev. ed. Chicago, University of Chicago Press, 1952.

Ley, Willy. *Gas Giants: The Largest Planets.* New York, McGraw-Hill Book Company, 1969.

—— and von Braun, Wernher, *The Exploration of Mars.* New York, The Viking Press, Inc., 1956.

Moore, Patrick. *Guide to Mars.* New York, The Macmillan Company, 1958.

——. *The Planet Venus,* 3d ed. New York, The Macmillan Company, 1960.

NASA. *Mariner-Mars 1969: A Preliminary Report.* Washington, D.C., National Aeronautics and Space Administration, 1969.

Peek, Bertrand Meigh. *The Planet Jupiter.* London, Faber & Faber, Ltd., 1958.

Sandner, Werner. *The Planet Mercury.* Translated from the German by Alex Helm. London, Faber & Faber, Ltd., 1963.

——. *Satellites of the Solar System.* Trans. by Alex Helm. London, The Scientific Book Club, 1965.

Smith, Alex. G., and Carr, T. D. *Radio Exploration of the Planetary System.* Princeton, N.J., D. Van Nostrand Company, Inc., 1964.

Wheelock, Harold J., compiler. *Mariner Mission to Venus; Prepared for the National Aeronautics and Space Administration by the staff, Jet Propulsion Laboratory, California Institute of Technology.* New York, McGraw-Hill Book Company, Inc., 1963.

Life on Other Planets

Anderson, Poul. *Is There Life on Other Worlds?* New York, Crowell-Collier Press, 1963.

Cade, C. Maxwell. *Other Worlds Than Ours.* New York, Taplinger Publishing Company, 1966.

Cameron, A. G. W., ed. *Interstellar Communication.* New York, W. A. Benjamin, Inc., 1963.

Drake, Frank D. *Intelligent Life in Space.* New York, The Macmillan Company, 1962.

Firsoff, V. A. *Life Beyond the Earth.* London, Hutchinson & Company, Ltd., 1963.

Gatland, Kenneth, and Dempster, Derek. *Inhabited Universe.* New York, Fawcett World Library, 1959.

Jones, Harold Spencer. *Life on Other Worlds,* rev. ed. London, Hodder & Stoughton, Ltd., 1959.

Lowell, Percival. *Mars As the Abode of Life.* New York, The Macmillan Company, 1908.

——. *Mars and Its Canals.* New York, The Macmillan Company, 1906.

MacGowan, Roger A., and Orway, Frederick I., III. *Intelligence in the Universe.* Englewood Cliffs, N.J., Prentice-Hall, Inc., 1966.

Macrey, John W. *Alone In the Universe?* New York, The Macmillan Company, 1963.

Moore, Patrick, and Jackson, Francis. *Life In the Universe.* New York, W. W. Norton & Company, Inc., 1962.

Oparin, Aleksandr Ivanovich. *Life—Its Nature, Origin and Development.* Translated from the Russian by Ann Synge. New York, Academic Press, Inc., 1961.

Oparin, Aleksandr Ivanovich, and Fesenkov, V. *Life in the Universe.* Translated from the Russian by David A. Myshne. New York, Twayne Publishers, Inc., 1961.

Ovenden, Michael W. *Life in the Universe; A Scientific Discussion.* Garden City, N.Y., Doubleday & Company, Inc., 1962.

Shapley, Harlow, *The View from a Distant Star.* New York, Basic Books, Inc., 1963.

Shklovskii, I. S., and Sagan, Carl. *Intelligent Life in the Universe.* San Francisco, Holden-Day, Inc., 1966.

Sullivan, Walter, *We Are Not Alone: The Search for Intelligent Life on Other Worlds.* New York, McGraw-Hill Book Company, 1964.

Young, Richard S. *Extraterrestrial Biology.* New York, Holt, Rinehart and Winston, Inc., 1966.

Rockets, Artificial Satellites, and Space Probes

Berman, Arthur I. *The Physical Principles of Astronautics; Fundamentals of Dynamical Astronomy and*

Corliss, William R. *Propulsion Systems for Space Flight*. New York, McGraw-Hill Book Company, Inc., 1960.

Deutsch, J., and Klemperer, Wolfgang B., eds. *Space Age Astronomy*. New York, Academic Press, 1962.

Ducrocq, Albert. *Victory Over Space*. Translated from the French by Oliver Stewart. Boston, Little, Brown & Company, 1961.

Eisner, Will. *America's Space Vehicles; A Pictorial Review*. New York, Sterling Publishing Company, Inc., 1962.

Feodosiev, V. I., and Siniarev, G. B. *Introduction to Rocket Technology*. Translated from the Russian by S. N. Samburoff. New York, Academic Press, Inc., 1959.

Hobbs, Marvin. *Fundamentals of Rockets, Missiles, and Spacecraft*. New York, John F. Rider Publisher, Inc., 1962.

Howard, N. E. *Handbook for Observing the Satellites*. New York, Thomas Y. Crowell Company, 1958.

King-Hele, Desmond. *Satellites and Scientific Research*, 2d ed. rev. New York, Dover Publications, Inc., 1962.

Levitt, I., and Cole, Dandridge M. *Exploring the Secrets of Space*. Englewood Cliffs, N.J., Prentice-Hall, Inc., 1963.

Ley, Willy. *Rockets, Missiles, and Space Travel*, rev. and enlarged ed. New York, The Viking Press, Inc., 1961.

Newell, Homer E., Jr., ed. *Sounding Rockets*. New York, McGraw-Hill Book Company, Inc., 1959.

Ordway, Frederick I., Gardner, James Patrick, and Sharpe, Mitchell R., Jr. *Basic Astronautics, An Introduction to Space Science, Engineering, and Medicine*. Englewood Cliffs, N.J., Prentice-Hall, Inc., 1962.

Parry, Albert. *Russia's Rockets and Missiles*. Garden City, N.Y., Doubleday & Company, Inc., 1960.

Puckett, Allen E., and Ramo, Simon, eds. *Guided Missile Engineering*. New York, McGraw-Hill Book Company, Inc., 1959.

Singer, S. F., ed. *Progress in the Astronautical Sciences*, vol. 1. New York, Interscience Publishers, Inc., 1962.

Von Braun, Wernher, and Ordway, Frederick I., III. *History of Rocketry and Space Travel*. New York, Thomas Y. Crowell Company, 1969.

Woodbury, David O. *Outward Bound for Space*. Boston, Little, Brown & Company, 1961.

Yates, Raymond Francis, and Russell, M. E. *Space Rockets and Missiles*. New York, Harper & Row, Publishers, 1960.

Zaehringer, Alfred J. *Soviet Space Technology*. New York, Harper & Row, Publishers, 1961.

Space Flight

Armstrong, H. G., ed. *Aerospace Medicine*. Baltimore, The Williams & Wilkins Company, 1961.

Benedict, Elliott T., ed. *Weightlessness—Physical Phenomena and Biological Effects*. New York, Plenum Press, Inc., 1961.

Brown, J. A. U. *Physiology of Man in Space*. New York, Academic Press, Inc., 1963.

Brown, Kenneth, and Ely, Lawrence D., eds. *Space Logistics Engineering*. New York, John Wiley & Sons, Inc., 1962.

Clarke, Arthur C. *Interplanetary Flight*, 2d ed. New York, Harper & Row, Publishers, 1960.

Cummings, Clifford I., and Lawrence, Harold R., eds. *Technology of Lunar Exploration*. New York, Academic Press, Inc., 1963.

Fogel, Lawrence J. *Biotechnology: Concepts and Applications*. Englewood Cliffs, N.J., Prentice-Hall, Inc., 1963.

Gatland, Kenneth W. *Astronautics in the Sixties; A Survey of Current Technology and Future Development*. New York, John Wiley & Sons, Inc., 1962.

Gerathewohl, Siegfried J. *Principles of Bioastronautics*. Englewood Cliffs, N.J., Prentice-Hall, Inc., 1963.

Godwin, Felix. *The Exploration of the Solar System*. New York, Plenum Press, Inc., 1960.

Hanrahan, James Stephen, and Bushnell, David. *Space Biology; The Human Factors in Space Flight*. New York, Basic Books, Inc., 1960.

Hess, Wilmot N., and Mead, Gilbert D., eds. *Introduction to Space Science*, 2d ed. New York, Gordon and Breach, Science Publishers, 1968.

Ramo, Simon. *Peacetime Uses of Outer Space*. New York, McGraw-Hill Book Company, Inc., 1961.

Sells, S. B., and Berry, C. A., eds. *Human Factors in Jet and Space Travel*. New York, The Ronald Press Company, 1961.

INDEX

ablation, 80, 203
acceleration in flight, 217-218
achondrites, 79
aerolites, 79
aeronomy, 48
airglow, 54
albedo, 38, 72, 90, 234
Alpha Centauri, 9, 229
Alphonsus, lunar crater, 134, 135, 155
Alter, Dinsmore, 108-111, 155, 156
Antoniadi, E. M., 94, 162, 163
aphelion of planets, 21, 232, 233
apogee, 200, 202
Apollo manned space flight, 212, 213
arcjet electric propulsion, 198
arcs, 4
argument of the perigee, 200
Aristarchus, 91, 138, 155, 156
Armstrong, Neil A., 196
artificial gravity, 220
artificial satellites, see satellites, artificial
ascending node, 200
aspect of moon, 28
asteroids:
    albedos of, 72
    Betulia, 127
    capture by earth, 214
    characteristics of, 72-73, 236, 237
    orbits of, 30, 71-72
    origin of, 74-75
    for space ships, 182
    Trojan, 29, 73-74
astronauts on moon, 116, 118-119, 158, 190, 212
astronomical satellites, 206-208
astronomical unit (a.u.), 27
atmosphere:
    condition of, for life, 179
    of earth, 48-56, 234
    of Jupiter, 101-102
    of Mars, 98-101
    of Mercury, 93-95, 182
    of moon, 90-93
    of Neptune, 104
    of Pluto, 104
    of Saturn, 102-104
    in space cabins, 220-221
    of sun, 12-13
    temperature variation in, 49, 55
    of Uranus, 104
    of Venus, 95-98, 182
autumnal equinox, 8

Baldwin, Ralph B., 153, 163-164
balloons, high-altitude, 207
Barnard's star, 188-189, 229
blink microscope, 17
blue haze of Mars, 101
Bode, Johann E., 27

Bradley, Donald A., 53
Brahe, Tycho, 20-21
breakoff phenomenon, 222
bright markings of Mars, 170-171

calcium metabolism, 219-220
canals, or channels, of Mars, 172, 183-184
carbonaceous chondrites, 80
celestial sphere, 7-8
chemosphere, 54
chlorophyll, 178
chondrites, 79-80, 82
chondrules, 26, 79
chromosphere, 12
circular orbits, 201
cislunar space, 59
clefts, lunar, 107
clouds:
    of Mars, 100-101, 171-172
    noctilucent, 54-55
    of Venus, 95-96
coesite, 81-82
comets:
    changes in, 85
    characteristics of, 84-85
    hydrogen, halos of, 89
    orbits of, 85-87
    origin of, 88-89
conic sections, 19-20
conjunction, 159, 209, 210
constellations, map of, 8
continental drift, 36, 40-41, 128-129
Cooper, Gordon, 39, 215, 218
Copernicus, Nicolaus, 5, 20
corona, 12, 68, 70, 125
cosmic rays, 66-67
craters:
    of maar type, 153-154
    of moon, 106-107, 134-135, 153-154
    of nuclear explosions, 130, 154
    of terrestrial meteorites, 81-82, 128, 129, 154, 239
Cygni (61), star, 189, 229

dark markings:
    of Mars, 170-171, 183-186
    of Mercury, 163-164
    oases in Martian, 183
declination, 8-9
density of planets, 33, 36, 232-233
descending node, 200
dog-leg injection into orbit, 204
Dollfus, Audouin, 31-32, 94, 104, 157-158, 164, 166, 168-169, 171-173
domes, lunar, 107
dust:
    atmospheric, 50
    interplanetary, 67-70, 97, 223-224

lunar, 154
terrestrial, 69

earth:
    age of, 34
    atmosphere of, 48-58, 234
    core of, 43, 44
    crust of, 42-43
    gravity field of, 36
    hydrogen corona of, 70
    magnetic field of, 45-47
    mantle of, 43-44
    phases of, 199
    polar axis of, 7
    rotation of, 2
    shape of, 34-36
earthlight, 28-29, 38
eclipse:
    lunar, 112
    solar, 12, 13, 90
ecliptic, 5
electric power generation, 197-198
electric propulsion rockets, 198-199
electromagnetic rocket propulsion, 199
electromagnetic spectrum, 61-63
electrostatic rocket propulsion, 198-199
electrostatic shielding, 228
ellipse, 19
elliptical orbits, 200
equatorial bulge, 35
equatorial orbits, 200
equinoxes, 8
erosion of spacecraft, 224
escape velocity:
    from Milky Way Galaxy, 9
    from planets, 209, 240
    of rockets, 209, 240
Europa, 45, 173
exosphere, 55-56
explosive decompression, 225

far side of moon, 117
fireballs, 75
flattening ratio of earth, 35-36
fluidization, 154
focus of ellipse, 19, 200
fossils:
    of great age, 34
    in meteorites, 176-177
    in salt deposits, 175
free fall, 22

"g" or gravity, 22, 217
gal, 36
galaxies, 8, 10, 67, 121, 230
Galileo, 15, 23-24, 113
gamma, 45
Ganymede, 103, 173, 187
gases:
    in comets, 87

in earth's atmosphere, 50-51
of Jupiter, 102
of Mars, 98-100
of Mercury, 94-95
of moon, 92, 180-181
of Neptune, 104
of Saturn, 103
in space, 63
of Titan, 104
of Venus, 97
gauss, 45
Gegenschein or counterglow, 68
geoid, 35-36
geomagnetic cavity, 56-58
Glenn, John, 37-38
granules of sun, 12
gravireceptors, 218
gravitation, 22
gravities at surfaces of planets, 36, 208, 240
gravity, artificial, 220
gravity, field of, 36
greenhouse effect, 50-51, 97-98

Hadley Rill, 116, 118-119
harmonic law, or law of squares, 21
heat balance, of earth, 52
Herschel, William, 16
heterosphere, 48
Hohmann transfer orbits, 210, 240
homosphere, 48
Hoyle, Fred, 66-67
Humason, Milton L., 85
hydrocarbons:
    as condition for life, 179
    in meteorites, 174, 176
hydrosphere, 42
hyperbola, 20

impact theory of lunar surface, 116-120, 153-154
inclination, angle of, 200, 203-204
inclined orbits, 200, 203-204
inertial guidance, 195
inferior conjunction, see conjunction
infrared waves, 61
injection into orbit, 201-202, 204
interplanetary space, 59-70, 223-225
interstellar space, 61
Io, 45, 102, 173
ionosphere, 56

Jeans, James H., 24
Jupiter:
    atmosphere of, 101-102
    belts and zones of, 101-102
    interior of, 45
    magnetic field of, 102
    satellites of, 14, 127
    surface of, 172-173
    Van Allen region of, 102

Kepler, Johannes, 20-21
Kozyrev, N. A., 94, 155, 157
Kuiper, Gerard P., 18, 99, 120-121

Lagrangian positions:
    of Jupiter, 29, 73-74
    of moon, 29-30, 73
latitude, 9
launch vehicles, see rockets
libration of moon, 28
life:
    characteristics of, 174, 177-180
    conditions conducive to, 178-180
    on Mars, 144-145, 182-187
    on Mercury, 181-182
    in meteorites, 174-175
    on moon, 180-181
    origin of, 177-179
    on outer planets, 187
    in universe, 187-189
    on Venus, 182
life-support systems in space, 220-222
light-year, 9, 229, 230
lithosphere, 42
Local Group of galaxies, 10, 122, 123, 230
longitude, 9
Lowell, Percival, 95, 162-163, 165, 172
Lunar Module, 92-196
Lunar Rover, 119
Lyot, Bernard, 157, 164
Lyra, constellation, 9

maar craters, 153-154
Mädler's Square, lunar, 180
Maffe: galaxy, 123, 230
magnetic fields:
    of earth, 45-47
    of Jupiter, 102
    of space, 63
    in space flights, 228
    of sun, 64-65
    of Venus, 97
magnetic reversals, 46
magnetic shielding, 228
magnetic storms, 65-66
magnetograms, 64
magnetosphere, 57
magnitude:
    absolute, 229, 230
    unit and mean opposition, 38, 234
    visual, 16-17
manned orbital flights, 203-205, 212
manned space flights:
    acceleration in, 217-218
    artificial gravity for, 220
    atmosphere in, 220-221
    calcium metabolism in, 219
    dangers of, 223-228
    group reactions in, 223
    life-support systems, 220-222
    lunar exploration, 118-119, 140, 141, 213
    noise in, 215-216
    orbits for lunar landings, 212-213
    orthostatic hypotension from, 219

radiations in, 225-227
sensory deprivation in, 222
shielding in, 225, 227-228
spacesickness in, 218-219
stress tolerance in, 222-223
vibrations in, 216-217
weightlessness, 218-220
maps:
    of constellations, 8
    of Mars, 144-145, 168-169
    of Mercury, 162-164
    of moon, 108-111, 113, 114, 117
    of satellite orbits, 204
    of sky, 6
    of stars, 8
maria, or seas, of moon, 106, 110, 112-113, 115, 116-121
Mars:
    atmosphere of, 98-101
    bright markings of, 170
    canals, or channels, of, 172, 183-184
    canyons of, 144-145, 149, 212
    dark markings of, 148, 170-171, 184-187
    flyby of spacecraft, 211-212
    interior of, 44
    life on, 144-145, 182-187
    maps of, 144-145, 168-169
    Mariner 9 orbit of, 211
    polar caps of, 150, 170, 212
    retrograde motion of, 5-6
    rotation of, 146
    satellites of, 30, 31, 231
    seasons of, 170-171, 183
    surface of, 144-145, 147, 167-172
    temperature of, 167, 170
    vegetation theory of, 171
    volcanoes on, 144-145, 147
    volcanic theory of, 171, 185
    water on, 149, 150, 185
mascons, 153
mass of planets, 232-233
McLaughlin, Dean B., 185
Mercury:
    atmosphere of, 93-95, 182
    dark markings of, 163-164
    drawings of, 163-164
    interior of, 44
    life on, 181-182
    maps of, 162, 164
    rotation of, 162, 181
    solar transit of, 31
    surface of, 159-165
Mercury manned flights, 37-38
mesosphere, 54-55
meteor craters, 81, 239
meteor showers, 78, 238
meteorites:
    ablation of, 80-81
    fossils in, 176-177
    hydrocarbons in, 174, 176
    lunar, 117-120, 153-154
    organisms in, 174-177
    temperature of, 80

terrestrial, 79-80, 128, 129
types of, 79-80
velocities of, 79
meteoritics, 76-77
meteoroids, 75-76, 80, 223-225
meteors, 75, 77-78
micrometeorites, 78
Milky Way Galaxy, 8, 34, 122-123, 187, 230
Millman, Peter M., 77
minor planets, *see* asteroids
missile ranges:
  Atlantic, 203-205
  Pacific, 203-205
  Russian, 205
missiles, 190
mists or hazes:
  of Mercury, 94-95, 164-165
  of moon, 91, 155
Mohorovičić discontinuity, 43
molecules in space, 63
moon:
  atmosphere of, 90-93, 112
  clefts and rills of, 107, 139
  craters of, 106-107, 134-135, 137, 138, 153-154
  crust of, 116
  domes on, 107
  dust on, 92, 154, 158
  far side of, 117
  fluidization on, 154
  formation of, 120, 153
  life on, possibility of, 180-181
  magnetic field of, 47
  maps of, 108-111, 113, 114, 117
  maria, or seas, of, 106, 112-113, 115, 116-121
  mascons on, 153
  meteoritic impact on, 117, 120, 153-154
  motion of, 3-4, 27-29
  mountains of, 106, 116, 118-119, 142, 143
  orbital velocity of, 4, 27, 240
  phases of, 28, 53-54, 106
  red flashes on, 157
  soil of, 116, 158, 181
  surface composition of, 116, 157-158
  surface gravity of, 208, 240
  surface temperatures of, 111-113
  topography of, 105-112, 139, 140, 143, 152
  volcanoes on, 116-121, 143
  weather effects of, 52-54
motion:
  apparent, 3-5
  of earth, 6-8
  laws of, 22, 192-193
  retrograde, 5-6
  of sun, 7-9
Mount Hadley, moon, 118-119

natural satellites, *see* satellites, natural

nebula, 124
Neptune:
  atmosphere of, 104
  interior of, 45
  satellites of, 18
  surface of, 172
Newton, Isaac, 21-22, 192-193
Nicholson, Seth B., 17, 73, 231
noctilucent clouds, 54-55
nodes of orbits, 200
nova, 125
nuclear electric power generation, 197
nuclear explosion craters, 130, 153-154
nuclear propulsion rockets, 196-198
nuclear reactors for space, 198
nucleus, of comets, 84
nystagmus, 219

oases, Martian, 183
oblateness, 32, 232-233
obliquity of the ecliptic, 8
Occam's Razor, 180
occultation, 31, 90-91
opposition, 159, 209, 210
orbital velocities of planets, 209-210, 240
Orbiting Astronomical Observatories (OAO), 89, 207
Orbiting Solar Observatories (OSO), 207
orbits:
  angle of inclination of, 200
  circular, 201
  of comets, 21
  of earth satellites, 200-203, 240
  eccentricity of, 20
  elements of, 20-22, 200-201
  elliptical, 200
  equatorial, 200
  inclined, 200, 203-204, 210
  injection into, 201-202, 204
  manned flights in, 203-205, 212-214
  nodes of, 200
  of planets, 18-20, 210, 240
  polar, 200
  reentry from, 202-203
  synchronous, 201
  transfer, 210, 240
origin:
  of life, 177-179
  of moon, 26
  of planets, 23-26
orthostatic hypotension, 219
ozone, in earth's atmosphere, 51, 54

paludes, or marshes, of moon, 106
parabola, 20
payloads of rockets, 196
penumbra, of sunspot, 13
perigee, 200-202
perihelion of planets, 21, 232-233
phases, 28, 199

photosphere, 12
photosynthesis, 178
planets:
  albedos of, 234
  atmospheres of, 93-104
  density of, 32, 41, 232-233
  distances from sun, 26-27
  distances from earth, 240
  escape velocities from, 209, 240
  inclination of equators of, 232-233
  inclination of orbits of, 240
  magnetic field of, 47
  mass of, 232-233
  oblateness of, 232-233
  orbital velocities of, 209-210, 240
  orbits of, 232-233
  revolution of, 232-233
  rotation of, 32, 232-233
  sizes and shapes, 30-32, 232-233
  surface gravities of, 208, 240
  surfaces of, 159-173
Plato, lunar walled-plain, 91
Pluto, 17, 45, 104
polar caps of Mars, 170
polar orbits of earth, 200
Priamus, 74
propellants, rocket:
  liquid, 193-194
  solid, 193-194
proprioreceptors, 218
propulsion, *see* rocket engines
Ptolemaic system, 4-5
pulsar, 124

quadrature, 28
quasars, 62
quiet sun, 13

radar waves, 61
radiant, of meteor, 78
radiation doses, 225-226
radiations in space flight, 225-226
radio waves, 61, 125
radius vector, 21
Ranger, *see* space probes
Red Spot of Jupiter, 101-103
reentry of spacecraft, 202-203
regression of plane of orbits, 200
rem, 225
resolution:
  of Mars, 159
  of moon, 105
  of Venus, 159
retrograde motion, 5-6
revolution:
  of earth, 8
  of planets, 232-233
rift system, 36, 40, 41
Right Ascension, 8-9
rills or rilles, lunar, 107, 116, 118-119, 141, 142
rocket engines:
  booster, 190
  combustion in, 192

components of, 190-191
electric propulsion, 198-199
fuel of, 192
nuclear propulsion, 196-198
sustainer, 190
rocket guidance, 194-195
rocket systems, 193-195
rockets:
    launching of, 192
    payloads of, 195-196
    principles of, 192-193
    problem areas of, 216
    sounding, 131, 190, 207
    stages of, 190-191
    types of, United States, 151, 196
Roemer, Elizabeth, 14, 16, 74, 85,
    87, 88
roentgen, 225
rotation:
    of earth, 2, 7-8
    of planets, 32-33, 182, 232-233
    of sun, 11-12

satellites, artificial:
    asteroids as, 213-214
    astronomical, 206-208
    Explorer, 56
    Imp, 57, 58
    Lunar Orbiter, 116, 120, 138,
        139, 212
    manned, 151, 203-205
    Nimbus, 40, 52
    OAO, 89, 207
    OGO, 89
    orbits of, 200-203, 211, 240
    OSO, 207
    Polyot, 204
    reentry of, 202-203
    Sputnik, 72, 195
    Vanguard, 35-36, 56
    weather, 38, 40, 131-133
satellites, natural, 11, 14-16, 18, 30,
    45, 126, 127, 173, 231
Saturn:
    atmosphere of, 102-104
    interior of, 45
    satellites of, 16, 126
    rings of, 103-104
    surface of, 172-173
Schiaparelli, Giovanni V., 33, 162,
    172
seasons of Mars, 170-171, 183
seismic waves, 42-43
sensory deprivation, 222
shatter cones, 81
shielding of spacecraft, 225, 227
Shoemaker, Eugene M., 153-154
siderites, 79
siderolites, 79
Sinton, William M., 111, 186
sinuses, or gulfs, of moon, 106
Slipher, E. C., 183
solar cells, 197
solar constant, 51-52, 234
solar cycle, 13

solar electric power generation, 197-
    198
solar flares, 13, 64-66, 164-165, 207,
    226-227
solar prominences, 12-13
solar system, 13-17, 26
solar transit, 31, 32, 93
solar wind, 63-64, 126
sound, speed of, 234
sounding rockets, 131, 207
space exploration, 2, 208-214
space flights, manned, see manned
    space flights
space medicine, 215-220
space probes:
    aiming zones of, 211
    cometary, 89
    interplanetary, 208-212, 240
    lunar, 208-212, 240
    Lunik, 28, 60, 113, 115-116, 195,
        196, 212
    Mariner 2, 60, 97-98
    Mariner 9, 6, 32, 101, 171, 182,
        210
    Mars 1, 224
    Pioneer, 1, 60, 102, 152
    Ranger, 105, 115, 134-135, 212
    solar, 13
    Surveyor, 2, 136-137, 141, 192,
        199, 212
    Viking, 145, 211
space regions, 59-61
space station, 191, 208
spacecraft, 191, 196
spacesickness, 218-219
spicules of sun, 12
spores in meteorites, 175
stars near sun, 8-9, 189, 229
Stratoscope balloons, 207
stratosphere, 48-49
stress tolerance, 222-223
sun:
    characteristics of, 11-13
    motion of, 7-9
    polar caps of, 66
sunspots, 13, 65
superior conjunction, see conjunction
surface:
    of Jupiter's satellites, 173
    of Mars, 167-172
    of Mercury, 159-165
    of moon, 105-120, 153-158
    of Venus, 165-167
surface gravities of planets, 208, 220,
    240
synchronous orbit, 201

tektites, 82-84
temperature:
    as condition for life, 179
    of earth's atmosphere, 49-52, 54,
        55
    of lunar surface, 111-113
    of Mars' surface, 167, 170
    of Mercury's surface, 162

of meteorites, 80
of meteoroids, 80
in space, 70
in space flight, 221
of sun, 11-12
of Venus' surface, 97-98
terminator, 28, 91, 95
terrestrial space, 60
thermocouples, 197
thermosphere, 55-56
Titan, 45, 173, 187
Titov, Gherman, 215, 219
Tombaugh, Clyde W., 17
Tovo, minor planet, 30
torr, 223
tracking networks, 205-206
transfer orbits, Hohmann, 210, 240
Trojan asteroids, 29, 73-74
troposphere, 48-49

umbra, of sunspot, 13
Uranus:
    atmosphere of, 104
    interior of, 45
    satellites of, 18
    surface of, 172
Urey, Harold C., 105, 176

vacuum in space, 63, 70, 223
Van Allen regions, 56-58, 102, 205,
    225, 226
vegetation theory of Mars, 171-172,
    186-187
Venus:
    atmosphere of, 95-98, 182
    drawings of, 166
    interior of, 44
    life on, 182
    magnetic field of, 97
    mass of, 23
    radial markings of, 165-166
    rotation of, 33, 182
    surface of, 165-167
    temperature of, 166
    views of, 160
vernal equinox, 8
visual magnitude, 16-17
volcanic theory:
    of lunar surface, 116-117, 120,
        153
    of Martian surface, 147, 171, 185-
        186

water as condition for life, 180
weather effects of moon, 52-54
weightlessness, 218-220
Whipple, Fred L., 89, 90
Wiechert-Gutenberg discontinuity,
    43

yellow veils, or clouds, of Mars, 171-
    172

zodiac, 68
zodiacal light, 67-68